Nanotechnology
for
Photovoltaics

Nanotechnology
for
Photovoltaics

Edited by
Loucas Tsakalakos

CRC Press
Taylor & Francis Group
Boca Raton London New York

CRC Press is an imprint of the
Taylor & Francis Group, an **informa** business

CRC Press
Taylor & Francis Group
6000 Broken Sound Parkway NW, Suite 300
Boca Raton, FL 33487-2742

First issued in paperback 2019

ISBN-13: 978-1-4200-7674-5 (hbk)
ISBN-13: 978-0-367-38435-7 (pbk)

Library of Congress Cataloging-in-Publication Data

Nanotechnology for photovoltaics / editor, Loucas Tsakalakos.
    p. cm.
    Includes bibliographical references and index.
    ISBN 978-1-4200-7674-5 (hardcover : alk. paper)
    1. Photovoltaic power generation. 2. Photovoltatic cells. 3. Semiconductor films. I. Tsakalakos, Loucas. II. Title.

TK1087.N36 2010
621.31'244--dc22   2009046419

For Olga, George, and Maria

Contents

Preface

In recent years there has been a significant, resurgent interest in renewable energy systems due to concerns regarding greenhouse gas-related environmental effects, energy security, and the rising costs of fossil-fuel-based energy. Solar energy conversion is of particular interest owing to the abundance of the source. The vast majority of today's commercial solar cells are based on crystalline silicon (Si), despite the fact that they are relatively expensive to produce. The rest of the market in photovoltaics (PV) is based on thin films of cadmium telluride, and to a lesser extent, copper indium gallium diselenide (CIGS) and amorphous Si, which promise lower cost, albeit with lower module efficiencies to date. Silicon is not the ideal semiconducting material for solar energy conversion owing to its indirect band gap, which makes optical absorption inefficient. Yet it is the material of choice in the PV industry today because it is the second most abundant element in the earth's crust, making it a relatively inexpensive semiconductor. Silicon PV also benefits from the tremendous technological base developed for Si by the electronics industry. The above discussion highlights three key questions facing the PV field: (1) How can the efficiency of solar modules be increased to competitive levels with other energy sources? (2) How can the cost of solar modules be decreased to a level suitable for primary power generation? (3) How can both of these goals be achieved in a single PV device and related manufacturing process, thus breaking the conventional paradigm of high efficiency at high cost? These questions lead to yet another question that is the central theme of this book: Can nanotechnology be used to address the above three questions, and if so, how? While thin film PV are helping to reduce the cost of solar energy, the community is well aware that revolutionary approaches are required.

This book, which focuses on the application of nanotechnology to PV technologies, targets the broad community interested in PV. This group includes industrial scientists and engineers working in the field, as well as professors and graduate students performing basic PV research. It is assumed that the reader has a fundamental working knowledge of the solid-state physics of semiconductors, as well as optoelectronic device physics. Additional physics concepts introduced in the book are directly related to phenomena that arise when materials are scaled to dimensions approximately below 100 nm.

The goal of the book is to inform the reader of state-of-the-art developments in the nanophotovoltaics field, primarily from a practical point of view, at the same time providing sufficient fundamental background such that the various research approaches are placed within the proper physical context as related to PV performance enhancement. The book is not, however, written as a textbook, although it could be used to supplement more basic texts on photovoltaics.

While early pioneering work may be found in the literature, it is only very recently that the application of nanotechnology to PV has seen an intense level of interest from the scientific community. It is therefore timely to have a book that describes the major efforts in applying nanotechnology to PV with an emphasis on practical demonstrations and discussions of the technical challenges associated with the application of nanostructures. Although (happily) advances will be made after publication of the book, the core material dealing with basic physics and properties of nanostructures should provide the reader with the background needed to understand basic strategies employed by future works in this field.

The book begins with a review of the applications of PV devices and the performance requirements followed by a brief discussion of thin films. Various so-called generation III, high-efficiency solar cell concepts are currently being explored, some of which require the use of nanoscale quantum structures. A review of the key advanced band structure concepts for obtaining efficiencies above the Shockley–Queisser (detailed balance limit) single bandgap efficiency limit of ~31% is provided. This includes the use of multiple bandgaps, intermediate bands, upconversion, downconversion, and carrier multiplication.

The basic optical properties of nanostructured materials as related to PV applications are discussed in detail, followed by a detailed description of nanoscale optoelectronic device physics related to performance.

The book then explores the recent literature in the application of various classes of nanostructures to photovoltaics. These are classified as (1) nanocomposites and nanostructured materials (focus on nanostructured organic cells), (2) quantum wells, (3) nanowires and nanotubes, and (4) nanoparticles and quantum dots. With regard to the last category, we describe solar cells that are based on quantum dots, as well as on the use of nanoparticles and quantum dots to enhance the performance of conventional solar cells via spectral conversion or plasmonics. Recent advances in luminescent solar concentrators employing nanoparticles are also provided. Each chapter contains a brief review of the historical development for the nanostructure class under consideration, the main applications beyond PV, and the major synthetic methods, followed by a critical review of the leading works that have utilized the particular nanostructure type in a solar cell or in which fundamental measurements of key parameters of interest to PV have been performed. Both the potential advantages of each nanostructure approach and the remaining technical challenges are discussed, with an emphasis on possible future areas of research interest. The book concludes with a summary of the major processing approaches and challenges faced by applying the various nanostructures to PV applications, with particular attention to future scale-up and nanomanufacturing issues (Epilogue).

I gratefully acknowledge my colleagues at GE Global Research and beyond for support of this work, in particular G. Trant, Dr. D. Merfeld, Dr. M. L. Blohm, Dr. J. LeBlanc, Dr. T. Feist, K. Fletcher, E. Butterfield, B. Norman, M. Idelchik, Dr. C. Lavan, M. Beck, Dr. S. Rawal, J. Likar, P. Rosecrans, and Dr. C. Korman. I am also grateful to the engineers and scientists with whom I have had the pleasure to work in the area of nano-PV, including Dr. B. A. Korevaar, J. Balch, J. Fronheiser, Dr. O. Sulima, Dr. J. Rand, R. Wortman, Dr. R. Rodrigues, Dr. U. Rapol, and Dr. J. D. Michael. The support of Ms. Luna Han and the staff at Taylor & Francis in preparation of this work is kindly appreciated. I also thank all the authors of the contributed chapters to this book for their insights into recent advances and remaining challenges of applying nanostructures and nanotechnology to photovoltaics. Finally, I express my deepest appreciation and gratitude to my wife and children for their patience and support in the preparation and editing of the manuscript.

Loucas Tsakalakos, PhD
Niskayuna, New York

The Editor

Loucas Tsakalakos, PhD, is a senior scientist and program leader in the Thin Films Laboratory, Micro & Nano Structures Technologies Organization at the General Electric Global Research Center in Niskayuna, New York. He received his BS degree (1995) from Rutgers University, and his MS (1998) and PhD (2000) degrees in materials science and engineering from the University of California–Berkeley. His expertise is in the integration of heterogeneous thin-film and nanostructured materials systems for micro- and nano-device applications; he also has extensive experience in the characterization of materials. Since joining GE Global Research in 2000, Dr. Tsakalakos has designed and implemented integrated electronic and sensor systems for defense applications, studied cathode materials for lighting applications, and is a founding team member of GE's Nanotechnology Program. His major area of research has been the development of nanostructured materials and devices, primarily using nanowires/tubes, working with multidisciplinary teams both within GE and in collaboration with external partners. He is currently leading the advanced/next-generation PV efforts within GE Global Research's Solar Energy Platform, which includes research to apply nanostructures to photovoltaics. Dr. Tsakalakos is a member of Tau Beta Pi (the National Engineering Honor Society), the author or co-author of over thirty journals, conference proceedings, and book chapter publications, and holds five U.S. patents.

The Contributors

Sheila Bailey
Photovoltaic and Space
    Environments Branch/RPV
NASA Glenn Research Center
Cleveland, Ohio

I. M. Ballard
Quantum Photovoltaics Group
Physics Department
Imperial College London
London, United Kingdom

Keith W. J. Barnham
Quantum Photovoltaics Group
Physics Department
Imperial College London
London, United Kingdom

Rahul Bose
Quantum Photovoltaics Group
Physics Department
Imperial College London
London, United Kingdom

B. C. Browne
Quantum Photovoltaics Group
Physics Department
Imperial College London
London, United Kingdom

Andreas Büchtemann
Fraunhofer Institute for Applied
    Polymer Research
Golm, Germany

D. B. Bushnell
QuantaSol Ltd.
Richmond upon Thames, United
    Kingdom

Kylie Catchpole
Centre for Sustainable Energy
    Systems
Department of Engineering
Australian National University
Canberra, Australia

Ngai L. A. Chan
Quantum Photovoltaics Group
Physics Department
Imperial College London
London, United Kingdom

Amanda J. Chatten
Quantum Photovoltaics Group
Physics Department
Imperial College London
London, United Kingdom

J. P. Connolly
Quantum Photovoltaics Group
Physics Department
Imperial College London
London, United Kingdom

Michael G. Debije
Chemical Engineering and
    Chemistry
Eindhoven University of
    Technology
Eindhoven, The Netherlands

N. J. Ekins-Daukes
Quantum Photovoltaics Group
Physics Department
Imperial College London
London, United Kingdom

Daniel J. Farrell
Quantum Photovoltaics Group
Physics Department
Imperial College London
London, United Kingdom

M. Fuhrer
Quantum Photovoltaics Group
Physics Department
Imperial College London
London, United Kingdom

R. Ginige
Quantum Photovoltaics Group
Physics Department
Imperial College London
London, United Kingdom

G. Hill
EPSRC National Centre for III-V
    Technologies
Sheffield, United Kingdom

Seth M. Hubbard
Department of Physics
Graduate Faculty Microsystems
    Engineering
Graduate Faculty Golisano
    Institute for Sustainability
Rochester Institute of Technology
Rochester, New York

A. Ioannides
Quantum Photovoltaics Group
Physics Department
Imperial College London
London, United Kingdom

D. C. Johnson
Quantum Photovoltaics Group
Physics Department
Imperial College London
London, United Kingdom

D. König
ARC Photovoltaics Centre of
    Excellence
University of New South Wales
Sydney, Australia

Bas A. Korevaar
Thin Films Laboratory
Micro & Nano Structures
    Technologies
General Electric—Global Research
    Center
Niskayuna, New York

En-Shao Liu
Department of Electrical and
    Computer Engineering
Microelectronics Research Center
University of Texas
Austin, Texas

M. C. Lynch
Quantum Photovoltaics Group
Physics Department
Imperial College London
London, United Kingdom

Liberato Manna
NNL—National Nanotechnology
    Laboratory of CNR-INFM
IIT Research Unit
Lecce, Italy

M. Mazzer
CNR-IMEM
University of Parma
Parma, Italy

James T. McLeskey Jr.
Department of Mechanical
    Engineering
Virginia Commonwealth
    University
Richmond, Virginia

A. Meijerink
Debye Institute for NanoMaterials
    Science, Condensed Matter and
    Interfaces
Utrecht University
Utrecht, The Netherlands

Qiquan Qiao
Department of Electrical
    Engineering
South Dakota State University
Brookings, South Dakota

Jana Quilitz
Fraunhofer Institute for Applied
    Polymer Research
Golm, Germany

Ryne Raffaelle
Director, National Center for
    Photovoltaics
National Renewable Energy
    Laboratory
Golden, Colorado

J. S. Roberts
EPSRC National Centre for III-V
    Technologies
Sheffield, United Kingdom

C. Rohr
Quantum Photovoltaics Group
Physics Department
Imperial College London
London, United Kingdom

R. E. I. Schropp
Debye Institute for NanoMaterials
    Science, Surfaces, Interfaces and
    Devices
Utrecht University
Utrecht, The Netherlands

T. N. D. Tibbits
QuantaSol Ltd.
Richmond upon Thames, United
  Kingdom

Loucas Tsakalakos
Thin Films Laboratory
Micro & Nano Structures
  Technologies
General Electric—Global Research
  Center
Niskayuna, New York

Emanuel Tutuc
Department of Electrical and
  Computer Engineering
Microelectronics Research Center
University of Texas
Austin, Texas

W. G. J. H. M. van Sark
Copernicus Institute for
  Sustainable Development and
  Innovation, Science, Technology
  and Society
Utrecht University
Utrecht, The Netherlands

Ye Xiao
Quantum Photovoltaics Group
Physics Department
Imperial College London
London, United Kingdom

E. T. Yu
Department of Electrical and
  Computer Engineering
Microelectronics Research Center
University of Texas
Austin, Texas

1

Introduction to Photovoltaic Physics, Applications, and Technologies

Loucas Tsakalakos

1.1 INTRODUCTION TO PHOTOVOLTAICS

Photovoltaics (PV) is defined as the science and technology of converting light to electricity, the most common form being the utilization of light from the sun. The concept was first demonstrated by French physicist Edmond Becquerel in 1839 via electrochemical studies of AgCl-coated Pt electrodes (Becquerel 1839). It took another 115 years for the demonstration of the first semiconductor solar cell using silicon (Si) to occur by researchers from AT&T Bell Labs (Chapin et al. 1954), with an initial power conversion efficiency of ~6%. In the intervening years there were various photovoltaic developments, the most important related to the use of a solid substance, selenium (Se), as an active layer (Adams and Day 1876) and the use of Se in a crude module (Fritts 1883). However, it was the application of a known semiconductor and the accompanying complete picture of the band structure of such materials (which was developed in the first part of the twentieth century) that enabled rapid developments in the field of PV in the last 50+ years. It has taken 40 to 50 years to develop a detailed understanding of the physics, materials science, optics, and chemistry of PV materials/devices, and the methods to manipulate them, such that record efficiencies close to that of the single-bandgap entitlement could be developed. Reasons for this will be described below and in more detail in subsequent chapters of this book.

Concurrent with the development of various PV technologies in the last 50 or so years, in the last 10 to 20 years there has also been a strong general trend in the broad science and engineering community of attempting to control the structure of matter at the sub-100 nm length scale in order to fashion novel physical, chemical, and biological properties. This is a mesoscopic length scale, in which matter is below the realm where bulk properties prevail, yet slightly larger than atomic dimensions. This field has collectively been referred to as nanoscience or nanotechnology, and has truly touched upon all technical fields in various ways (Contescu et al. 2008). We will use the term *nanotechnology* in this book to emphasize the applied nature of the field, though of course much basic scientific work has been and continues to be performed. Nanotechnology is by no means a unified, homogeneous field. Rather, it is a collection of many topics in multiple domains that deal with the structure of organic, inorganic, and biological materials and devices (and combinations thereof) with a critical length of approximately <100 nm in order to provide unique properties and functionality.

Nanotechnology has not only impacted nearly all fields of science and engineering, but also contributed to and facilitated the breakdown of the barriers that have traditionally existed between the various science and engineering disciplines. It is not uncommon in recent years for graduate students in mechanical engineering to perform experiments typically associated with biochemistry, or for biologists to use nanofabricated devices to advance proteomics and genomics. Areas of research and development in which nanotechnology has had a strong impact include materials science (in a broad manner, which in itself has traditionally been multidisciplinary), chemistry, physics, electronics, photonics, biochemistry, proteomics, genomics, mechanics, composites and other load-bearing materials/structures, pharmacy/medicine, biomedical engineering, surgery, pathology, microelectomechanical systems (MEMS), sensors (chemical, biological, and physical), actuators, thermoelectrics, fuel cells, and many others.

The intersection of photovoltaics and nanotechnology, i.e., nanophotovoltaics (nano-PV) is a relatively recent phenomenon. While some nanostructure-related concepts were discussed in the 1980s and early 1990s, particularly as related to nanocrystalline silicon thin films (often referred to as microcrystalline Si in the PV literature—see Shah et al. [2003] and references therein)—it is only in this decade that the use of various emerging classes of nanostructures has been applied to PV. An early example of using nanostructured materials was shown by Grätzel and coworkers, who used compacted and annealed nano-sized titania powders to demonstrate a dye-sensitized solar cell (O'Regan and Grätzel 1991). Perhaps one of the first attempts at utilizing distinct emerging nanostructures in PV was by Alivisatos and coworkers, who used CdSe nanocrystals in combination with organic hole conducting layers to make a novel solar cell with a power conversion efficiency of 1.7% (Huynh et al. 2002). Other developments have occurred since this work, and it is only in the last four to five years that a concerted effort has been made to utilize nanostructures for PV in various formats, and hence "crystallize" this field of nanophotovoltaics. It is thus timely to present a comprehensive book that outlines the major development of various nanostructures for photovoltaics. A more detailed description of the source of electricity we are primarily interested in, i.e., the solar spectrum, is provided below, followed by the conventional semiconductor mechanism used to create useful power from the sun.

The power source we are interested in is radiation from the sun. The nuclear reactions that occur on the sun produce a broad radiation spectrum with an integrated spectral irradiance (power per unit area) of 1,366 W/m^2 in outer space, as measured by earth-orbiting research satellites (Green 1982). The measured solar spectrum on the earth's surface is dependent on various factors, such as latitude and the angle of incidence, which impact the total air mass (AM) density of the atmosphere through which the sunlight has passed. These factors have a marked effect on the integrated spectral irradiance as well as the shape of the spectrum (Gueymard et al. 2002; Gueymard 2004). More specifically, $AM = 1/\cos\theta$, where θ is the angle of incidence relative to radiation directly normal to the earth's surface. The intensity is typically higher in locations such as the southwestern United States and equatorial countries than in the northeastern United Sates or northern Europe (e.g., Germany).

Figure 1.1 shows the so-called extraterrestrial AM0 (air mass 0, i.e., no atmosphere) spectrum that one measures in space.[*] This represents the solar energy arriving at the earth prior to interaction with the atmosphere. Due to the variability of the solar spectrum on the earth, and to facilitate direct comparison of photovoltaic technologies, two other standard terrestrial spectra have been defined, the global and direct AM1.5 spectra, i.e., AM1.5G and AM1.5D, respectively.[†] The AM1.5D spectrum represents the direct solar component of the AM1.5G spectrum, which also includes the indirect hemispherical component of sunlight encompassing 2π steradians within the field of view of the tilted plane. These American Society for Testing and Materials (ASTM)–defined spectra represent the

---

[*] http://rredc.nrel.gov/solar/spectra/am0/
[†] http://rredc.nrel.gov/solar/spectra/am1.5/

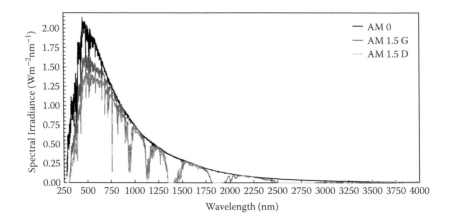

FIGURE 1.1

ASTM standard AM0, AM1.5G, and AM1.5D solar spectra. (Adapted from NREL http://rredc.nrel.gov/solar/spectra/am0/ and http://rredc.nrel.gov/solar/spectra/am1.5/)

mean spectra encountered in the continental United States throughout the course of a year and assume a tilt angle of 37° toward the equator (from the North Pole), which is approximately the mean latitude of the United States. The AM1.5G and D spectra are also plotted in Figure 1.1 for comparison. It is evident that the integrated spectral irradiance is less than AM0, i.e., ~1,000 W/m², and that distinct absorption bands exist at the longer wavelengths due to absorption by various chemical species in the atmosphere (e.g., H_2O, O_2, CO_2, etc.) as well as UV absorption due to ozone and other species. Both AM0 and AM1.5 spectra will be utilized throughout this book.

Photovoltaics is concerned with how the above standard spectra, as well as real spectra encountered throughout the day in specific locations, can be effectively harnessed to produce as much power as possible. The maximum optical power available for terrestrial applications for a 1-m² module is ~1,000 W (~700 W electrical power), though with practical efficiencies this is typically closer to between 50 and 200 W. The detailed reasons for this will be described both below and in Chapter 3; however, in order to begin to understand these PV efficiency losses better, it is helpful to briefly review the basic mechanism for solar energy conversion using a semiconductor.

The conventional means of converting solar energy to electrical energy with a semiconducting material is accomplished by forming a junction between a region in the semiconductor that has been doped with a controlled impurity level to produce an excess of electrons (e.g., phosphorus in silicon, n-region) and a region that is doped to produce an excess of holes (e.g., boron in silicon, p-region). The so-called p-n junction between these two regions produces a depletion width on either side of the junction (W_{Dp} on the p-side and W_{Dn} on the n-side) due to diffusion of charge carriers across the junction leaving space charges from ionized dopant ions, and a corresponding electric field gradient that is capable of separating photogenerated electron-hole pairs that diffuse to this junction. We can analyze this in more detail by considering the semifree carrier charge density in the n- and p-type region as follows (Sze and Ng 2007):

$$n = N_C e^{(E_F - E_C)/kT} \tag{1.1}$$

$$p = N_V e^{(E_V - E_F)/kT} \tag{1.2}$$

where N_C and N_V are the effective density of states in the conduction and valence bands, respectively, and E_F, E_C, and E_V are the energy at the Fermi level, conduction band edge, and valence band edge, respectively. When the p-n junction is formed, the Fermi level, E_F, must be constant across the junction in equilibrium. If we assume an abrupt change in doping at the p-n junction (abrupt junction), the depletion approximation (box-like dopant ionization distribution), and by solving the Poisson's equation in one dimension (x direction), one can derive the following expression for the electric field:

$$\mathrm{E}(x) = -\frac{qN_A(x + W_{Dp})}{\varepsilon_S} \quad \text{for } W_{Dp} \leq x \leq 0 \tag{1.3}$$

$$\mathrm{E}(x) = -E_m + \frac{qN_D x}{\varepsilon_S} = -\frac{qN_D(W_{Dn} - x)}{\varepsilon_S} \quad \text{for } 0 \leq x \leq W_{Dn} \tag{1.4}$$

where E_m is the maximum electric field at $x = 0$:

$$|E_m| = \frac{qN_D x}{\varepsilon_S} = \frac{qN_D W_{Dn}}{\varepsilon_S} = \frac{qN_A W_{Dp}}{\varepsilon_S} \tag{1.5}$$

in which q is the elementary charge and ε_s is the dielectric constant of the semiconductor. Integrating Equations 1.4 and 1.5 produces the potential distribution in each region. The depletion widths are given by the following expressions:

$$W_{Dp} = \sqrt{\frac{2\varepsilon_s \psi_{bi}}{q} \frac{N_D}{N_A(N_A + N_D)}} \tag{1.6}$$

$$W_{Dn} = \sqrt{\frac{2\varepsilon_s \psi_{bi}}{q} \frac{N_A}{N_D(N_A + N_D)}} \tag{1.7}$$

$$W_{Dn} + W_{Dp} = \sqrt{\frac{2\varepsilon_s}{q}\left(\frac{N_A + N_D}{N_A N_D}\right)\psi_{bi}} \tag{1.8}$$

in which ψ_{bi} is the built-in potential (or maximum/total). The depletion width for the case of a one-sided abrupt junction—in which one side is highly/degenerately doped compared to the other one (e.g., p⁺-n or n⁺-p) and which is typical of standard silicon solar cells—reduces from Equation 1.8 to the following:

$$W_D = \sqrt{\frac{2\varepsilon_s \psi_{bi}}{qN}} \tag{1.9}$$

A graphical representation of the space charge/depletion regions, electric field distribution, potential distribution, and corresponding band structure for the one-dimensional case is shown in Figure 1.2.

It is of interest to determine the current-voltage characteristics of the p-n junction in both the absence and presence of light. We begin with the current densities of the former, J_p and J_n, on either side of the junction as (Sze and Ng 2007)

$$J_n = \mu_n n \nabla E_{Fn} \tag{1.10}$$

$$J_p = \mu_p p \nabla E_{Fp} \tag{1.11}$$

where μ is the charge carrier mobility and E_{Fn} and E_{Fp} are the quasi-Fermi levels in the n- and p-regions, respectively. If the quasi-Fermi levels are

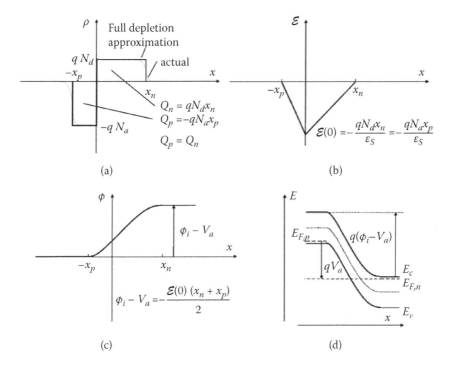

FIGURE 1.2

(a) Charge density distribution. (b) Electric field distribution. (c) Potential energy distribution. (d) Band structure for a typical p-n junction with an abrupt dopant profile. (With permission from *Principles of Semiconductor Devices*, 2007, by Bart Van Zeghbroeck, http://ecee.colorado.edu/~bart/book/Figure 4.3.1)

constant, which is the case in thermal equilibrium, then the n and p current densities are zero, i.e., there is no current flow. Upon application of a forward bias voltage (V) to the junction, the potential barrier at the junction is lowered, hence increasing the current density, whereas under a reverse bias the potential barrier is increased, hence decreasing the current density. The quasi-Fermi levels remain nearly constant within the depletion regions due to the high carrier concentrations there, and it can be shown that $qV = E_{Fn} + E_{Fp}$. By taking into account the continuity equations and the Einstein relation for the diffusion coefficient ($D = kT\mu/q$), the minority carrier current densities at the depletion width edges can be deduced as follows:

$$J_p = -qD_p \left. \frac{dp_n}{dx} \right|_{W_{Dn}} = \frac{qD_p p_{no}}{L_p} \left[\exp\left(\frac{qV}{kT}\right) - 1 \right] \tag{1.12}$$

which is the hole diffusion current on the n-side. L_p is the minority carrier diffusion length (holes on n-side), which is defined as $L_p \equiv \sqrt{D_p \tau_p}$, in which τ_p is the minority carrier lifetime (in this case holes in the n-region) associated with recombination in the semiconductor crystal. The minority carrier diffusion current on the p-side is similarly deduced as

$$J_n = qD_n \frac{dn_p}{dx}\bigg|_{W_{Dp}} = \frac{qD_n n_{po}}{L_n}\left[\exp\left(\frac{qV}{kT}\right) - 1\right] \quad (1.13)$$

Equations 1.12 and 1.13 can be combined to give the total current as follows:

$$J = J_p + J_n = \left[\frac{qD_p p_{no}}{L_p} + \frac{qD_n n_{po}}{L_n}\right]\left[\exp\left(\frac{qV}{kT}\right) - 1\right] = J_o\left[\exp\left(\frac{qV}{kT}\right) - 1\right] \quad (1.14)$$

This is the well-known Shockley or ideal diode equation, which assumes only diffusion current. If we also assume there is recombination in the depletion region, then the equation becomes

$$J = J_o\left[\exp\left(\frac{qV}{\eta kT}\right) - 1\right] \quad (1.15)$$

The ideality factor, η, is equal to 1 when the junction is diffusion current dominated (ideal) and equal to 2 when the junction is dominated by recombination current (poor-quality junction material), and can take on values in between when neither process dominates. The I-V characteristics of the p-n junction are well known and show a rectifying behavior (Figure 1.3).

When light is applied to the junction, one can model the solar cell as an ideal diode in parallel with a current source, J_L, that accounts for the photogenerated electron-hole pairs (excitons) in the semiconductor, and a resistive load. This can be combined with the standard diode equation to yield a general expression for a semiconducting p-n diode-based photovoltaic cell:

$$J = J_o \left[\exp\left(\frac{qV}{\eta kT} \right) - 1 \right] - J_L \tag{1.16}$$

where the photogenerated current density in a semiconductor of bandgap E_g is related to the applied light spectrum ϕ_{ph} as

$$J_L = q \int\limits_{h\nu=E_g}^{\infty} \frac{d\phi_{ph}}{dh\nu} d(h\nu) \tag{1.17}$$

Based on this expression, one can see that application of light to a PV device leads to a shift of the diode current-voltage (I-V) or current density-voltage (J-V) equation into the fourth quadrant (Figure 1.3), and thus also leads to power generation that can be applied to a load (the area under the curve). The J-V plot contains several critical parameters that directly relate to the efficiency of the solar cell. The short-circuit current density (or current), J_{sc}, provides the maximum current density that is obtainable when the two leads of the device are connected, i.e., under no load. The maximum J_{sc} available from the AM1.5 solar spectrum to a single-bandgap solar cell (which is directly related to the integrated solar photon flux)

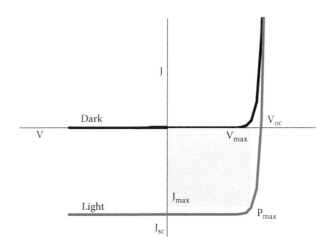

FIGURE 1.3
Schematic of the current density-voltage plot for a p-n junction diode in the dark and under illumination. The critical device performance parameters for the photovoltaic effect are shown.

is ~62 mA/cm^2, and for a Si solar cell ($E_g = 1.1$ eV), the maximum potential current density is ~40 mA/cm^2. In a well-functioning silicon solar cell this value is ~30 mA/cm^2. On the other hand, the open-circuit voltage, V_{oc}, provides the maximum voltage one would measure across the two leads of the device should they be disconnected. The maximum V_{oc} that may be obtained by the device is equal to the difference in the quasi-Fermi levels (the built-in potential), and in a well-performing single-bandgap device this would be in the range of 0.6 to 0.8 V. The fact that these values are lower than the maximum possible for typical solar cell materials (Si, CdTe, CuInGaSe$_2$) can be seen by the expression for V_{oc} derived from the ideal diode law:

$$V_{OC} = \frac{kT}{q} \ln\left(\frac{I_L}{I_o} + 1\right) \approx \frac{kT}{q} \ln\left(\frac{I_L}{I_o}\right) \tag{1.18}$$

The logarithmic dependence of V_{oc} on the saturation current thus leads to the less than expected voltage since real materials have higher than ideal saturation currents due to recombination and other device loss mechanisms. It should be noted that indeed the ideal saturation current is related to the minority carrier lifetimes in an inverse square root fashion, and is directly proportional to $\exp(-E_g/kT)$ (see Sze and Ng 2007).

Another important feature of the J-V curve is the so-called fill factor (FF). This quantifies the "squareness" of the J-V curve under solar illumination, and can be defined as the ratio of the area of the curve under the maximum power point of the cell (see Figure 1.3), i.e., $P_{max} = I_{max}V_{max}$, to the area associated with open and closed circuit ($P = I_{sc}V_{oc}$). These parameters directly relate to the power conversion efficiency of the device, η, under direct sunlight as

$$\eta = \frac{I_m V_m}{P_o} = \frac{AJ_{sc}V_{oc}FF}{P_o} \tag{1.19}$$

in which P_o is the applied power density to the device (i.e., 100 mW/cm^2 for AM1.5 global sunlight).

In addition to measuring the J-V characteristics of a photovoltaic device, another important characterization that must be performed is the quantum efficiency of the device as a function of wavelength. The quantum efficiency can be defined as the probability of collecting an electron-hole pair (exciton) that can be used by the device for power generation. The external quantum efficiency (EQE) is based on the J_{sc} measured for the PV device as a function of wavelength (the probability of collecting incident photons, whereas the internal quantum efficiency (IQE) takes into account the reflectance loss of the device and provides a measure of the QE for absorbed photons (A):

$$IQE = EQE/A = EQE/(1 - R - T) \cong EQE/(1 - R) \qquad (1.20)$$

QE data can provide valuable insights into the loss mechanisms of the solar cell, including potential for front surface recombination (at short wavelengths), effectiveness in harnessing longer-wavelength photons (e.g., in which the absorption coefficient is typically low for indirect-bandgap inorganic semiconductors), or regions of the spectrum in which the solar device cannot absorb light (e.g., for some polymer-based PV devices). Figure 1.4 shows a typical EQE plot for a silicon solar cell with a power conversion efficiency of ~13%.

Another important feature of semiconductor-based PV is the fact that the efficiency of the solar energy converter increases with increasing concentration of sunlight falling upon it. While the detailed physics of this are beyond the scope of this chapter, it can be shown directly by consideration of Equations 1.18 and 1.19. The photogenerated current is expected to increase monotonically with concentration, whereas V_{oc} increases logarithmically. Therefore, it is clear that efficiency will increase with concentration of sunlight. This is important to keep in mind for

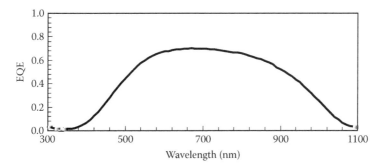

FIGURE 1.4

Example of the external quantum efficiency (EQE) of a Si solar cell with a power conversion efficiency of 13%. Losses are observed at short wavelengths due to front surface recombination and long wavelengths due to the low absorption coefficient of Si in the near IR.

applications such as terrestrial concentrated photovoltaics and related world-record solar cell devices based on multiple bandgaps (see below), since these record efficiencies are always at relatively high concentrations of sunlight (>200 suns).

It is important to note that while the majority of PV technologies are based on the p-n junction, i.e., the p-n homojunction found in silicon photovoltaics that make up ~85% of the PV market in 2008, there are other mechanisms for solar energy conversion. Thin-film technologies based on CdTe and $Cu(In,Ga)Se_2$ (CIGS), which will be discussed in more detail below, are based on heterojunction concepts, in which the charge-separating junction is formed by depositing a second (higher-bandgap) material with band offset that also leads to the formation of a charge-separating field gradient. CdS is the standard material used to form this junction in CdTe and CIGS, though efforts to develop new materials are under investigation and, in some cases, are in production (for CIGS) (Bhattacharya and Ramanathan 2004). There are three types of heterojunctions that are possible (Figure 1.5) (Pallab 1997). Type I heterojunctions are fashioned such that the higher-bandgap conduction and valence bands straddle those of the lower-bandgap material. Type II heterojunctions are formed in a manner such that the bandgaps are staggered; i.e., there is a downward energy offset for both the conduction and valence bands. Finally, type III heterojunctions contain a so-called broken gap; i.e., the valence band of one material overlaps with the conductions band of the other material. The most widely used heterojunction is type II, which describes the band alignment for CdS/CdTe (see Figure 1.5) (Tomita et al. 1993). Other examples of heterojunction-based solar cells include quantum well solar cells, which are described in detail in Chapter 5, and hydrogenated amorphous silicon (a-Si:H)/crystalline silicon solar cells, which have been commercialized under the title of HIT (heterojunction with intrinsic thin layer) cells by Sanyo (Taguchi et al. 2000). Schottky junctions between a semiconductor and a non-ohmic metal can also provide a photovoltaic effect (Xing et al. 2008).

The focus of this book is on the use of nanostructures and nanotechnology at the solar cell level. However, a commercial PV system involves a collection of series-connected solar cells that form a module, with multiple modules connected to balance of system (BOS) components that in most applications convert direct current (DC) electricity to alternating current (AC) electricity for practical use. There are many applications of solar technologies, each with their own requirements. It is therefore instructive

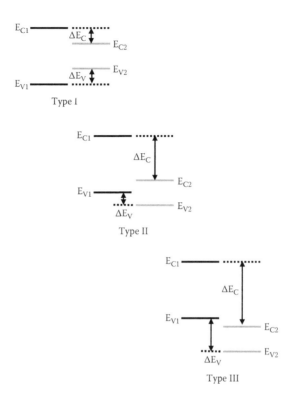

FIGURE 1.5
Schematic of type I, II, and III heterojunction band structure alignment showing band offset energies.

to review these various markets for PV in order to better understand how various performance features may be beneficial.

1.2 SOLAR CELL MARKETS

The overall market for photovoltaics has grown tremendously since approximately 2000, with compound annual growth rates (CAGRs) of 30 to 40% (Figure 1.6), and the industry generated ~$37B in worldwide revenues in 2008 (see Marketbuzz™ 2009: Solarbuzz™ Annual World Solar Photovoltaic Industry Report). The impact of the economic downturn in late 2008 and early 2009 on the PV market remains to be seen. This growth has been primarily driven by incentive programs implemented by several countries, such as Germany, Japan, and Spain, and to a lesser extent programs introduced by states in the United States, such as California and New Jersey. Recent

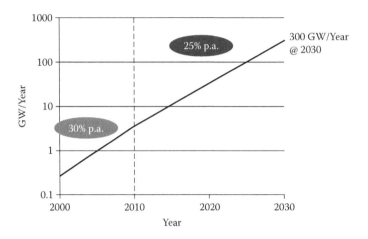

FIGURE 1.6

2006 prediction for future growth of the PV industry assuming a growth rate of 30% per year and a slower rate of 25% per year after 2010. The worldwide recession of 2008–2009 may have an impact on these growth predictions in the near term. (With permission from Hoffman, W., "PV Solar Electricity Industry: Market Growth and Perspective," 2006. 3285–3311. Copyright of Elsevier.)

federal legislation in the United States is expected to ignite a strong market upswing for photovoltaics. Growth in China and Korea is also occurring, and the markets in Australia, Brazil, India, Italy, France, Greece, and other countries may grow in the coming years. The market is by no means homogeneous. There are numerous segments that are of interest for PV, with each requiring varying performance features, cost structures, value chains, and channels to market. The market-dependent performance features of interest for a PV system include the following parameters: (1) system cost, (2) installed cost per peak power ($/W_p$), (3) total system power output (W_p), (4) annual energy yield (kWh/kW_p), (5) performance ratio ($W_{achieved}/W_{rated}$), (6) module lifetime (years), (7) rated module power (W), (8) module efficiency, (9) inverter efficiency, (10) total system efficiency, (11) inverter lifetime (years), (12) module/system temperature coefficient ($°C^{-1}$), (13) investment payback time (years), and ultimately (14) levelized cost of electricity (LCOE) ($/kWh). Prior to discussing the technical details of typical PV systems, the major solar energy market segments are first summarized below.

1.2.1 Residential and Commercial Building Rooftop

The rooftop market represents a major fraction of today's solar market, approximately 70 to 80% of the market (Hoffman 2006). It is estimated that the total area of available residential and commercial rooftops in the United States alone is enough to generate approximately 710,000 MW of power.* In Germany, where ~46% of the total world PV market is found, residential and commercial rooftops make up nearly 89% of the market. A small fraction of the world residential component is off-grid at present, though there is great potential in developing countries; however, the vast majority is the grid-connected market. Under some national and regional incentive programs, e.g., in the United States, it is possible for individuals and companies to sell unused power back to the grid at above-market rates, whereas in Germany one can sell all PV power to the grid (up to a nationwide total MW cap), making the investment payback time relatively short and allowing for a good business proposition.

There are several system-level technical features that are important for this market. Of particular importance is the method of mounting the modules to the rooftop (Figure 1.7). Methods that do not require drilling into the roof are desirable, as are suitable thermal management approaches. In the case of large commercial rooftops, which are typically flat, nonpenetrating systems are of particular interest. Since tracking is typically not possible, proper mounting of the module array to achieve as high power output as possible over the course of the day is critical. Technologies that can provide omnidirectional solar energy absorption are therefore of great interest for such applications. Since rooftops are subject to varying environmental conditions such as wind, rain, snow, and hail, the module packaging and electrical interconnects must be robust. Mounting and packaging approaches that can protect modules against standing water and other environmental effects on commercial rooftops are desired. Historically, PV systems in this market segment are warranted for a period of 20 to 30 years. Hence, this places challenging requirements on the module and BOS lifetimes (or mean time to failure [MTTF]).

1.2.2 Building-Integrated Photovoltaics

Building-integrated photovoltaics (BIPV) applications are those in which the PV modules are intimately integrated with the building structure. Examples include building facades and windows that contain PV elements

---

* http://www.ef.org/documents/PV_pressrelease.pdf

FIGURE 1.7
Picture of a roof-mounted PV system. (Otani, K. et al. *Field Experience with Large-Scale Implementation of Domestic PV Systems and with Large PV Systems on Buildings in Japan.* 2004. 449–459. John Wiley & Sons. With permission.)

for power generation (Figure 1.8). The BIPV market is relatively small at present (<5%), though the potential market size is quite large given the general growth in large building construction worldwide.

This market contains several specific requirements. Often it is desirable to use solar cell modules that are semitransparent. This may include the use of thin solar cells that have a relatively small total absorption (and hence low efficiency), or it may contain smaller solar cells mounted within a package with enough space between cells to allow some sunlight into the building. Often the modules must be mechanically robust, since they may be subject to mechanical stresses caused by wind, hail, natural expansion/contraction of the building structure, and thermal expansion.

1.2.3 Ground-Mounted Systems

Another important market segment is ground-mounted centralized PV systems. These are generally based on large area arrays of nontracking rails/fixtures with mounted PV modules rated for 50 to 250 W that can produce powers in the kilowatt to megawatt range by laying out the modules over many acres of land (Figure 1.9). Some of the largest ground-mounted systems may be found in the south of Spain, as well as in Germany. For

FIGURE 1.8
Pictures of building-integrated PV systems in Hannover and Bamberg, Germany. (With permission from Hagemann, I., "Architectural Considerations for Building-Integrated Photovoltaics," *Prog. Photovolt. Res. Appl.* 4 [1996]: 247–258. Copyright of John Wiley & Sons.)

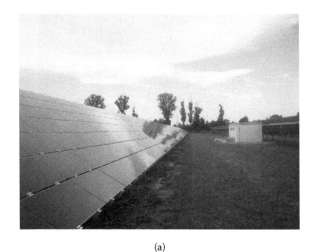

(a)

(b)

FIGURE 1.9

Images of ground-mounted PV systems in (a) Halbergmoss, Germany (Courtesy of O. Stern and O. Mayer) and (b) Aichi Airport, Japan (From Ichiro Araki et al. 2009. "Bifacial PV system in Aichi Airport-site demonstration research plant for new energy power generation," *Solar Energy Materials & Solar Cells*. Elsevier. 911–916.) (See color insert following page 206.)

example, one of the largest planned solar power plants in the world is currently being developed in Brandenburg, Germany, with an 80-MW total capacity produced on 460 ha of land (Wagner 2008). Tracking (usually single-axis) is increasingly being used for this market segment.

Practical issues that must be considered for ground-mounted systems include the impact of animals, such as droppings, and damage to insulation, as well as maintaining the field to minimize shading by vegetation. The environmental impact of such systems is currently the subject of debate.

1.2.4 Concentrator PV

Concentrator photovoltaics (CPV) is another market segment that at present is very small (~1% of the total PV market) but has strong future potential. While there are some efforts to create low-profile roof-mounted low-level concentrator systems (Bowden et al. 1994), the major application of CPV is in centralized PV power stations, similar to the ground-mounted systems discussed above (McConnell et al. 2004). CPV systems typically differ from ground-mounted PV systems in that they provide for a medium to high level of concentration using reflective or refractive optics in order to achieve high efficiencies, and also typically provide dual-axis tracking capability. Since concentration can only be achieved over relatively small areas using direct sunlight, it thus becomes feasible to use relatively expensive, high-efficiency solar cells such as multijunction solar cells (see below), since the total cost of the cells will be low (McConnell and Symko-Davies 2006). The majority of the system cost is thus transferred to the concentrating optics and tracking system.

A typical design of a CPV dish is shown in Figure 1.10, in which it is clear that the large dish is used to collect the sunlight, concentrate it to a level of between 100 and 1,500 suns, and focus the light on the solar cells mounted on the end of a receiver plate at the focal spot of the mirror. While solar cells utilized in such applications are often based on high-efficiency multijunctions, research in using relatively lower-cost high-efficiency silicon solar cells is also being pursued (Slade et al. 2005). It is typically necessary to use a suitable thermal management strategy to minimize heating of the solar cell in the receiver so that the efficiency of the solar cell, as well as performance of other components, is not degraded. Recently it has been shown that by cell miniaturization and proper mounting of the solar cell to a heat sink, one can improve passive heat rejection as well as increase the concentration at which the

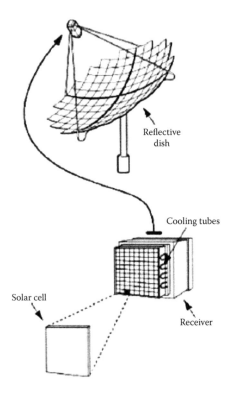

FIGURE 1.10
Schematic of a concentrator PV dish. (From Luque, A. et al. *Photovoltaic Concentration at the Onset of Its Commercial Deployment.* 2006. Prog. *Photovolt. Res. Appl.* 14:413–28. With permission.)

efficiency peaks, and it is possible to maintain a near-ambient temperature at the heat sink–substrate interface (at most ~10 K above) without active cooling at concentrations up to 5,000 suns (Korech et al. 2007; Gordon et al. 2004). Thermal management strategies employed in the electronics industry can also be applied to CPV applications up to 2,000 suns, in particular by using liquid-metal interfaces in a passive scheme (van Kessel et al. 2008). Key additional factors in design of solar cells for CPV include temperature coefficient and metallization to minimize resistive losses in the finger contacts and bus bars.

1.2.5 Space

Solar cells have been used for power generation on space-based systems since the 1950s. The space market is small, though very important for multiple applications, including scientific instruments, commercial communications satellites, etc. Key requirements are that the cell technologies achieve high efficiency under AM0 solar irradiation, have a low temperature coefficient, and have good radiation hardness. This latter requirement is due to the fact that the PV systems may be subject to harsh radiation environments, including high-energy electrons, protons, alpha particles, gamma rays, and heavy ions. The system design and module packaging must also be robust to minimize the effect of micro-meteorite impacts, and it must be amenable to facile deployment once in space. The standard solar cell used is the multijunction solar cell (Fatemi et al. 2000; Karam et al. 1998). Such cells have also been used in low-concentration space modules (Stavrides et al. 2002). There is, however, increasing interest in the use of thin-film solar cells for space applications, since the thin absorber layer allows for enhanced radiation hardness, and the high specific power (W/kg) of such cells on thin flexible foils provides additional advantages to space applications (Liu et al. 2005; Dhere et al. 2002).

1.2.6 Consumer Electronics

The market for powering consumer products by photovoltaics is relatively small, but it is expected that as the demand for portable electronics grows, the consumer market will also grow. One of the first applications of PV in consumer products was the use of amorphous silicon thin-film solar cells (see below) for calculators. PV-powered calculators were first introduced to the market in 1980 by Sanyo, with efficiencies in the range of 4 to 6% (Arya 2004). Other applications include mobile phones, portable MP3 players, etc. Indeed, solar-powered cell phone chargers may be found on the market. Due to the required form factors for these applications, a particularly attractive technology is the use of flexible PV devices such as thin-film amorphous silicon, CIGS, and organic photovoltaics.

Having discussed the major markets for photovoltaics, it is instructive to provide a more detailed discussion of PV systems. This will then be followed by detailed descriptions of conventional cell technologies and emerging technologies that have the potential to provide enhanced efficiency at low cost.

1.3 PV SYSTEMS

By its very nature, electrical power generation using photovoltaics leads to a direct current (DC) form of electricity. Since the electrical grid is

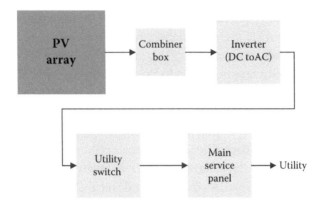

FIGURE 1.11

High-level system diagram of a grid-connected PV system. (Adapted from California Energy Commission, "A Guide to Photovoltaic (PV) System Design and Installation," http://www.energy.ca.gov/reports/2001-09-04_500-01-020.PDF)

based on alternating current (AC) electricity, as are most major electrical components/appliances, it is thus necessary to convert PV-generated DC power to AC. As has been alluded to above, there are additional components (BOS) in a PV system that must be considered to provide high efficiency and AC power output.

Figure 1.11 shows a schematic of the main components in a typical grid-connected PV system. The design changes if the system is backed by a battery system, or if the system is off the grid. Nevertheless, a typical PV system will essentially consist of a combiner box that integrates the power generated from each module, followed by an inverter that converts DC power to AC. For a grid-connected system a utility switch may be included, which is capable of placing either all or any unused power generated by the system onto the grid. Wiring for these components, as well as bypass diodes, is also typical. Other components that may be part of a system include ground fault protectors, metering, as well as batteries for systems that have integrated storage.

While some systems can have single- or dual-axis tracking with significant benefits in energy yield output of about 20 to 30% (Moore et al. 2005), most are nontracking since this adds cost and may lead to additional long-term reliability challenges (Maish 1999). As tracking costs are reduced and reliability enhanced, more new installations are using such systems. Nevertheless, due to multiple factors, including the fact that most systems are nontracking, shading effects, temperature coefficient effects, dirt/debris buildup on the modules, etc., the actual power that is produced is typically less than what the system is rated for. A useful metric of the usefulness of the PV system is the performance ratio (PR), which is the ratio of the total power produced to that for which the system is rated (Marion et al. 2005). For typical systems, this value can range from 50 to 90%, and for small-scale systems this can be less than 70% if various environmental factors (e.g., shading) are impacting the system. However, for large-scale systems, the PR is typically greater than 75% due to improved inverter efficiencies and improved power losses. The total efficiency of the system, which accounts for the efficiency of not only the modules but also the inverters and other components in the system (resistance losses, diode losses, etc.), is also a useful metric. It has been shown that for modules based on polycrystalline Si, the total efficiency ranges from 4 to 12% over the course of the year, even though the modules are rated at a higher efficiency (So et al. 2006). PV array efficiency is typically reduced at low irradiance due to nonlinear current-voltage output of the modules, though the main factor impacting efficiency is the surface temperature of the modules. The inverter efficiency is also impacted by low irradiance, though usually inverters perform at nearly the specified performance parameters.

All these components must be as low in cost as possible, of high reliability, and in the case of the inverters, highly efficient. Typical inverters attain efficiencies greater than 90%, yet they often suffer from reliability problems. Inverter lifetimes are typically on the order of ten years, though the panels may be warrantied at the 20- to 25-year level. Having briefly discussed the system-level concerns of photovoltaics, we next discuss the major solar cell technologies that are currently utilized, followed by advanced concepts that are the focus of this book. For more details on PV systems engineering, see the works by Messenger and Ventre (2003) and Buresch (1983).

1.4 SOLAR CELL TECHNOLOGIES

Solar cell technologies have been classified into three generations: silicon technologies (generation I), thin-film technologies (generation II), and generation III technologies (Green 2003). These are best visualized by plotting module power conversion efficiency as a function of module cost ($/m^2) (Figure 1.12). Lines of constant $/W are observed in the figure. The regions shown delineating the three generations are intended to

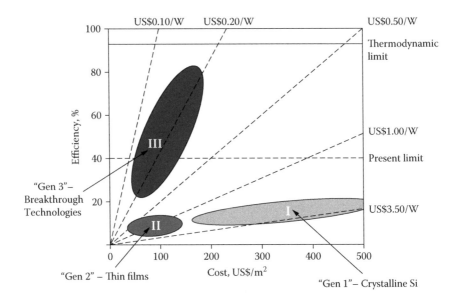

FIGURE 1.12
Solar module landscape as defined by M. Green showing module efficiency as a function of cost ($/m^2). Lines of constant $/W are evident. (Adapted from Green, M. *Third Generation Photvoltaics: Advanced Solar Energy Conversion.* 2003. Springer-Verlag.)

provide a guideline, though an exact boundary for each generation is difficult to define with accuracy. Silicon helped to spark the PV industry and has dominated the market for the last ~50 years. Si technologies provide a medium efficiency at relatively high costs. Thin-film technologies reduce costs by applying a relatively thin absorber layer to a low-cost substrate (glass, metal foil, or polymer), though the module efficiency is often lower than generation I technologies for reasons discussed below. Generation II technologies have been known for greater than three decades; however, only in the 2000s has widespread commercial adoption of thin films by the market occurred. Commercial thin-film PV (TFPV) modules with costs less than $1.00/W are now available. Generation III technologies hold the promise of providing low costs (less than $1/W) yet high efficiencies (greater than ~20%). The exact technologies that will provide these features are yet to be established. The major generation III concepts that are currently being explored in academic and industrial laboratories will be discussed below, and there is a strong possibility that nanostructured devices may play a major role in the emergence of such concepts in the medium to long term. A more detailed description of the three generations of photovoltaics is provided.

1.4.1 Silicon (Generation I)

1.4.1.1 Single-Crystalline Si

Silicon is the most widely used PV technology. The first viable semiconductor-based solar cell was shown using Si single crystals, and the first commercial applications of PV used Si. Until about 2004, Si made up ~95% of the PV market, though in recent years this has dropped to ~85% due to the rapid growth of thin-film technologies. Interestingly, from a physics perspective Si is not the ideal material for PV, since its indirect bandgap means that the absorption coefficient is relatively low. A high Si thickness of ~125 μm is required to absorb >90% of sunlight with energy above the bandgap, compared to the direct-bandgap semiconductor GaAs that only requires a thickness of ~0.9 μm. Si is favored for PV because it is relatively inexpensive compared to other materials; it is the second most abundant material in the earth's crust (Skinner 1979). Si PV technology also benefits from the tremendous Si technological base developed in the last few decades during the emergence and establishment of the electronics industry. This includes methods of growing high-quality single crystals using the czochralsi (CZ), float zone (FZ), and related methods, as well as means of controlling the purity of Si (achieved with FZ, the Siemens process, and other methods) (Woditsch and Koch 2002). A voluminous and detailed scientific understanding of the materials science, physics, and chemistry of Si exists, and it has been argued that Si is the most studied material in the history of humanity. Advances in materials science and engineering have provided techniques to remove or passivate bulk crystal defects in Si, and the surface of Si can be passivated to yield surface recombination velocities near zero.

Most Si PV technologies are based on crystalline p-n junction devices (Green 2003b). The absorber is typically p-type due to the lower cost of such wafers compared to n-type wafers, though n-type absorbers are preferable from a performance perspective due to the higher charge carrier mobilities and minority carrier lifetimes. Until recently, most Si wafers were obtained by the PV industry from the electronic industry. However, the rise of the PV industry in the 2000s led to an shortage of Si for PV, and hence a rise in Si materials costs. This problem has only recently started to be alleviated with the fabrication of Si materials factories that have added solar-grade silicon capacity. It is noted that the purity requirement for solar-grade silicon is not as stringent as for electronic-grade wafers (Woditsch and Koch 2002).

Doping of the emitter layer is achieved by vapor-phase diffusion (e.g., $POCl_3$), or to a lesser extent by applying a phosphorous- or boron-containing solution (e.g., phosphoric acid) or glass (e.g., borosilicate glass) to the top surface, followed by subsequent high-temperature diffusion drive-in and cleaning steps. Texturing of the cell may be implemented by a wet etch (e.g., dilute NaOH aqueous solution) to impart antireflective properties to the cell, and a thin silicon nitride (or oxide) layer may be used to passivate the top surface. Such a layer can also act as a quarter-wavelength antireflective layer. A blanket back contact and front grid contact may be applied by screen printing of an aluminum or silver-containing (front) paste, followed by an anneal to create a near-ohmic contact to the Si crystal.

The typical efficiency of commercially available Si solar modules is ~14 to 17%. Companies such as Sunpower provide higher module efficiencies (>19%) by using higher-quality silicon in combination with all back-contact schemes (Van Kerschaver and Beaucarne 2006). Among the highest commercially available efficiencies are provided by the amorphous silicon (a-Si)–based HIT cell, with cell-level efficiencies reported to be on the order of 22.3% (Tsunomura et al. 2008) and module efficiencies of ~18 to 20%. The major loss mechanisms in Si devices that limit efficiency are bulk recombination, depletion region recombination, front and back surface/contact recombination, contact resistance losses, front grid shading, series resistance loss, and reflection losses. The world record Si solar cell was demonstrated by Green and coworkers with an efficiency of nearly 25% by using high-quality float-zone Si, all back-contacts, and an optimized antireflection scheme employing inverted pyramidal structures (Zhao et al. 1998). Most Si cell technologies have Si wafer thicknesses of ~200 to 500 μm, though recently available modules use thicknesses of 150 to 200 μm. At these thicknesses special handling is required during manufacture to avoid fracture.

1.4.1.2 Polycrystalline Si

The manufacturing cost of single-crystalline Si solar cells can be relatively high due to the requisite single-crystal growth, dicing/wafering, and other processes. Use of silicon in a polycrystalline form (poly-Si) leads to lower costs since directional solidification, a much faster process, is utilized. The efficiency is usually lower owing to the presence of a high density of recombination centers at the grain boundaries, as well as more defects (dislocations, stacking faults, etc.) within the grains. Dopant segregation is also a concern. It is possible to fashion poly-Si into continuously formed ribbons, as is done by companies such as Evergreen Solar, Inc. (Janoch et al. 1997). Poly-Si can also be made in a so-called continuous molded wafer process, as was developed by Astropower, Inc. (Bai et al. 1997). The record power conversion efficiency for a poly-Si solar cell is 20% (Shultz et al. 2004).

1.4.2 Thin Films

Thin-film, or so-called generation II, photovoltaic technologies offer the promise of lower-cost modules by allowing for direct deposition of thin absorber layers on low-cost substrates such as glass, metal foils, and polymers. The cost ($/m^2) can be reduced by a factor of 2 to 5 times compared to bulk Si PV modules. While champion cells with efficiencies similar to those of conventional Si technologies have been demonstrated (16 to 20%), commercial thin-film modules yield efficiencies less than ~12%. A brief review of the major thin-film technologies is given below.

1.4.2.1 Crystalline Si

There has been a strong interest in developing Si thin-film solar cells in the last two to three decades. The primary reason is to combine the relatively low cost and abundance of Si raw materials with the process cost reductions provided by thin films. The main technical challenges with crystalline Si thin films are associated with (1) the significant loss of light absorption due to the low absorption coefficient of crystalline Si, and (2) loss of efficiency due to recombination at grain boundaries and crystal defects in the bulk of the grains.

The first challenge has been addressed by light trapping schemes that increase the path length of light through the solar cell. Light trapping can be implemented by either texturing of the glass substrate on which the Si film is deposited, or by depositing an antireflective film on top of the cell. Aberle and coworkers (Widenborg and Aberle 2007) demonstrated a new glass texturing process applied to Si thin-film devices with a potential for large-scale integration. They showed efficiencies on the order of 7%, with the texturing providing an enhancement in J_{sc} of 8 to 19% (relative) and in some cases also a slight V_{oc} enhancement. The optical absorption was increased over most of the spectrum compared to planar samples, save for in the near-UV portion, though the EQE only showed enhancements in the near infrared (IR) with total enhancements of 5 to 17%. This was explained by a minority carrier diffusion length in the absorber region that was significantly smaller than the width of that region, leading to a small impact from light absorption. It should be noted that backside reflection was also found to be important in improving the device performance, with the most promising backside reflector being a white broadband reflector material. The use of highly textured transparent conducting oxides (TCOs) such as aluminum-doped zinc oxide (Zno:Al) and fluorine-doped tin oxide (SnO_2: F) have been shown to increase J_{sc} by increasing the absorption and EQE in the near-IR portion of the solar spectrum (Müller et al. 2004).

The problem of recombination loss in polycrystalline Si thin-film solar cells has been addressed by efforts to increase the grain size of the thin films. One means to achieve this is by depositing an amorphous or nano-scale grain-size thin film, and subsequently annealing the film to achieve grain growth (Aberle 2006a). Grain sizes on the order of 1 to 10 μm can be achieved by aluminum-induced crystallization at temperatures close to 600°C (Nast et al. 1998). Grain sizes in the range of hundreds of micrometers to a few millimeters have been achieved by liquid phase epitaxy (LPE) on single-crystal Si and polycrystalline Si and alumina substrates (Wagner et al. 1993; Wang et al. 1996). Another means of minimizing recombination losses is to use rapid thermal annealing or plasma hydrogenation to reduce point defects densities, passivate surface states, and passivate deep-level states associated with dislocations and stacking faults in the grains.

Commercialization of crystalline Si thin-film solar cells has not been widespread. One of the first to commercialize polycrystalline Si solar cells was introduced by CSG Solar, which emanated from the University of New South Wales (Green et al. 2004; Basore 2006a,b; Aberle 2006b). CSG refers to crystalline silicon on glass. Texturing is imparted by depositing the amorphous Si film on a layer of half-micrometer Silicon oxide sphere. Metallization and interconnects for monolithic integration on glass substrates are fashioned using unique ink-jet printing and laser scribing processes.

Another type of crystalline silicon thin-film technology is based on so-called nanocrystalline silicon (nc-Si) (Rath 2003; Schropp et al. 2009; Zhao et al. 2005). These are fabricated by hot-wire CVD or plasma-enhanced chemical vapor deposition (PECVD) at relatively high plasma power, substrate temperature, and pressures in order to nucleate fine-grained (sub-

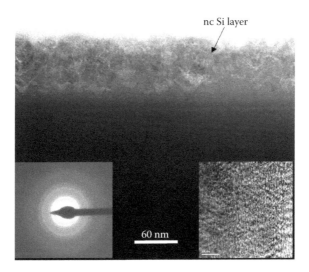

FIGURE 1.13

Transmission electron microscope image of a nanocrystalline silicon (nc-Si) thin film deposited on a glass substrate by PECVD. The left inset shows the corresponding electron diffraction pattern showing the expected polycrystalline ring pattern, and the right inset shows a high-resolution transmission electron microscopy (TEM) micrograph. (TEM data by M. Larsen.)

100 nm) silicon thin-film materials (Figure 1.13). Due to the high density of grain boundaries, the minority carrier lifetimes are quite low, and a p-i-n (electric field drift-based) device structure is usually employed to allow for efficient collection of photogenerated carriers. The efficiency of nc-Si solar cells is relatively low, i.e., <6%, and they have not found widespread commercial use as stand-alone devices. As with most Si thin-film deposition technologies, the deposition rates are relatively low (≤2 nm/s) but can be increased to between 10 and 20 nm/s by various means (e.g., ion beams, etc.).

1.4.2.2 Amorphous Si and Tandems

Unlike nc-Si solar cells, hydrogenated amorphous silicon (a-Si:H)-based thin-film solar cells have obtained relatively large-scale market adoption and have a long history of scientific and technological development. Amorphous Si is typically deposited by PECVD at relatively low substrate temperatures (<200°C) with low power and pressures. As a result, Si films are formed in an amorphous phase that has a higher (~1.7 eV vs. 1.1 eV) and direct bandgap. Charge carrier mobilities and carrier lifetimes are significantly reduced compared to crystalline Si, with μ on the order of 1 to 20 cm²/V-s (compared to 50 to 1300 cm²/V-s for typical c-Si, depending on doping type and concentration) and τ less than 1 µs (compared to 1 µs to 1 ms for typical c-Si).

The relatively poor charge transport properties of a-Si necessitates the use of an n-i-p solar cell structure to allow for electric-field-assisted transport of charge carriers to the charge-separating junctions (Carlson and Wronski 1976). The typical efficiency of commercial a-Si modules is in the range of 5 to 6%, making them useful for market segments where low cost and low efficiency are acceptable. Large-area flexible a-Si-based solar cells are also available for military and related applications.

While solar cells based only on a-Si:H provide a relatively low efficiency, a-Si has been incorporated into higher-efficiency devices using the concept of a heterojunction device (Xu et al. 2006), or by stacking multiple-bandgap materials to improve the absorption characteristics of the cell (a tandem cell). The former has already been discussed, and is available commercially under the trademark HIT cell device (Taguchi et al. 2000). The latter devices are also commercially available, with United Solar Ovonics Corporation among the leaders in this technology. This technology employs a-Si as the top cell of a three-bandgap structure in which the bottom two cells are composed of a-SiGe with varying Ge content to reduce the bandgap. Other variants of this technology include the use of a-SiC as the top cell material and the use of nc-Si as one of the tandem cells. Efficiencies as high as 15% have been reported for laboratory cells, with stabilized efficiencies on the order of 13% (Yan et al. 2006). This technology is also readily applicable to flexible modules (Deng 2005), though large-scale production of such cells has just begun. Tandem structures may also combine poly/nano-crystalline Si films with a-Si-based cell layers. Turnkey tandem a-Si/nc-Si production systems are available on the market (from Applied Materials, Oerlikon, and others) (Meier et al. 2007; von Roedern and Ullal 2008), and several companies are producing modules based on these, including Signet, Sharp, Mitsubishi Heavy Industries, Topray Solar, EPV Solar, Sontar (Q-Cells), and others.

1.4.2.3 CdTe

CdTe thin-film modules have achieved the highest market penetration owing to the relatively high module efficiencies (about 10%) and low cost

($<\$1/W_p$). Indeed, CdTe is the first thin-film technology to achieve large-scale manufacturing and installation, with First Solar currently leading the market at production levels of >75 MW in 2006 (Meyers 2006), and currently at >500 MW, and companies such as Primestar Solar (GE majority stake investor), Abound Solar, Calyxo, and Arendi also developing CdTe modules. The standard CdTe device structure is based on a superstrate architecture, in which a TCO is first deposited on a low-cost soda lime glass by sputtering, followed by deposition of a thin CdS window layer (~100 to 300 nm), a thick CdTe absorber layer (1 to 8 μm), and a back metal contact. Sun illumination takes place from the backside of the original glass deposition substrate. Scribing steps are used to allow for a monolithically integrated module, and the module is laminated on the backside with a low-cost glass sheet. The CdTe layer can be deposited by low-cost deposition methods such as closed-space sublimation and sputtering.

The key technological challenges for CdTe PV technology are associated with improving device performance and environmental and health concerns associated with Cd in manufacturing and at the end of module life. The record efficiency for CdTe was obtained by the U.S. National Renewable Energy Laboratory (NREL) at ~16.5% (Wu et al. 2001). Module-level efficiencies are lower for several reasons, with the main contribution being from short-wavelength (<500 nm) absorption in the CdS window layer. It has been shown that reducing the CdS thickness improves the short-circuit current density, though typically at a loss of V_{oc} and FF (Meyers 2006). Losses associated with grain boundary space charge and related effects also strongly contribute to reduction in efficiency. Interestingly, the efficiency of polycrystalline CdTe solar cells is equal to or higher than that of single-crystal devices, which have been argued to be related to doping type inversion in the region near the heterojunction (Bosio et al. 2005). The purity of grains compared to bulk materials may also play a role in this observation. Concerns about the hazardous nature of Cd are addressed at the end of module life by robust buy-back and recycling programs, and it is well known that Cd within the CdTe crystal phase (and embedded in glass sheets) does not pose the health hazards associated with pure Cd (Moskowitz et al. 1994). Appropriate environmental health and safety (EHS) measures must be in place, however, during manufacture of CdTe PV modules. The CdTe layer can be deposited by low-cost deposition methods such as sublimation-based deposition and sputtering.

1.4.2.4 CIGS

The $Cu(In,Ga)Se_2$ (CIGS) are other materials systems of great interest in thin-film form for photovoltaics. These materials, which can also substitute sulfur for selenium (so-called CIGSSe), have been studied for the past couple of decades (Rau and Schock 1999), and record efficiencies of nearly 20% have been shown by NREL (Repins et al. 2008). As a result of this high record efficiency and the potential for low-cost processing, numerous companies are attempting to scale up production of CIGS, including Showa Shell, Nanosolar, Solyndra, Solybro, Miasole, Global Solar, Solopower, Solarion, Würth Solar, ISET, and others. Unlike CdTe, CIGS solar cells are fashioned in a standard substrate configuration, and it is also possible to deposit CIGS at relatively low temperatures (approximately 500°C) on metal or polymer substrates to enable flexible solar products (Otte et al. 2006). CIGS thin films are primarily deposited using co-evaporation/ evaporation or sputtering, and to a lesser extent electrochemical deposition (Guimard et al. 2003), or ion-beam-assisted deposition (Lippold et al. 2001). Since they are quaternary compounds, control of thin-film stoichiometry during manufacture is critical. There are also efforts to fabricate all or partly solution-deposited CIGS solar cells (Mitzi et al. 2009; Li et al. 2008), with some predicting these could provide the ultimate path to ultra-low-cost ($<\$1/W_p$), roll-to-roll, and flexible PV modules. CIGS module power conversion efficiencies typically range from 7 to 12%, with small area modules yielding 14%. The problem of CdS UV absorption is also of concern for CIGS, though suitable replacements to CdS have been shown in R&D and production CIGS technologies (Matsunaga et al. 2009).

1.4.2.5 Organic and Dye-Sensitized Solar Cells

Thin-film solar cells based on organic materials such as polymers and dyes have also been the subject of intense research in recent years as a potential means of achieving very low PV module costs ($<\$1/W_p$). The benefits of organic-based solar cells include the ability to employ solution-based processes that may also lead to roll-to-roll manufacturing in a manner similar to that of newspapers. Organic solar cells are fundamentally different from inorganic semiconductor-based devices in that they are based on exiton transport rather than nearly free charge carrier. Electron-hole pairs in polymer semiconductors are tightly bound (high binding energy) and have short diffusion lengths, a few to a few tens of nanometers, relative to the optical absorption depths (Gregg 2005). Carrier generation occurs only in the vicinity of the charge-separating junction at which excitons are dissociated by the strong local electric field, leading to electron generation on one side and hole generation on the other. Organic semiconductors also suffer from relatively narrow absorption spectra, which tends to limit the overall available photocurrent.

Early organic solar cell devices were based on planar junctions (Tang 1986); however, it is now generally accepted that a higher surface area junction device architecture, typically referred to as a bulk heterojunction organic PV device, is required for improved performance (Forrest 2005). In practice this often leads to intermixed phases without clear separation, which therefore may produce significant shunting in the devices that limits efficiency. The maximum possible power conversion efficiency is difficult to obtain from first principles due to the difficulty in modeling complex junction geometries and the corresponding exciton dissociation dynamics; however, an upper bound of ~20% has been estimated (Forrest 2005). There is strong interest in utilizing nanostructures to improve the performance of organic solar cells, as discussed later in Chapter 4. Among the highest power conversion efficiencies obtained for a polymer solar cell is ~6% (under low-level concentration, 200 mW/cm^2, AM1.5G illumination) by using a tandem design (Kim et al. 2007), and more recently at ~7.9% (Solarmer).

Another organic-based concept is the dye-sensitized solar cell, first demonstrated by Grätzel and coworkers (O'Regan and Grätzel 1991; Wang et al. 2005). This is a hybrid organic-inorganic design, in which a porous, nanocrystalline titanium oxide film is used as the electron conductor, which is in contact with an electrolyte solution that also contains organic light-absorbing dyes near the interfaces. Charge transfer occurs at the interface, such that holes are transported in the electrolyte. Power conversion efficiencies of ~11% have been demonstrated and commercialization of dye-sensitized PV modules is under way (Han et al. 2006). Technological challenges include lifetime (also of major concern for polymer-based PV cells) and packaging of the solution-containing PV cells. As a result of the latter issue, there are efforts to develop solid-state equivalents of dye-sensitized solar cells (Yum et al. 2008).

Thus far, generation I and II technologies have been described. Below the major high-efficiency concepts are reviewed, with more details provided in Chapter 3.

1.4.3 Generation III

Generation III photovoltaic technologies, as defined by M. Green (2003a), are those that provide both a relatively high efficiency (approximately ≥20%) and low cost (approximately ≤\$200/m^2 or ≤\$1/W_p). The cost requirement almost necessarily implies a cost structure and manufacturing strategy similar to that for thin films, i.e., deposition of low-cost, thin absorber layers on a low-cost substrate (glass, metal, polymer). It is also generally accepted that in order for the efficiency requirement to be established, the generation III technology should provide a limiting efficiency greater than the single-bandgap Shockley–Queisser limit of ~31% (Shockley and Queisser 1961). Therefore, the generation III concepts currently being explored provide this fundamental efficiency enhancement by mechanisms that attempt to overcome the efficiency-limiting features of single-bandgap solar cells: (1) absorption and conversion of a broader portion of the solar spectrum, primarily low-energy photons that are not absorbed by a single-bandgap semiconductor; (2) better utilization of high-energy photons in the solar spectrum that are typically thermalized by interaction with lattice phonon or absorption in upper layers of a solar cell structure; and (3) creation of more than one electron or photon per given absorbed photon to improve the J_{sc}. The major known concepts are now reviewed, followed by a discussion of how nanostructures may be of use in generation I, II, and III photovoltaics.

1.4.3.1 Multijunction Solar Cells

Multijunction, or multiple-bandgap, solar cells (MJSCs) utilize two or more semiconductor p-n junctions of different bandgaps in order to absorb a greater fraction of the solar spectrum. The cells are typically monolithically integrated and series connected using a tunnel junction, with current matching between the cells provided by adjusting the bandgap and thickness of each cell. The theoretical entitlement of multiple bandgaps has been analyzed and shown to be 44% for two bandgaps, 49% for three bandgaps, 54% for four bandgaps, and 66% for an infinite number of bandgaps (Marti and Araujo 1996).

MJSCs are the only advanced PV concept that has been demonstrated at the laboratory level and is also in commercial production for space and terrestrial concentrator applications. The world record efficiency has been shown by a triple-junction device at just over 41% under 454 suns concentrated sunlight* (King et al. 2007), and most recently at 41.6% (364 suns) by Boeing Spectrolab. At 1 sun under AM0 or AM1.5 illumination the efficiency of MJSCs is on the order of 30 to 32%. These devices are typically heteroeptaxially or metamorphically integrated compound semiconductors based on InGaAs/InGaP on low-bandgap Ge substrates. Their cost per unit area is orders of magnitude higher than that of Si and thin-film solar cells due to the expensive substrates and many layers grown at low deposition rates to minimize crystal defects. As such, while mechanistically MJSCs can be considered generation III devices, based on their cost this is not the case. They are thus used in the space market, in which cost is less of a technology concern than supplying ample power to space systems, and in terrestrial concentrator applications small areas are needed, and hence the high solar cell cost per unit area does not significantly impact the overall system cost. In summary, the key advantage of MJSCs is high efficiency, and the main disadvantages are the requirement for current matching between the cells within the monolithic device and the high cost per unit area.

1.4.3.2 Intermediate Band Solar Cells

Another proposed concept for capturing low-energy light is to introduce energy levels within the bandgap of a semiconductor that can provide lower-energy transitions, and hence the ability to capture more IR photons within the solar spectrum. Theoretical analysis shows that these so-called intermediate band solar cells (IBSCs) have the potential for a maximum efficiency of ~63% (Luque and Marti 1997). In practice, there are severe requirements on the transition rates and excited state lifetimes, and no full IBSC has been shown. Recently, however, a key operating mechanism was demonstrated, i.e., the demonstration of photocurrent due to the transition of carriers from the intermediate band to the conduction band (Marti et al. 2006). IBSCs are being explored in bulk semiconductors (Yu et al. 2003; Wahnón et al. 2006) and in quantum dot assemblies (Luque and Marti 2006).

1.4.3.3 Hot Electron Solar Cells

One of the fundamental loss mechanisms in single-bandgap solar cells is the thermalization of charge carriers promoted to levels above the

---

* http://www.ise.fraunhofer.de/press-and-media/pdfs-zu-presseinfos-englisch/2009/press-release-world-record-41.1-efficiency-reached-for-multi-junction-solar-cells-at-fraunhofer-ise-pdf-file.

bandgap by interaction with phonons (quantized vibrations of the crystal lattice). If these "hot electrons" could be extracted prior to the thermalization process, which typically occurs in the picosecond timeframe, the limiting efficiency could be increased significantly. Since impact ionization, or the generation of additional charge carriers per photogenerated carrier, also occurs high in the conduction band (and low in the valence band), the limiting efficiency increases to ~53% (Wurfel 1997). The key challenges for demonstrating a practical hot electron solar cell are (1) slowing the thermalization rate of hot electrons (Conibeer et al. 2008b) and (2) creating energy-selective contacts that allow for extraction of the hot electrons within a narrow energy band without cooling them (Conibeer et al. 2008a).

1.4.3.4 Multiple-Exciton Generation

Multiple-exciton generation (MEG) involves the formation of more than one electron-hole pair per given photon, thus increasing the short-circuit current density of the solar cell. MEG occurs in all semiconductors at a very low efficiency; however, it has recently been shown that greater than two excitons per input photon can be generated in quantum-confined semiconductor wells and dots (Nozik 2001; Ellingson et al. 2005), with some groups reporting up to seven excitons per photon (Schaller et al. 2006). Other groups have argued the effect is not more efficient in nanocrystals than in the bulk (Trinh et al. 2008), or in some systems is not present (Ben-Lulu et al. 2008). Scientific challenges for MEG-based solar cells include charge separation and contact to quantum structures, though several approaches to quantum dot solar cells have been proposed (Nozik 2002). It has been shown that nonradiative energy transfer is a mechanism that can be used to extract excited carriers from quantum dot assemblies to a p-i-n junction device (Chanyawadee et al. 2009).

1.4.3.5 Upconversion

Another major loss mechanism for single-junction solar cells is related to the lack of absorption of solar photons with energies below the semiconductor bandgap. While intermediate bands are one means of harnessing those photons by changing the band structure of the semiconductor, it has also been proposed to achieve this by attaching an absorber layer below the solar cell that can convert the low-energy photons to higher-energy photons that can be absorbed by the PV cell. The maximum quantum efficiency of this upconversion process is 50%, since at best two low-energy photons are required to produce one high-energy photon. Trupke et al. (2002b) analyzed the limiting efficiency of a solar cell with an attached upconverter and showed it could increase from 31 to 44%. A first practical implementation of upconversion was shown by Shalav et al. (2005, 2007) with a Si solar cell EQE of ~10^{-4}% in the wavelength range of 1,480 to 1,580 nm, corresponding to absorption from the upconverting NaYF4:Er^{3+} phosphor. Further challenges for upconversion include increasing the upconversion quantum efficiency and making the upconverting layer more broadband (Strumpel et al. 2007).

1.4.3.6 Downconversion

The opposite of upconversion is the process of converting one high-energy photon to two low-energy photons. This downconversion process, also referred to as quantum splitting or quantum cutting in the literature, was analyzed with respect to increasing the limiting efficiency of single-band-gap solar cells and found to do so from ~31 to ~ 39% (Trupke et al. 2002a). Quantum splitting has been shown for lighting applications, in which the photon energies available are higher than those in the solar spectrum (Piper et al. 1974). To date no materials that can effectively quantum split UV photons in the solar spectrum (300 to 500 nm) have been demonstrated—hence significant basic materials research is required. However, downshifting phosphor single crystals employing photoluminescence (one low-energy photon out for one high-energy photon in) has been applied to thin-film solar cells with absolute efficiency enhancements of 0.5 to 0.8% (Hong and Kawano 2003).

1.4.4 Nanostructured Concepts

The above discussion on generation III PV concepts highlights the strong potential role of nanostructures, nanoscience, and nanotechnology to contribute to achievement of ultra-low-cost, high-efficiency solar energy systems. Here we review the major classes of nanostructures that will be described in more detail in subsequent chapters, highlighting their major structural and physical characteristics, nonphotovoltaic applications, and methods of fabrication/synthesis. The nanostructures are classified by dimensionality, i.e., nanostructured bulk materials and composites (3D), quantum wells (2D), nanowires and nanotubes (1D), and nanoparticles and quantum dots (0D).

1.4.4.1 Nanostructured Bulk Materials and Composites

Nanostructured bulk materials were among the first nanoscale materials to be studied. These have been primarily studied for their mechanical properties, since the well-known Hall-Petch relationship predicts that the strength of material is inversely proportional to the grain size. Nanostructured metals (Al, Ni, etc.) and ceramics (oxides, nitrides, carbides, etc.) have been fabricated by various methods, including ball milling, cryo-milling, and direct synthesis of nanopowders by chemical methods or laser ablation, followed by various sintering methods (e.g., hot pressing). Key scientific challenges include the maintenance of nanoscale grain sizes upon sintering, since sintering by its nature is driven by reduction in surface area, as well as minimizing grain boundary sliding for fine-grained materials. Numerous nanocomposites have been demonstrated for mechanical and electrical applications, with hybrid organic-inorganic composites of particular interest. Inorganic nanocomposite solar cells have been demonstrated with promising efficiency (Nanu et al. 2005), and nanostructured organic solar cells will be explored in Chapter 4.

1.4.4.2 Quantum Well

Quantum wells were among the first structures in which quantum confinement effects were systematically explored, and hence can be viewed as prototypical nanostructures for the study of such effects in PV. Quantum wells are fabricated by epitaxial growth of bandgap and (usually) lattice-mismatched semiconductors by various means, including metal-organic chemical vapor deposition (MOCVD), molecular beam epitaxy (MBE), and related methods. It is possible to control the deposition of individual layers in a multiquantum well (MQW) structure to the submonlayer level, allowing for the creation of artificial lattices (e.g., superlattices) that yield novel electrical characteristics. Applications of quantum wells, which have been studied since the 1980s, include high-speed electronics (e.g., high electron mobility transistors [HEMTs]), lasers, light-emitting diodes (LEDs), and sensors.

1.4.4.3 Nanowire/Tube

Nanowires and nanotubes represent a class of structures with unique properties that are currently being explored for photovoltaic applications. The synthesis of nanowires and nanorods, solid (usually single-crystalline) elongated structures with submicron diameters, can be traced to work in the 1960s by Wagner and Ellis on vapor-liquid-solid (VLS) growth (Wagner and Ellis 1964), which was furthered in the 1970s by various workers in efforts to grow single-crystal whisker structures. In the 1980s quantum wires fabricated by top-down lithography and etching methods were shown in the literature, and in the 1990s early efforts to grow whiskers in the subwavelength regime appear. In 1998 a seminal paper by Lieber and coworkers on growth of nanowire semiconductors by laser ablation was published, thus reigniting interest in the field (Morales and Lieber 1998). The first silicon nanowire-based transistors were shown in 2000 by Lieber and coworkers and Heath and coworkers (Cui et al. 2000; Chung et al. 2000), and since then there has been a tremendous interest in applications of nanowire materials. These include for electronics, biological and chemical sensors, thermoelectrics, field emission displays, LEDs, lasers, photodetectors, mechanical composites, and superconductors. Nanowire structures can be synthesized by chemical vapor deposition, electrochemical deposition, wet etching, solution-based chemical synthesis, critical-point synthesis, and other methods.

Carbon nanotubes (CNTs) were first discovered in 1991 by Iijima and coworkers, and have been considered the prototypical nanostructure for the study of electrical and optical phenomena in one-dimensional quantum structures. Numerous applications in the areas of electronics, photonics, sensing, mechanics, and other fields have been explored. CNTs are essentially single-layer carbon sheets (graphene) that have been rolled and joined along a particular direction within the two-dimensional lattice (chirality), giving rise to multiple types of CNTs that may be either semiconducting or metallic in nature (Saito et al. 1992). About two-thirds of CNTs in a typical CNT array are semiconducting in nature (with various bandgaps), with the rest being metallic (Wilder et al. 1998). To date, it has not been possible to deterministically control the chirality of nanotubes, though recent works have shown it is possible to control the chirality distribution favoring semiconducting CNTs over metallic ones (Li et al. 2004). Nanotubes composed of inorganic materials have also been demonstrated, including GaN (Goldberger et al. 2003),

Pb(Zr,Ti)O$_3$ (Min and Lee 2006), titania (Kasuga et al. 1999), and have alumina (Mei et al. 2003).

1.4.4.4 Nanoparticle/Quantum Dots

Nanoparticles and quantum dots, or zero-dimensional nanostructures, are also of great interest for various applications in PV. Quantum dots (QDs) and nanoparticles (NPs) have been synthesized in numerous materials systems, including semiconductors and metals, and it is possible to create core-shell structures also employing dielectrics. Quantum dots have been synthesized by solution-based chemical methods, chemical vapor methods, and physical vapor deposition. Among the first works in this field were by Bawendi and coworkers (Murray et al. 1993) and Alivisatos and coworkers (Colvin et al. 1991). Quantum dots and nanoparticle applications beyond PV include biological labeling (Alivisatos et al. 2005), nanoelectronics (Cui et al. 2004), and many others.

1.4.4.5 PV Materials Tetrahedron

The above discussion highlights the broad potential for application of nanostructures to PV science and technology. While there is great potential in nanostructures having an impact on producing low-cost, high-efficiency photovoltaics, one should look to existing PV technologies as an indication of the timeline for making a significant technological impact. For example, in the case of silicon PV, it took approximately 40 years from the invention of the first Si solar cell to achieve record efficiencies of nearly 25% for champion cells and ~20% at the module level. The PV market only reached significant market size and growth rates in the 2000s. Thin-film photovoltaics have been known for more than 30 years, but have only reached significant champion cell levels in the last decade, and promising market penetration within the last few years. Commercial CdTe thin-film modules still provide significantly less efficiencies (~11%) than champion devices (~16.5%), and champion devices have more room for improvement. Furthermore, CIGS technologies are still relatively immature from the perspective of meeting the demands of large-scale PV applications, though rapid development progress is being made.

This relatively long timescale for PV technology maturation is fundamentally related to the challenging nature of photovoltaic materials science, physics, chemistry, and engineering. In addition to the fundamental

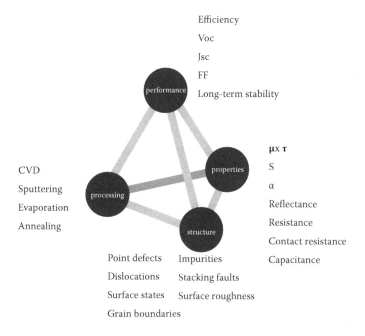

FIGURE 1.14

PV materials tetrahedron showing the relationship between processing, materials structure, materials properties, and ultimately device/system level performance that must be optimized in a low-cost manufacturing process. The detailed parameters are noted. Application of nanostructures to PV poses additional complications due to emerging physics, surface-related recombination, etc.

loss mechanisms discussed above, a PV module requires the simultaneous optimization of many parameters. This is best visualized by a so-called PV materials tetrahedron (Figure 1.14), in which these parameters are identified with respect to (1) processes used for form materials in usable form factors, (2) the impact of these processes on materials defects and structure (usually in a nonequilibrium state), (3) basic materials properties of interest to PV (mobility, minority carrier lifetime, surface recombination velocity), and (4) the final PV performance parameters that lead to system-level output (V_{oc}, J_{sc}, FF, P_{mp}, efficiency, long-term reliability, etc.). There is a strong interrelation between these features within the materials tetrahedron, and it can be argued that the fundamental challenge with PV technology development is the requirement that these parameters must be simultaneously optimized, whether in a laboratory environment or in a low-cost, high-speed manufacturing process.

The application of nanostructures to PV thus provides additional complications to this tetrahedron, and hence to ultimate development/commercialization of nano-PV technologies. New physics such as multiple-exciton generation or upconversion is not fully understood, whereas nanostructures inherently provide significantly larger surface areas, leading to challenges associated with surface traps and related effects. The field of nanophotovoltaics is relatively new, and hence we can expect many basic science and engineering breakthroughs to occur in the coming years that will lead to promising paths for low-cost, high-efficiency PV systems. The current state of the art in this field for the various classes of nanostructures is reviewed in this book. We begin with a discussion of the basic electrical and optical properties of nanostructures, a discussion of new device physics associated with nanoscale PV devices, and detailed reviews of the application of bulk nanostructures to organic PV, quantum wells, nanowires/tubes, and nanoparticles/quantum dots to photovoltaics. We will end with an overview of the known potential manufacturing processes that nanotechnology provides and prospects for these to yield low-cost, scalable PV manufacturing.

REFERENCES

Aberle, A. G. 2006a. Progress with polycrystalline silicon thin-film solar cells on glass at UNSW. *J. Crystal Growth* 287:386–90.

Aberle, A. G. 2006b. Fabrication and characterization of crystalline silicon thin-film materials for solar cells. *Thin Solid Films* 511–512:26–34.

Adams, W. G., and Day, R. E. 1876. The action of light on selenium. *Proc. R. Soc. Lond.* 25A:113–17.

Alivisatos, A. P., Gu, W., and Larabell, C. 2005. Quantum dots as cellular probes. *Annu. Rev. Biomed. Eng.* 7:55.

Arya, R. R. 2004. Amorphous silicon based solar cell technologies: Status, challenges, and opportunities. *Mater. Res. Soc. Symp. Proc.* 808:A7.5.

Bai, Y., Ford, D. H., Rand, J. A. Hall, R. B., and Barnett, A. M. 1997. 16.6% efficient Silicon-Film™ polycrystalline siliconsolar cells. In *Proceedings of the 26th IEEE Photovoltaic Specialists Conference*, Anaheim, CA. pp. 35–38.

Basore, P. A. 2006a. CSG-1: Manufacturing a new polycrystalline silicon PV technology. In *Proceedings of the IEEE 4th World Conference of Photovoltaic Energy Conversion*, pp. Waikoloa, HI, May 7–12, 2089–93.

Basore, P. A. 2006b. CSG-2: Expading the production of a new polycrystalline silicon PV technology. In *Proceedings of the 21st European Photovoltaic Solar Energy Conference*, Barcelona, Spain, September 4–9, pp. 544–48.

Becquerel, A. E. 1839. 'Recherches sur les effets de la radiation chimique de la lumière solaire au moyen des courants électriques' and 'Mémoire sur les effets électriques produit sous l'influence des rayons solaires.' *C. R. Acad. Sci.* 9:145–49, 561–67.

Ben-Lulu, M., Mocatta, D., Bonn, M., Banin, U., and Ruhman, S. 2008. On the absence of detectable carrier multiplication in a transient absorption study of InAs/CdSe/ZnSe Core/Shell1/Shell2 quantum dots. *Nano Lett.* 8:1207–11.

Bhattacharya, R. N., and Ramanathan, K. 2004. Cu(In,Ga)Se$_2$ thin film solar cells with buffer layer alternative to CdS. *Solar Energy* 77:679–83.

Bosio, A., Romeo, N., Podestà, A., Mazzamuto, S., and Canevari, V. 2005. Why CuInGaSe2 and CdTe polycrystalline thin film solar cells are more efficient than the corresponding single crystal? *Cryst. Res. Technol.* 40:1048–53.

Bowden, S., Wenham, S. R., Dickinson, M. R., and Green, M. A. 1994. High efficiency photovoltaics roof tiles with static concentrators. In *Proceedings of the 1st IEEE World Conference Photovoltaic Energy Conversion*, Vol. 1, Waikoloa, HI, December, pp. 774–77.

Buresch, M. 1983. *Photovoltaic energy systems: Design and installation*. New York: McGraw-Hill.

Carlson, D. E., and Wronski, C. R. 1976. Amorphous silicon solar cell. *Appl. Phys. Lett.* 28:671.

Chanyawadee, S., Harley, R. T., Henini, M., Talapin, D. V., and Lagoudakis, P. G. 2009. Photocurrent enhancement in hybrid nanocrystal quantum-dot *p-i-n* photovoltaic devices. *Phys. Rev. Lett.* 102:077402.

Chapin, D. M., Fuller, C. S., and Pearson, G. L. 1954. A new silicon p-n junction photocell for converting solar radiation into electrical power. *J. Appl. Phys.* 25:676–77.

Chung, S. W., Yu, J.-Y., and Heath, J. R. 2000. Silicon nanowire devices. *Appl. Phys. Lett.* 76:2068.

Colvin, V. L., Alivisatos, A. P., and Tobin, J. G. 1991. Valence band photoemission from a quantum dot system. *Phys. Rev. Lett.* 66:2786.

Conibeer, G. J., Jiang, C.-W., König, D., Shrestha, S., Walsh, T., and Green, M. A. 2008a. Selective energy contacts for hot carrier solar cells. *Thin Solid Films* 516:6968–73.

Conibeer, G. J., König, D., Green M. A., and Guillemoles, J. F. 2008b. Slowing of carrier cooling in hot carrier solar cells. *Thin Solid Films* 516:6948–53.

Contescu, C., Schwarz, J. A., and Putyera, K., eds. 2008. *Dekker encyclopedia of nanoscience and nanotechnology*. 2nd ed. Boca Raton, FL: Taylor & Francis.

Cui, Y., Duan, X., Hu, J., and Lieber, C. M. 2000. Doping and electrical transport in silicon nanowires. *J. Phys. Chem.* 104B:5213.

Cui, Y., Bjork, M. T., Liddle, J. A., Sonnichsen, C., Boussert, B., and Alivisatos, A. P. 2004. Integration of colloidal nanocrystals into lithographically patterned devices. *Nano Lett.* 4:1093.

Deng, X. 2005. Optimization of a-SiGe based triple, tandem and single junction solar cells. In *Proceedings of the 31st IEEE Photovoltaic Specialists Conference*, Orlando, FL, January 3–5, pp. 1365–70.

Dhere, N. G., Ghongadi S. R., Pandit, M. B., Jahagirdar, A. H., and Scheiman, D. 2002. CIGS$_2$ thin-film solar cells on flexible foils for space power. *Prog. Photovolt. Res. Appl.* 10:407–16.

Ellingson, R. J., Beard, M. C., and Johnson, J. C. 2005. Highly efficient multiple exciton generation in colloidal PbSe and PbS quantum dots. *Nano Lett.* 5:865.

Fatemi, N. S., Pollard, H. E., Hou, H. Q., and Sharps, P. R. 2000. Solar array trades between very high efficiency multi-junction and Si space solar cells. In *Proceedings of the 28th IEEE Photovoltaic Specialists Conference*, Anchorage, AK, September, pp. 1083–86.

Forrest, S. D. 2005. The limits to organic photovoltaic cell efficiency. *MRS Bull.* 30:28–21.

Fritts, C. E. 1883. A new form of selenium cell. *Am. J. Sci.* 26:465–72.

Goldberger, J., He, R., Zhang, Y., et al. 2003. Crystal gallium nitride nanotubes. *Nature* 422:599.

Gordon, J. M., Katz, E. A., Feuermann, D., and Huleihil, D. 2004. Toward ultra-high-flux photovoltaic concentration. *Appl. Phys. Lett.* 84:3642–44.

Green, M. A. 1998. *Solar cells*. Kensington, Australia: University of New South Wales.

Green, M. A. 2003a. *Third generation photovoltaics: Advanced solar energy conversion.* Berlin: Springer-Verlag.

Green, M. A. 2003b. Crystalline and thin-film silicon solar cells: State of the art and future potential. *Solar Energy* 74:181–92.

Green, M. A., Basore, P. A., Chang, N., et al. 2004. Crystalline silicon on glass (CSG) thin-film solar cell modules. *Solar Energy* 77:857–63.

Gregg, B. A. 2005. The photoconversion mechanism of excitonic solar cells. *MRS Bull.* 30:20–22.

Gueymard, C. 2004. The sun's total and spectral irradiance for solar energy applications and solar radiation models. *Solar Energy* 76:423–53.

Gueymard, C., Myers, D., and Emery, K. 2002. Proposed reference irradiance spectra for solar energy systems testing. *Solar Energy* 73:443–67.

Guimard, D., Bodereau, N., Kurdi, J., et al. 2003. Efficient $Cu(In,Ga)Se_2$ based solar cells prepared by electrodeposition. *Mater. Res. Soc. Symp. Proc.* 763: B6.9.1–6.

Hagemann, I. 1996. Architectural considerations for building-integrated photovoltaics. *Prog. Photovolt. Res. Appl.* 4:247–58.

Han, L., Fukui, A., Fuke, N., Koide, N., and Yamanaka, R. 2006. High efficiency of dye-sensitized solar cell and module. In *Proceedings of the IEEE 4th World Conference on Photovoltaic Energy Conversion*, Vol. 1, Waikoloa, HI, May 7–12, p. 179.

Hoffman, W. 2006. PV solar electricity industry: Market growth and perspective. *Solar Energy Mater. Solar Cells* 90:3285.

Hong, B.-C. and Kawano, K. 2003. PL and PLE studies of $kMgF_3$:Sm crystal and the effect of its wavelength conversion on CdS/Cdte solar cells. *Solar Energy Mater. Solar Cells.* 80:417–32.

Huynh, W. U., Dittmer, J. J., and Alivisatos, A. P. 2002. Hybrid nanorod-polymer solar cells. *Science* 295:2425–27.

Janoch, R., Wallace, R., and Hanoka, J. I. 1997. Commercialization of silicon sheet via the string ribbon crystal growth technique. In *Proceedings of the 26th IEEE Photovoltaic Specialists Conference*, Anaheim, CA, September, pp. 95–98.

Karam, N. H., et al. 1998. High efficiency GaInP2/GaAs/Ge dual and triple junction solar cells for space applications. In *2nd World Conference and Exhibition on Photovoltaic Solar Energy Conversion*, Vienna, Austria, July 6–10, p. 3534.

Kasuga, T., Hiramatsu, M., Hoson, A., Sekino, T., and Niihara, K. 1999. Titania nanotubes prepared by chemical processing. *Adv. Mater.* 11:1307.

Kim, J. Y., Lee, K., Coates, N. E., et al. 2007. Efficient tandem polymer solar cells fabricated by all-solution processing. *Science* 317:222.

King, R. R., Law, D. C., Edmondson, K. M., et al. 2007. 40% efficient metamorphic GaInP/GaInAs/Ge multijunction solar cells. *Appl. Phys. Lett.* 90:183–516.

Korech, O., Hirsch, B., Katz, E. A., and Gordon, J. M. 2007. High-flux characterization of ultrasmall multijunction concentrator solar cells. *Appl. Phys. Lett.* 91:064101.

Li, X. C., Soltesz, I., Wu, M., Ziobro, F., Amidon, R., and Kiss, Z. 2008. A nanoparticle ink printing process for all printed thin film copper-indium-selenide (CIS) solar cells. In *Proceedings of the SPIE: Nanoscale Photonic and Cell Technologies for Photovoltaics*, San Diego, CA, August, 7047: 70470E-70470E-9.

Li, Y., Mann, D., Rolandi, M., et al. 2004. Preferential growth of semiconducting single-walled carbon nanotubes by a plasma enhanced CVD method. *Nano Lett.* 4:317–21.

Lippold, G., Neumann, H., and Schindler, A. 2001. Ion beam assisted deposition of Cu(In,Ga) Se_2 films for thin film solar cells. *Mater. Res. Soc. Symp. Proc.* 668:H3.9.1–6.

Liu, S. H., Simberger, E. J., Matsumoto, J., Garcia, A., III, Ross, J., and Nocerino, J. 2005. Evaluation of thin-film solar cell temperature coefficients for space applications. *Prog. Photovolt. Res. Appl.* 13:149–56.

Luque, A., and Marti, A. 1997. Increasing the efficiency of ideal solar cells by photon induced transitions at intermediate levels. *Phys. Rev. Lett.* 78:5014.

Luque, A., and Marti, A. 2006. Recent progress in intermediate band solar cells. In *Proceedings of the 4th World Conference on Photovoltaic Energy Conversion*, Waikoloa, HI, May 7–12, p. 49.

Luque, A., Sala, G., and Luque-Heredia, I. 2006. Photovoltaic concentration at the onset of its commercial deployment. *Prog. Photovolt. Res. Appl.* 14:413–28.

Maish, A. 1999. Defining requirements for improved photovoltaic system reliability. *Prog. Photovolt. Res. Appl.* 7:165–73.

Marion, B. Adelstein, J., Boyle, K., et al. 2005. Performance parameters for grid-connected PV systems. In *Proceedings of the 31st IEEE Photovoltaics Specialists Conference and Exhibition*, Orlando, FL, January 3–7, pp. 1601–6.

Marti, A., Antol, E., Stanley, C. R., et al. 2006. Production of photocurrent due to intermediate-to-conduction-band transitions: A demonstration of a key operating principle of the intermediate-band solar cell. *Phys. Rev. Lett.* 97:247701.

Marti, A., and Araujo, G. L. 1996. Limiting efficiencies for photovoltaic energy conversion in multigap systems. *Solar Energy Mater. Solar Cells* 43:203.

Matsunaga, K., Komaru, T., Nakayama, Y., Kume, T., and Suzuki, Y. 2009. Mass-production technology for CIGS modules. *Solar Energy Mater. Solar Cells*, 93:1134–8.

McConnell, R., Ji, L., Lasich, J., and Mansfield, R. 2004. Concentrator photovoltaic qualification standards for systems using refractive and reflective optics. In *Proceedings of the 19th European PV Solar Energy Conference and Exhibition,* Paris, France, June 7–11.

McConnell, R., and Symko-Davies, M. 2006. Multijunction photovoltaic technologies for high performance concentrators. In *Proceedings of the 4th World Conference on Photovoltaic Energy Conversion*, Vol. 1, Waikoloa, HI, May 7–12, pp. 733–36.

Mei, Y. F., Wu, X. L., Shao, X. F., Siu, G. G., and Bao, X. M. 2003. Formation of an array of isolated alumina nanotubes. *Europhys. Lett.* 62:595.

Meier, J., Kroll, U., Benagli, S., et al. 2007. Recent progress in up-scaling of amorphous and micromorph thin film silicon solar cells to 1.4 m^2 modules. *Mater. Res. Soc. Symp. Proc.* 989:0989-A24-01-12.

Messenger, R. A., and Ventre, J. 2003. *Photovoltaic systems engineering*. 2nd ed. Boca Raton, FL: CRC Press.

Meyers, P. V. 2006. First solar polycrystalline CdTe thin film PV. In *Proceedings of the IEEE 4th World Conference on Photovoltaic Energy Conversion*, pp. 2024–27.

Min, H.-S., and Lee, J.-K. 2006. Ferroelectric nanotube array growth in anodic porous alumina nanoholes on silicon. *Ferroelectrics* 336:231.

Mitzi, D. B., Yuan, M., Liu, W., et al. 2009. Hydrazine-based deposition route for device-quality CIGS films. *Thin Solid Films* 517:2158–62.

Moore, L., Post, H., Hayden, H., Canada, S., and Narang, D. 2005. Photovoltaic power plant experience at Arizona public service: A 5-year assessment. *Prog. Photovolt. Res. Appl.* 13:353–63.

Morales, A. M., and Lieber, C. M. 1998. A laser ablation method for the synthesis of crystalline semiconductor nanowires. *Science* 279:20811.

Moskowitz, P. D., Steinberger, H., and Thumm, W. 1994. Health and environmental hazards of CdTe photovoltaic module production, use, and decommissioning. In *Proceedings of the IEEE 1st World Conference on Photovoltaic Energy Conversion*, pp. 115–18.

Müller, J., Rech, B., Springer, J., and Vanecek, M. 2004. TCO and light trapping in silicon thin film solar cells. *Solar Energy* 77:917–30.

Murray, C. B., Norris, D. J., and Bawendi, M. G. 1993. Synthesis and characterization of nearly monodisperse CdE (E = sulfur, selenium, tellurium) semiconductor nanocrystallites. *J. Am. Chem. Soc.* 115:8706.

Nanu, M., Schoonman, J., and Goossens, A. 2005. Nanocomposite three-dimensional solar cells obtained by chemical spray deposition. *Nano Lett.* 5:1716–19.

Nast, O., Puzzer, T., Koshier, L. M., Sproul, A. B., and Wenham, S. R. 1998. Aluminum-induced crystallization of amorphous silicon on glass substrates above and below the eutectic temperature. *Appl. Phys. Lett.* 73:3214.

Nozik, A. J. 2001. Spectroscopy and hot electron relaxation dynamics in semiconductor quantum wells and quantum dots. *Annu. Rev. Phys. Chem.* 52:193.

Nozik, A. J. 2002. Quantum dot solar cells. *Physica* 14E:115.

O'Regan, B., and Grätzel, M. 1991. A low-cost, high-efficiency solar cell based on dye-sensitized colloidal TiO_2 films. *Nature* 353:737.

Otani, K., Kato, K., Takashima, T. et al. 2004. Field experience with large-scale implementation of domestic PV systems and large PV systems on buildings in Japan. *Prog. Photovolt. Res. Appl.* 12:449–59.

Otte, K., Makhova, L., Braun, A., and Konovalov, I. 2006. Flexible $Cu(In,Ga)Se_2$ thin-film solar cells for space application. *Thin Solid Films* 511–12:613–22.

Pallab, B. 1997. *Semiconductor optoelectronic devices*. Englewood Cliffs, NJ: Prentice Hall.

Piper, W. W., DeLuca, J. A., and Ham, F. S. 1974. Cascade fluorescent decay in Pr^{3+}-doped fluorides: Achievement of a quantum yield greater than unity for emission of visible light. *J. Lumin.* 8:344.

Rath, J. K. 2003. Low temperature polycrystalline silicon: A review on deposition, physical properties and solar cell applications. *Solar Energy Mater. Solar Cells* 76:431–87.

Rau, U., and Schock, H. W. 1999. Electronic properties of $Cu(In,Ga)Se_2$ heterojunction solar cells—Recent achievements, current understanding, and future challenges. *Appl. Phys. A Mater. Sci. Process.* 69:131.

Repins, I., Contreras, M. A., Egaas, B., et al. 2008. 19.9%-efficient $ZnO/CdS/CuInGaSe^2$ solar cell with 81.2% fill factor. *Prog. Photovolt. Res. Appl.* 16:235–39.

Saito, R., Fujita, M., Dresselhaus, G., and Dresselhaus, M. S. 1992. Electronic-structure of chiral graphene tubules. *Appl. Phys. Lett.* 60:2204–6.

Schaller, R. D., Sykora, M., Pietryga, J. M., and Klimov, V. I. 2006. Seven excitons at a cost of one: Redefining the limits for conversion efficiency of photons into charge carriers. *Nano Lett.* 6:424.

Schropp, R. E. I., Rath, J. K., and Li, H. 2009. Growth mechanism of nanocrystalline silicon at the phase transition and its application in thin film solar cells. *J. Crystal Growth* 311:760–64.

Schultz, O., Glunz, S. W., and Willeke, G. P. 2004. Multicrystalline silicon solar cells exceeding 20% efficiency. *Prog. Photovolt. Res. Appl.* 12:553–58.

Shah, A. V., Meier, J., Vallat-Sauvain, E. et al. 2003. Material and solar cell research in microcrystalline silicon. *Solar Energy Mater. Solar Cells* 78:469–91.

Shalav, A., Richards, B. S., and Green, M. A. 2005. Application of $NaYF4:Er^{3+}$ up-converting phosphors for enhanced near-infrared silicon solar cell response. *Appl. Phys. Lett.* 86:013505.

Shalav, A., Richards, B. S., and Green, M. A. 2007. Luminescent layers for enhanced silicon cell performance: Up conversion. *Solar Energy Mater. Solar Cells* 91:829–42.

Skinner, B. J. 1979. Earth resources. *Proc. Natl. Acad. Sci. USA* 76:4212–17.

Slade, A., Stone, K. W., Gordon R., and Garboushian, V. 2005. High efficiency solar cells for concentrator systems: Silicon or multi-junction? *Proc. SPIE* 5942:59420O.

So, J. H., Jung, Y. S., Yu, B. G., Hwang, H. M., and Yu, G. J. 2006. Performance results and analysis of large scale PV system. In *Proceedings of the IEEE 4th World Conference on Photovoltaic Energy Conversion*, Vol. 1, pp. 2375–78.

Stavrides, A., King, R. R., Colter, P., Kinsey, G., McDanal, A. J., O'Neil, M. J., and Karam, N. H. 2002. Fabrication of high efficiency, III-V multi-junction solar cells for space concentrators. In *Proceedings of the 29th IEEE Photovoltaic Specialists Conference*, New Orleans, LA, May 17–24, pp. 920–22.

Strumpel, C., McCann, M., Beaucarne, G., et al. 2007. Modifying the solar spectrum to enhance silicon solar cell efficiency—An overview of available materials. *Solar Energy Mater. Solar Cells* 91:238.

Sze, S. M., and Ng, K. K. 2007. *Physics of semiconductor devices*. 3rd ed. Hoboken, NJ: Wiley-Interscience.

Taguchi, M., Kawamoto, K., Tsuge, S., et al. 2000. HIT™ cells—High efficiency crystalline Si cells with novel structure. *Prog. Photovolt. Res. Appl.* 8:503–13.

Tang, C. W. 1986. Two-layer organic photovoltaic cell. *Appl. Phys. Lett.* 48:183.

Tomita, Y., Kawai, T., and Hatanaka, Y. 1993. Carrier transport properties of sputter-deposited CdS/CdTe heterojunction. *Jpn. J. Appl. Phys.* 32:1923–28.

Trinh, M. T., Houtepen, A. J., Schins, J. M., et al. 2008. In spite of recent doubts carrier multiplication does occur in PbSe nanocrystals. *Nano Lett.* 8:1713–18.

Trupke, T., Green, M., and Würfel, P. 2002a. Improving solar cell efficiencies by down-conversion of high energy photons. *J. Appl. Phys.* 92:1668.

Trupke, T., Green, M., and Würfel, P. 2002b. Improving solar cell efficiencies by up-conversion of sub-band-gap light. *J. Appl. Phys.* 92:4117.

Tsunomura, Y., Yoshimine, Y., and Taguchi, M. 2009. Twenty-two percent efficiency HIT solar cell. *Solar Energy Mater. Solar Cells.* 93:670–73.

Van Kerschaver, E., and Beaucarne, G. 2006. Back contact solar cells: A review. *Prog. Photovolt. Res. Appl.* 14:107–23.

van Kessel, T. G., Martin, Y. C., Sandstrom, R. L., and Guha, S. 2008. Extending photovoltaic operation beyond 2000 suns using a liquid metal thermal interface with passive cooling. In *Proceedings of the 33rd IEEE Photovoltaic Specialists Conference*, San Diego, CA, May 11–16, p. 405.

von Roedern, B., and Ullal, H. A. 2008. Critical issues for commercialization of thin-film PV technologies. *Solid State Technology*, February. Available from http://www.solid-state.com/display_article/319149/5/none/none/Feat/Critical-issues-for-commercialization-of-thin-film-PV-technologie.

Wagner, B. F., Schetter, Ch., Sulima O. V., and Bett, A. 1993. 15.9% efficiency for Si thin film concentrator solar cell grown by LPE. In *23rd IEEE Photovoltaic Specialists Conference*, Louisville, KY, May, p. 356.

Wagner, H. 2008. Just a few more hurdles to 80 MW. *PHOTON International*, September. Available from http://www.photon-magazine.com/news_archiv/details.aspx?cat=News_PI&sub=europe&pub=4&parent=1267.

Wagner, R. S., and Ellis, W. C. 1964. Vapor-liquid-solid mechanism of single crystal growth. *Appl. Phys. Lett.* 4:89.

Wahnón, P., Palacios, P., Sánchez, K., Aguilera, I., and Conesa, J. C. 2006. Ab-initio modeling of intermediate band materials based on metal doped chalcopyrite compounds. In *Proceedings of the 4th World Conference on Photovoltaic Energy Conversion*, Waikoloa, HI, May 7–12, p. 63.

Wang, P., Klein, C., Humphry-Baker, R., Zakeeruddin, S. M., and Grätzel, M. 2005. Stable ≥8% efficient nanocrystalline dye-sensitized solar cell based on an electrolyte of low volatility. *Appl. Phys. Lett.* 86:123508.

Wang, T. H., Ciszek, T. F., Schwerdtfeger, C. R., Moutinho, H., and Matson, R. 1996. Growth of silicon thin layers on cast MG-Si from metal solutions for solar cells. *Solar Energy Mater. Solar Cells* 41/42:19–30.

Widenborg, P. I., and Aberle, A. G. 2007. Polycrystalline silicon thin-film solar cells on AIT-textured glass superstrates. Vol. 2007, *Adv. Optoelectr.* 24584.

Wilder, J. W. G., Venema, L. C., Rinzler, A. G., Smalley, R. E., and Dekker, C. 1998. Electronic structure of atomically resolved carbon nanotubes. *Nature* 391:59–62.

Woditsch, P., and Koch, W. 2002. Solar grade silicon feedstock supply for PV industry. *Solar Energy Mater. Solar Cells* 72:11–26.

Wu, X., Keane, J. C., Dhere, R. G., et al. 2001. In *Proceedings of the 17th European Photovoltaic Solar Energy Conference*, Vol. 2, Munich, Germany, October 22–26, p. 995.

Würfel, P. 1997. Solar energy conversion with hot electrons from impact ionization. *Solar Energy Mater. Solar Cells* 46:43–52.

Xing, J., Jin, K., He, M., Lu, H., Liu, G., and Yang, G. 2008. Ultrafast and high-sensitivity photovoltaic effects in TiN/Si Schottky junction. *J. Phys. D Appl. Phys.* 41:195103.

Xu, Y., Hu, Z., Diao, H., et al. 2006. Heterojunction solar cells with n-type nanocrystalline silicon emitters on p-type c-Si wafers. *J. Non-Cryst. Solids* 352:1972–75.

Yan, B., Yue, G., Owens, J. M., Yang, J., and Guha, S. 2006. Over 15% efficient hydrogenated amorphous silicon based triple-junction solar cells incorporating nanocrystalline silicon. In *Proceedings of the IEEE 4th World Conference on Photovoltaic Energy Conversion*, Vol. 2, Waikoloa, HI, May 7–12, p. 1477.

Yu, K. M., Walukiewicz, W., Wu, J., et al. 2003. Diluted II-VI oxide semiconductors with multiple band gaps. *Phys. Rev. Lett.* 91: 246403.

Yum, J.-H., Chen, P., Grätzel, M., and Nazeeruddin, M. K. 2008. Recent developments in solid-state dye-sensitized solar cells. *ChemSusChem* 1:699–707.

Zhao, J., Wang, A., Green, M. A., and Ferrazza, F. 1998. 19.8% efficient "honeycomb" textured multicrystalline and 24.4% monocrystalline silicon solar cells. *Appl. Phys. Lett.* 73:1991.

Zhao, Z. X., Cui, R. Q., Meng, F. Y., Zhou, Z. B., Yu, H. C., and Sun, T. T. 2005. Nanocrystalline silicon thin films deposited by high-frequency sputtering at low temperature. *Solar Energy Mater. Solar Cells* 86:135–44.

2

Optical Properties of Nanostructures

Kylie Catchpole

2.1 INTRODUCTION

Optical nanostructures are everywhere in the world around us. To give just a few examples, the bright white color of paint, sunscreen, and toothpaste, the black of soot, and the brilliant colors of butterfly wings and holograms on credit cards are all due to optical nanostructures. To broaden the field even more, everything that you see that has a matte appearance, that is, is not shiny, appears that way because it is scattering light and has a nanoscale optical structure.

In this chapter, I will discuss how optical nanostructures can lead to changes in the reflection, scattering, and absorption properties of a material and how these effects can be used to improve the efficiency of solar cells. I will also discuss photoluminescence, since this has an important relation with solar cell efficiency. Photoluminescence is less ubiquitous than other nanoscale optical effects, but it can be seen in fluorescent road signs and safety vests, as well as the striking colors of gemstones when illuminated by ultraviolet light in geological displays.

It is important to understand that when we discuss nanoscale optical phenomena, we are far away from the region of geometrical optics, where light can be viewed as a ray that reflects or refracts from interfaces. We are in the region of wave optics, where a particle can scatter more light than is incident on it, where light can be completely blocked from propagating through a material whose components are completely transparent, and where many other weird and wonderful things can occur, as we shall see below.

The length scales involved in optical nanostructures are generally much larger than those involved in nanostructures that have electronic effects. This is because optical nanoscale effects generally occur on the scale of

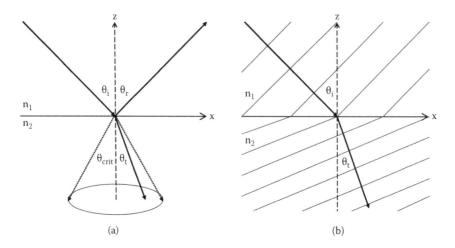

FIGURE 2.1

(a) Reflection and refraction at an interface according to Snell's law showing the incident, reflected, and transmitted angles, along with the critical angle and the escape cone. (b) Refraction showing the wavefronts in each medium. The number of wavefronts per unit length in the *x*-direction is unchanged.

the wavelength of light, which is of the order of hundreds of nanometers, whereas electronic nanoscale effects occur on the scale of the wavelength of electrons, which is of the order of 1 nm.

2.2 BACKGROUND

This section contains some basic background necessary for understanding optical phenomena. For more detail, see the very readable text by Hecht (2001).

2.2.1 Snell's Law and the Momentum of a Photon

A good starting point is the familiar situation of refraction at an interface according to Snell's law. A plane wave is incident at an angle θ_i on an interface between two materials with refractive indices n_1 and n_2 (Figure 2.1a). This results in reflection with an angle θ_r, $\theta_r = \theta_i$ and transmission with an angle θ_t that satisfies

$$n_1 \sin\left(\theta_i\right) = n_2 \sin\left(\theta_t\right) \tag{2.1}$$

For light incident from a lower index medium (for convenience let us say that $n_1 < n_2$), the maximum refracted angle, or critical angle, θ_{crit}, occurs for light incident with an angle of close to 90°. This is important for photovoltaics because it defines an escape cone. If light is to be trapped within a solar cell, it needs to be propagating in the semiconductor (medium 2) with an angle beyond the escape cone.

It is sometimes not realized that Snell's law can be related directly to conservation of momentum. The energy of a photon is given by $E = hf$, where f is the frequency of the radiation and h is Planck's constant. This means that a photon in a vacuum carries momentum $p = E/c = h/\lambda$ since $c = \lambda f$. (Here c is the speed of light in a vacuum and λ is the wavelength of the light.) The momentum of a photon traveling in the xz-plane can be separated into an x-component given by $p \sin(\theta)$ and a z-component given by $p \cos(\theta)$. An interface defined by $z = $ constant can change the z-component of the momentum, but not the x-component. This means that we must have $p_{x1} = p_{x2}$. The momentum p is directly proportional to the wavevector $k = 2\pi/\lambda$. Thus, the conservation of the x-component of the momentum is often written as the conservation of the x-component of the wavevector:

$$k_{x1} = 2\pi/\lambda_1 \sin\left(\theta_i\right) = 2\pi/\lambda_2 \sin\left(\theta_t\right) = k_{x2} \qquad (2.2)$$

We can divide this equation by 2π and multiply by c/f (making use of the definition of refractive index as the ratio of the speed of light in vacuum to the speed of light in a medium, $n_1 = c/(\lambda_1 f)$) to arrive at Snell's law. We can see from Figure 2.1b that the number of wavefronts per unit length in the x-direction, which is proportional to k_x and to p_x, is unchanged by refraction.

A physical feeling for the x-component (also called the in-plane component) of the wavevector is very useful in understanding nanoscale optical phenomena. For example, as we have seen above, when light incident from air passes through a planar interface, the x-component of its momentum is unchanged. This means that it can also escape from a planar thin-film structure by transmission through the rear or by reflection at the rear and transmission through the front surface. In order to trap light within a solar cell structure, the in-plane component of its momentum must be changed.

2.2.2 Polarizability, Permittivity, and Refractive Index

In this section we explore the relationship between the familiar concept of refractive index and the less familiar microscopic optical responses of materials. This is important because when designing nanoscale optical structures, it is necessary to be aware of the origin of the optical properties of materials, as well as how the different optical properties are interconnected, and cannot be engineered in isolation.

When Maxwell's equations are solved in free space, traveling wave solutions are found that have a velocity $c = (\varepsilon_0\mu_0)^{-1/2}$, where $\varepsilon_0 = 8.85 \times 10^{-12}$ Fm^{-1} and $\mu_0 = 4\pi \times 10^{-7}$ NA^{-2} are the electric permittivity and magnetic permeability of free space, respectively. Similarly, electromagnetic waves in a medium with relative permittivity ε and relative permeability μ propagate with velocity $v = (\varepsilon\varepsilon_0\mu\mu_0)^{-1/2}$. The definition of refractive index is $n = c/v$, so the relationship between refractive index and permittivity is

$$n = \sqrt{\varepsilon\mu} \cong \sqrt{\varepsilon} \qquad (2.3)$$

since $\mu \cong 1$ in the optical part of the electromagnetic spectrum. Note that Equation 2.3 also holds for complex values of the refractive index and dielectric function, i.e., $\tilde{n} = n + ik$ and $\varepsilon = \varepsilon' + i\varepsilon''$, although the above argument must be modified in that case. In the case of complex $\tilde{n}$ and ε, the imaginary parts are related to absorption in the material. Although the permittivity, ε, is a less familiar concept than the refractive index, it is more appropriate when describing the microscopic optical behavior of materials, so we will use it for the rest of this section, along with illustrations of the resulting effect on n and k. The extinction coefficient, k, may also be less familiar than the absorption coefficient, α, which is used to describe the exponential attenuation of intensity, I, with distance, d, according to the Beer-Lambert law, $I = I_0 \exp(-\alpha d)$. The absorption coefficient and the extinction coefficient are related by $\alpha = 4\pi k / \lambda$, which can be derived in a straightforward way by considering the intensity of an electromagnetic wave at two points (Pankove 1971).

In the following section we will derive an approximate expression for ε in order to gain insight into how the electronic structure of a material determines ε, and hence n, k, and α.

Maxwell's equations in a vacuum relate the electric displacement, $\mathbf{D}$, to the electric field, $\mathbf{E}$, via $\mathbf{D} = \varepsilon_0\mathbf{E}$. The electric displacement relates to charge densities and is measured in Coulombs per square meter, while the electric field is related to forces and potential differences and is measured in volts per meter. When an electromagnetic field is applied to a material, the material polarizes. At optical frequencies we are mostly concerned with electronic polarization, which is a shift of the electron cloud relative to the nucleus. The polarization leads to an additional term in the expression for the electric displacement:

$$\mathbf{D} = \varepsilon_0 \mathbf{E} + \mathbf{P} \qquad (2.4)$$

Since, for many materials, the polarization is linearly dependent on the field, we can write this as

$$\mathbf{D} = \varepsilon_0 \mathbf{E} + \mathbf{P} = \varepsilon_0 (1 + \chi_e) \mathbf{E} = \varepsilon \varepsilon_0 \mathbf{E} \qquad (2.5)$$

where χ_e is the electric susceptibility of the material. Thus, if we could derive an equation for the polarization density, **P**, as a function of **E**, we could also obtain ε. In order to do this we assume that the movement of charges in the material can be described by classical harmonic oscillators, i.e., springs. This turns out to be an accurate description even when the system is described quantum mechanically, although the parameters need to be reinterpreted in a quantum mechanical sense. For the purposes of illustration we consider the force between an electron cloud and its nucleus to be represented by one harmonic oscillator. Using **F** = *m***a**, the equation of motion of the oscillator is given by

$$\mathbf{F} = q\mathbf{E} - K\mathbf{x} - b\frac{d\mathbf{x}}{dt} = m\frac{d^2\mathbf{x}}{dt^2} \qquad (2.6)$$

where $q\mathbf{E}$ is the driving force on an electron with charge q, $-K\mathbf{x}$ is the restoring force with spring constant K, $b\,d\mathbf{x}/dt$ is the damping force, which describes losses in the system, and m is the mass of the electron. Using the fact that **E** has a time dependence given by $\cos(\omega t)$, this can be solved to obtain

$$\mathbf{x} = \frac{(q/m)\mathbf{E}}{\omega_0^2 - \omega^2 - i\gamma\omega} \qquad (2.7)$$

where $\omega_0^2 = K/m$ is the resonant frequency of the oscillator and $\gamma = b/m$ is the decay rate. The induced dipole moment **p** of an electronic oscillator is given by $\mathbf{p} = q\mathbf{x}$. The polarization density **P** is then the dipole moment of an oscillator multiplied by the number of oscillators per unit volume N, i.e.,

$$\mathbf{P} = N\mathbf{p}$$

$$= \frac{(Nq^2/m)\mathbf{E}}{\omega_0^2 - \omega^2 - i\gamma\omega} \tag{2.8}$$

$$= \frac{\omega_p^2}{\omega_0^2 - \omega^2 - i\gamma\omega}\varepsilon_0\mathbf{E}$$

Here $\omega_p = q\sqrt{N/m\varepsilon_0}$ is the plasma frequency, which is especially important in the optical response of metals and heavily doped semiconductors.[*] The plasma frequency and plasmonics for solar cell applications are discussed in detail in Chapter 11.

From Equation 2.5 it then follows that

$$\varepsilon = 1 + \frac{\omega_p^2}{\omega_0^2 - \omega^2 - i\gamma\omega} \tag{2.9}$$

Plots of the real and imaginary parts of ε, along with n and k for a one-resonator system, are given in Figure 2.2. We can interpret these results in the following way. Far below a resonance, the oscillator responds only weakly to the driving field. As the frequency increases toward a resonance, the oscillator begins to resonate, and the amplitude and the damping of the oscillation increase. Far above all resonances, the oscillators can no longer respond fast enough to the driving field, and ε and n approach 1.

We can also see that between resonances, ε' and the refractive index increase with frequency, whereas at a resonance they decrease with frequency.

The fact that oscillators cannot respond immediately to the applied field leads to the Kramers–Kronig relations:

$$\varepsilon'(\omega) = 1 + \frac{2}{\pi}P_c\int_0^\infty \frac{\Omega\varepsilon''(\Omega)}{\Omega^2 - \omega^2}d\Omega \tag{2.10}$$

or equivalently,

---

[*] Equation 2.8 applies to isolated oscillators, e.g., gases. For dense materials there is a correction due to the induced field of surrounding oscillators (see Hecht 2001).

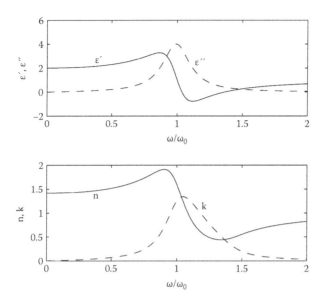

FIGURE 2.2

Plots of the real and imaginary parts of (a) ε and (b) n and k for a one-resonator system.

$$n(\omega) = 1 + \frac{2}{\pi} P_c \int_0^\infty \frac{\Omega k(\Omega)}{\Omega^2 - \omega^2} d\Omega \qquad (2.11)$$

where P_c is the Cauchy principal value of the integral and the integration is performed over angular frequency Ω. The derivation of the Kramers–Kronig relations is rather mathematical (see, for example, Stenzel 2005), but they are very useful for understanding the relationship between the real and imaginary parts of the dielectric function and the complex refractive index. By inspection of Equation 2.10, we can see that if a material has no absorption at all at any frequency ($\varepsilon'' = 0$), then ε' and hence n will be equal to 1 (here we are actually describing a vacuum). In contrast, if a material has strong absorption over a large wavelength range (for example, a semiconductor), it will have a high refractive index.

2.3 LAMBERTIAN SCATTERING

The interaction of light with materials structured on the scale of the wavelength is well understood only for a few limited cases, so we restrict the discussion here to isotropic scattering and, in the next sections, periodic structures and single particles. For more general structures, work to date has been largely empirical, although since computing power is now becoming available to calculate the behavior of such systems, we can expect progress to accelerate.

The benchmark for light trapping in solar cells is the Lambertian case, which assumes lossless, isotropic scattering of light. Isotropic scattering leads to a radiance (intensity per unit solid angle) that varies as $\cos(\theta)$. For a solar cell with refractive index n with a Lambertian front surface, at each scattering event, a fraction $1/n^2$ is scattered out of the solar cell into the air. Without a rear reflector, this leads to an average path length for light within the solar cell of $2n^2W$, where W is the thickness of the solar cell. That is, the path length is enhanced by a factor of $2n^2$, compared to a single pass across the device. The factor of 2 is due to the oblique propagation of light across the device. If the solar cell also has an ideal rear reflector, the path length enhancement is increased to $4n^2$. The derivation of the path length enhancement for Lambertian light trapping is readily available elsewhere, so it is not repeated here (Yablonovitch 1982; Green 1995).

The exact fraction of light that is coupled out for a real scattering structure will depend on the details of the structure, but we can expect that for a strongly scattering semiconductor structure the coupled out fraction will be small (though this does not mean it is negligible!). The reason for the small fraction of out-coupled light is the high refractive index, or equivalently, the high polarizability, of semiconductor materials. This can be understood through the use of a simple analogy. Imagine you are standing on a large plate of jelly (or Jell-O, if you are American), and jumping gently up and down. You can immediately see that most of your energy will be going into making the jelly move, and that only a small fraction will go into making the air move. The reason for this is the high polarizability of the jelly; i.e., it takes much more force to make jelly move than to make air move.

An alternative way of thinking about Lambertian light trapping, which is particularly useful for wavelength-scale structures, uses the concept of optical mode density. An optical mode is characterized by a frequency, ω, and an in-plane wavevector, k_x. For thick substrates such as wafers that are many wavelengths thick, k_x can take on a continuously varying range of values and there is a continuous distribution of modes, as illustrated in Figure 2.3a. For isotropic (Lambertian) radiation and for thick substrates with refractive index n, the density of optical modes is given by Boyd, 1983.

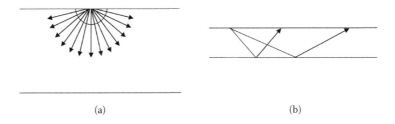

(a)   (b)

FIGURE 2.3
A schematic illustration of the photonic modes in (a) a thick substrate and (b) a thin (waveguiding) layer. In a thick substrate there is a continuous distribution of modes while in a thin layer only a few modes can propagate.

$$\rho = \frac{\omega^2 n^3}{\pi^2 c^3} \tag{2.12}$$

We can see from Equation 2.12 that the density of modes is much higher in a semiconductor than in air. Isotropic light scattering distributes light equally among all the optical modes available. There are more optical modes available in the semiconductor than the air, so more light is scattered into the semiconductor. The absorption in the semiconductor is enhanced by a factor proportional to n^2 rather than n^3 because of the reduced speed of propagation of light in the semiconductor (Brendel 2003).

For very thin solar cells with a thickness comparable to the wavelength of light, light that is transmitted into the semiconductor interferes with itself, either constructively or destructively, depending on the path length across the device. As a result, only certain values of k_x are possible for propagating waves. These values of k_x correspond to certain effective angles of propagation, as illustrated in Figure 2.3b, although strictly for such a thin layer we should imagine light as a field rather than as a ray. Thus, the solar cell acts as a waveguide along which only certain modes may propagate, and the number of optical modes is reduced compared to the case for a thick substrate. Because of this reduction in optical mode density within the semiconductor, the fraction of light coupled into the solar cell is reduced under isotropic scattering; i.e., the effectiveness of Lambertian light trapping is reduced for very thin solar cells (Stuart and Hall 1997). This reduction is not very severe for silicon and other high-index semiconductors. For example, for a structure supporting three modes (corresponding to a 165 nm thick silicon film on glass, for 1,000 nm wavelength light), the path length enhancement is already within 25% of the Lambertian limit valid for thick structures. However, for organic materials or other nanoengineered materials that have a lower refractive index and which also tend to be very thin, the reduction in optical modes available could have a significant effect. For a 200-nm thick solar cell with refractive index of 2 deposited on glass, the maximum path length enhancement for a 800-nm light will only be about half that of the Lambertian limit for thick structures.

2.4 PERIODIC PHOTONIC STRUCTURES

Periodic structures have interesting optical properties and can provide very effective light trapping and antireflection. Aside from the idealized Lambertian case, periodic structures are the only other case of large-scale structure for which the optical properties can be calculated relatively easily at present. As such, periodic structures are both important in and of themselves and a way to start understanding more complicated aperiodic structures.

There are three main types of periodic photonic structures that have been used to enhance the efficiency of solar cells: diffraction gratings, Bragg stacks (also called 1D photonic crystals), and 3D photonic crystals. Some of the properties of the three types of structures are illustrated in Figure 2.4. Diffraction gratings can trap light in the x-direction (for groove-type gratings) or in the x- and y-directions (for pillar-type gratings). Bragg stacks are relatively simple to fabricate and can give very high reflectance, but this is limited to a relatively narrow range of incident angles. Three-dimensional photonic crystals can give very high reflectance for the full range of incident angles but require a more complicated fabrication process. These properties of periodic photonic structures occur over a particular wavelength range; the wavelength range can be either broad or narrow, depending on the design of the structure. Periodic photonic structures are also responsible for the brilliant colors of butterfly wings and opals (http://www.webexhibits.org/causesofcolor/).

To understand the behavior of these photonic structures, it is useful to look at the optical modes that can propagate within them. We consider the Bragg stack first, since it is the simplest case.

When light is incident on a periodic photonic structure, it interferes with itself to create standing waves. There are only certain types of standing

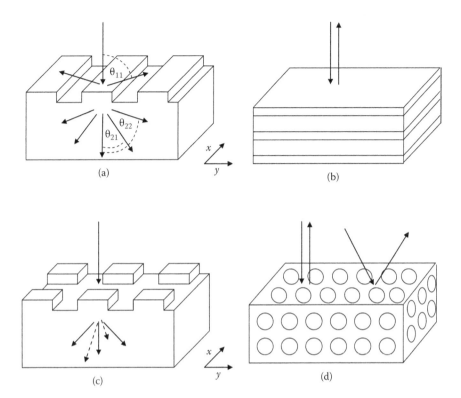

FIGURE 2.4
(a) Rectangular groove diffraction grating. (b) Bragg stack. (c) Pillar diffraction grating.
(d) 3D photonic crystal.

waves that are possible, and these are known as the modes of the structure. For example, for the two modes illustrated in the periodic structure
in Figure 2.5a, either the nodes or the antinodes of the standing wave
are in the high-index region (Joannopoulos et al. 1995). This leads to two
types of modes with two different frequency ranges. Between these two
frequency ranges is a photonic bandgap, where light interferes destructively with itself and hence cannot propagate through the material. Thus,
photonic crystals are highly reflective within the photonic bandgap. For
a 1D photonic crystal, or Bragg stack, this high reflectivity only occurs
for near-normal angles of incidence, but for 3D photonic crystals it can
occur for all angles of incidence. The optical behavior of photonic crystals is related to the electronic band structure of semiconductors such as
silicon. In semiconductors the wavefunctions of the electrons interfere to
give energies at which electrons may travel through the semiconductor
and energies at which they may not (the electronic bandgap).

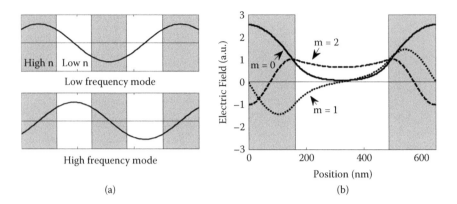

FIGURE 2.5

(a) Standing waves within a 1D photonic crystal. (b) The *x*-component of the electric field for modes within a rectangular groove diffraction grating for a period of 650 nm and wavelength of 1,000 nm. (After Catchpole, K. R., "A Conceptual Model of the Diffuse Transmittance of Lamellar Diffraction Gratings on Solar Cells," *J. Appl. Phys.* 102 [2007]: 013102.) The gray region is silicon and the white region is air.

The propagation of light within photonic crystals and the propagation of light in diffraction gratings are also closely related. Diffraction gratings can form a very effective structure for trapping light within solar cells, as well as leading to a great deal of insight into the optimization of light trapping structures. We will first discuss the angles at which light is diffracted and the constraints this places on the optimum period of the grating. This will be followed by a discussion of how the fraction of light coupled into the solar cell can be maximized and how this is related to the modes of a photonic crystal.

Figure 2.4a shows an example of the type of grating under consideration. The light is normally incident from a material with refractive index n_1 and enters a material with refractive index n_2. Diffracted reflected and transmitted light is present in regions 1 and 2 with diffracted angles θ_{1m} and θ_{2m}, respectively, where m is an integer. The diffracted angles are given by the well-known grating equation for normally incident light:

$$n_1 \sin\theta_{1m} = n_2 \sin\theta_{2m} = \frac{m\lambda}{L} \tag{2.13}$$

where L is the diffraction grating period. At these diffracted angles the wavefronts incident on each period of the grating interfere constructively because their path difference is an integral multiple of the wavelength.

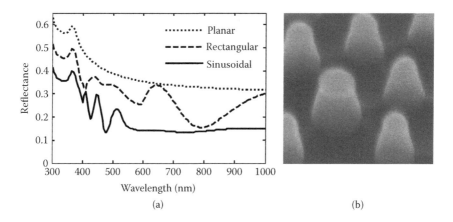

FIGURE 2.6

(a) The reflectance of a sinusoidal and a rectangular grating on a silicon substrate, both with period 200 nm and height 200 nm, showing the reduced reflectance obtained with a sinusoidal profile. In both cases, the reflectance could be reduced further with a low-index coating. The reflectance of planar silicon is shown for reference. (b) Nanopillars etched in silicon. (Reproduced from Inns, D., P. Campbell, et al., "Wafer Surface Charge Reversal as a Method of Simplifying Nanosphere Lithography for RIE Texturing of Solar Cells," *Adv. Optoelectronics* [2007]: 60:32707. With permission.) The cone shape of the pillars contributes to the antireflection effect. Here the feature size is larger, leading to scattering as well as antireflection. On the tops of some of the pillars the silica nanospheres used as an etching mask can be seen.

Equation 2.13 can be rearranged to give an expression for the in-plane wavevector of the diffracted wave. So in region 2, for example, we have

$$k_x = \frac{2\pi n_2}{\lambda} \sin \theta_{2m} = m \frac{2\pi}{L} \qquad (2.14)$$

Thus, another way of thinking of the action of a grating is that it can change the in-plane wavevector of incident light by an integral multiple of $2\pi/L$. The quantity $(2\pi/L)\hat{\mathbf{x}}$ is known as the grating vector, where $\hat{\mathbf{x}}$ is the unit vector in the x-direction.

From Equation 2.13, we can see that light that is diffracted into a semiconductor layer with refractive index n_2 will only be able to escape into air (with refractive index $n_1 = 1$) when $m\lambda/L < 1$. This is an important condition for the design of a grating on a solar cell because it is desirable that diffracted orders are trapped. Note that the above condition does not depend on the refractive index of the solar cell material (Eisele et al. 2001). We can also see that for $\lambda/n_2 < L < \lambda$ there will be diffracted orders that propagate within the semiconductor but no diffracted orders that propagate in air; i.e., a measurement of the reflectance and transmittance of the structure taken in air will show no diffraction, even though light is being diffracted within the semiconductor. (This is also an important point for scattering structures; measurements taken in air give only limited information about the scattering that occurs within the semiconductor [Krc et al. 2003; Stiebig et al. 2006]).

The concept of photonic modes within periodic structures is also useful for understanding diffraction gratings. In fact, the modes that propagate within a rectangular diffraction grating are the same modes that propagate in a 1D photonic crystal. The only difference is that the angle of incidence has changed by 90°. Figure 2.5b shows the modes that can propagate within the grating region of a rectangular air/silicon diffraction grating. Each mode has an effective refractive index that is related to the fraction of the energy of the mode that propagates within the high-index material. In Figure 2.5b, mode 0 has 97% of its energy in the silicon and has an effective refractive index of 3.4, close to the value for silicon of 3.7. Mode 2 has 48% of its energy in the silicon, and has an effective refractive index of 1.4, between that of silicon and air. Normally incident light can couple to both modes 0 and 2, but not to mode 1, due to its symmetry properties. Because modes 0 and 2 have different effective refractive indices, the modes propagate at different speeds, and this leads to interference. Transmission through the diffraction grating is maximized when the modes within the grating interfere constructively. This means that the ideal height for a rectangular diffraction grating can be quite easily calculated (Catchpole 2007). The effective refractive indices, n^{eff}, are solutions of the photonic crystal equation:

$$f(n^{\text{eff}}) = \cos(k_{xr}L_r)\cos(k_{xg}L_g) - \frac{1}{2}\left(\tau\frac{k_{xr}}{k_{xg}} + \frac{1}{\tau}\frac{k_{xg}}{k_{xr}}\right) \times$$

$$\sin(k_{xr}L_r)\sin(k_{xg}L_g) = \cos(k_x L) \tag{2.15}$$

This is a transcendental equation, but it can be easily evaluated numerically. Here k_0 is the wavevector of the incident wave, $k_x = k_0 \sin(\theta_{in}) = (2\pi/\lambda)n_1 \sin(\theta_{in})$ is the x-component of the wavevector of the incident wave, and

$$k_{xi} = k_0(n_i^2 - (n^{\text{eff}})^2)^{1/2}; \quad i = r, g \tag{2.16}$$

are the x-components of the wavevectors in the ridges (r) and grooves (g) of the grating, respectively. τ is equal to 1 for transverse-electric (TE) polarization and n_r^2 / n_g^2 for transverse-magnetic (TM) polarization, and $L = L_r + L_g$. Constructive interference and hence transmission through the grating is maximized when the grating height is equal to an odd integral multiple of

$$h = \frac{\lambda}{2 \left| n_2^{\text{eff}} - n_0^{\text{eff}} \right|} \tag{2.17}$$

This leads to an optimum height for a silicon grating in air of 200 nm. The zeroth diffracted order tends to couple most effectively to the grating modes when there are few grating modes present. Using this together with some impedance-matching considerations (Catchpole 2007) also allows us to predict the optimum period for a rectangular grating. For silicon in air the optimum period is 650 nm. The method can also be extended to rectangular pillar diffraction gratings (Catchpole and Green 2007). For pillar diffraction gratings an approximation is made to determine the values of the effective indices in the grating region since there is no known analytical solution for the modes in the grating region.

There have been a number of different types of solar cells fabricated incorporating photonic structures. These include blazed gratings (Heine and Morf 1995; Morf et al. 1997) and rectangular gratings (Eisele et al. 2001; Stiebig et al. 2006) as well as a combined grating/Bragg stack (Zeng et al. 2006). Blazed gratings have the advantage of an asymmetric structure, which can reduce out-coupling. Dye-sensitized cells showing absorption enhancement due to the presence of a photonic crystal have also been demonstrated (Nishimura et al. 2003). Fabrication methods include interference lithography (in which 3D structures such as photonic crystals have recently been demonstrated [Jang et al. 2007]), micromolding (Azzaroni et al. 2005), and nanoimprinting (Guo 2007; Heijna et al. 2008).

2.5 NANOSTRUCTURES FOR ANTIREFLECTION COATINGS

Antireflection coatings are an important way of increasing the absorption in solar cells, since the reflectance of a bare semiconductor substrate is around 30%. Standard antireflection coatings are formed of a single or double layer of materials such as Si_3N_4, TiO_2, and MgF_2. For a thin film of a homogeneous material such as these on a substrate, the incident and reflected beams will interfere. If the thin film has a refractive index equal to the square root of that of the substrate, and if the film is one-quarter of a wavelength thick, the incident and reflected beams will cancel exactly, leading to no reflection at that wavelength. Multiple thin films can extend the wavelength range over which low reflectance can be achieved. However, the use of this approach is limited by the availability of materials with the appropriate refractive index.

The lack of suitable materials can be circumvented by the use of optical nanostructures. As the period of a grating is reduced, the number of diffracted orders also reduces. For sufficiently small periods the grating can act as if it is an *effective medium*, which only refracts and does not diffract the incident light. In this way the refractive index of a nanostructured material can be engineered, allowing much greater flexibility in the design of antireflection coatings.

For a rectangular grating, such as that described in the previous section, there are three conditions that must be met in order for the grating to be able to be described as an effective medium (Lalanne and Hutley 2003). The first is that only the zeroth diffracted order is present, i.e., the period of the grating, L, satisfies

$$L < \frac{\lambda}{n} \tag{2.18}$$

where n is the refractive index of the substrate. This means that the grating will only refract rather than diffract the incident light. The second condition is that there is also only one mode (the zeroth grating mode) present in the grating region. If this condition is not met, interference between the different grating modes can occur, leading to additional features in the reflectance and transmission spectra (Lalanne et al. 2006). The index of the effective medium for this rectangular grating will be the effective index of the zeroth grating mode as described in the previous section. The minimum reflectance occurs at the quarter wavelength condition $h = \lambda/(4n^{eff})$. The third condition is that the grating is thick enough that evanescent waves cannot tunnel through the grating, which could be the case for grating thicknesses less than a quarter of a wavelength. As described in the previous section, for subwavelength pillar-type gratings (for example, a periodic array of nanowires), there is no exact result known for the effective refractive index of the mode within the grating region. There are a number of approximations available that are valid under different conditions (Lalanne and Lemercier 1996; Lalanne and Hutley 2003).

It is important to be careful when using approximations for the effective refractive index. Many approximations are strictly valid only in the long wavelength limit, where the wavelength of light is much larger than the period of the structure (Kikuta et al. 1995). This is usually not the case for nanostructures fabricated for the visible wavelengths of light, and yet effective medium approximations are frequently applied to these cases. For accurate results, Maxwell's equations should be solved for the system of interest where this is possible, for example, using rigorous coupled wave analysis (Moharam et al. 1995) for grating structures. Software is now commercially available to do this.

Reflectance can also be reduced by a nanostructure that leads to a gradual change in refractive index between the air and the substrate. An example of this is the moth eye structure, which has a period and height of around 200 nm with an approximately sinusoidal profile (Figure 2.6). Similar structures have also been etched in silicon, leading to very low reflectance (Kanamori et al. 1999). The optimum profile for antireflection has been determined to be a two-dimensional Klopfenstein taper (Grann et al. 1995).

The effective medium approach is an important conceptual tool for understanding the optical properties of subwavelength nanostructures. It is important to note, however, that the optimal surface structure for most designs of solar cells should have both antireflection *and* scattering properties.

2.6 SCATTERING FROM SINGLE PARTICLES

At the other extreme from diffraction from a periodic array is scattering from a single particle. This case is also of interest because it can be calculated without too much difficulty, and can lead to insights when studying more complicated optical structures. The scattering and absorption for a single sphere surrounded by an infinite medium were solved by Mie (1908). The results can be expressed as scattering and absorption cross-sections, C_{scat} and C_{abs}. These are defined as the power scattered or absorbed by a particle divided by the incident light intensity, and have units of area. Defining the extinction cross-section $C_{ext} = C_{scat} + C_{abs}$, we can interpret C_{ext} as the area over which the particle interacts with incident light. For particles comparable or smaller than the wavelength, C_{ext} can be much larger than the area of the particle. (This may appear to be paradoxical— see Bohren [1983] for a discussion of how a particle can absorb more than the light incident on it.)

It is useful to also define the scattering and absorption cross-sections normalized by the projected area of the particle, Q_{scat} and Q_{abs}. For a sphere, $Q_{scat} = C_{scat} / \pi r^2$ and $Q_{abs} = C_{abs} / \pi r^2$. (These are sometimes rather confusingly referred to as scattering and absorption efficiencies, which tends to reinforce the erroneous impression that they should have a maximum value of 1.) For a particle of dielectric function ε_1 that is small compared to the wavelength and embedded in a medium with dielectric function ε_m, we can make the quasi-static approximation that the electromagnetic field is uniform over the diameter of the particle at a given instant in time. This leads to the expressions

$$Q_{scat} = \frac{1}{6\pi}\left(\frac{2\pi}{\lambda}\right)^4 |\alpha|^2, \quad Q_{abs} = \frac{2\pi}{\lambda}\mathrm{Im}[\alpha] \tag{2.19}$$

where $\alpha = 3V(\varepsilon_1 - \varepsilon_m)/(\varepsilon_1 + 2\varepsilon_m)$ is the polarizability of the particle with volume V. Q_{scat} is proportional to the square of the polarizability because scattering can be viewed as a process of absorption followed by reradiation, both of which depend linearly on the polarizability. Because of this, larger particles scatter relatively more and absorb relatively less than small ones (up to the limit of the validity of the quasi-static approximation). We can also see from Equation 2.19 the well-known result that scattering increases as the wavelength of light decreases, leading to the red color of sunsets as blue light is scattered out of the direct beam from the sun.

A variety of calculation methods are available for particles of different shapes and sizes (MiePlot; Bohren and Huffman 1983; Mischenko et al. 2000). Some methods can also be used for particles on a substrate (Mischenko et al. 2004). For other cases it is necessary to use computationally intensive but flexible techniques such as finite element and finite difference methods (Taflove 1995; Volakis et al. 1998). It is important to note that results for single particles are likely to be significantly altered when interactions with neighboring particles are included. Nevertheless, the study of single particles is useful for obtaining an intuitive understanding, which may then be applied to multiparticle systems.

Examples of scattering structures currently used on solar cells are textured conducting oxides, textured metallic back reflectors, textured glass, and diffuse dielectric reflectors such as white paint (Cotter 1998; Krc et al. 2004; Springer 2005; Keevers et al. 2007). Other novel types of scattering structures being investigated include semiconductor nanowires, which can show high absorptance (Tsakalakos et al. 2007) and strong scattering (Muskens et al. 2008). Approaches for designing optimal scattering structures are also starting to emerge (Fahr et al. 2008).

2.7 ABSORPTION AND PHOTOLUMINESCENCE

For designing nanostructured solar cells, it is important to understand the relationship between absorption and photoluminescence, their impact on cell performance, and the scope for engineering these properties.

It is clearly desirable that a photovoltaic material should be a strong absorber. Strong absorbers reduce the amount of material required to make a solar cell, and also enable better carrier collection and increased voltages, due to decreased recombination. As an example, a strong absorber might have an absorption coefficient of $10^6\,cm^{-1}$. This leads to an absorption length of 10 nm (where the absorption length is defined as $1/\alpha$). For comparison, the absorption length of Si at a wavelength of 980 nm is 100 μm.

Absorption and photoluminescence are directly related since the absorptivity, A, of a body is equal to its emissivity, E, according to Kirchoff's Law. The absorptivity and emissivity measure how close a body comes to a black body with $A = E = 1$. The luminescence spectrum of a semiconductor is given by the generalized Planck equation (Wurfel 1982; Smestad and Ries 1992):

$$L(\hbar\omega) = E(\hbar\omega)\frac{2n^2}{h^3 c^2}\frac{(\hbar\omega)^3}{\exp[(\hbar\omega - \mu)/kT] - 1} \qquad (2.20)$$

where $\hbar\omega$ is the energy of the radiation and μ is the chemical potential. Here L is the spectral radiance, i.e., the power per unit area per projected solid angle per photon energy interval. From Equation 2.20 we can see that although luminescence may seem to be a loss, it is actually desirable that a photovoltaic material be highly luminescent. This is because high luminescence implies high emissivity (and hence high absorptivity) and high chemical potential. Chemical potential is equal to the difference in quasi-Fermi levels of electrons and holes and is related to the densities of electrons and holes via $np = n_i^2 \exp(\mu / kT)$, so a high chemical potential also implies high carrier densities and hence relatively low nonradiative recombination. This leads to high V_{oc}s. For maximum V_{oc} and hence maximum cell efficiency, the radiative recombination rate should be high compared to other recombination rates. The photoluminescence efficiency is defined as the radiative recombination rate divided by the total recombination rate under open-circuit conditions. High cell efficiencies have been achieved with silicon even though the photoluminescence efficiency is not high. The photoluminescence efficiency for a 200 μm thick c-Si solar with V_{oc} = 630 mV is about 10^{-4}. However, if the photoluminescence efficiency were 1, the V_{oc} would be about 850 mV (Smestad and Ries 1992). The effect of this can be seen in the V_{oc}s for GaAs solar cells, which are a larger fraction of the bandgap than for Si solar cells.

It is important to note that the radiative recombination rate is dependent on the optical environment. This leads to scope for manipulating the local density of optical states to increase the radiative recombination rate, for example, using photonic band structures or plasmonic nanostructures. This approach has been used in light-emitting diodes (Vuckovic et al. 2000) and in photoluminescence enhancement (Boroditsky et al. 1999; Biteen et al. 2005), and so from the above discussion, may also provide benefits in solar cells. However, it would be necessary to develop designs that can provide a significant enhancement across the solar spectrum and over a sufficiently large region within the device.

2.8 SUMMARY

To summarize, there is tremendous potential for the improvement of the optical properties of solar cells using nanostructures. Since optical nanostructures are already everywhere, it must be possible to make them cheaply. Using optical nanostructures, it is possible to manipulate the overall absorption of nanostructured materials and even fundamental properties like radiative recombination rates. Because of recent advances in fabrication and modeling, this is a very rapidly developing area of photovoltaics.

ACKNOWLEDGMENTS

I thank Prof. Albert Polman for his support for this work. I also acknowledge the financial support of an Australian Research Council fellowship (project DP0880017). This work is part of the research program of the Stichting voor Fundamenteel Onderzoek der Materie (FOM), which is financially supported by the Nederlandse organisatie voor Wetenschappelijk Onderzoek (NWO). It is part of an Industrial Partnership Program between FOM and the Foundation Shell Research.

REFERENCES

Azzaroni, O., P. L. Schilardi, et al. 2005. Surface-relief micropatterning of zinc oxide substrates by micromolding pulsed-laser-deposited films. *Appl. Phys. A Mater. Sci. Processing* 81:1113–16.

Biteen, J. S., D. Pacifici, et al. 2005. Enhanced radiative emission rate and quantum efficiency in coupled silicon nanocrystal-nanostructured gold emitters. *Nano Lett.* 5:1768.

Bohren, C. F. 1983. How can a particle absorb more than the light incident on it? *Am. J. Phys.* 51:323–27.

Bohren, C. F., and D. R. Huffman 1983. *Absorption and scattering of light by small particles.* New York: Wiley-Interscience.

Boroditsky, M., R. Vrijen, et al. 1999. Spontaneous emission extraction and Purcell enhancement from thin-film 2-D photonic crystals. *J. Lightwave Technol.* 17:2096.

Boyd, R. W. 1983. *Radiometry and the detection of optical radiation.* New York: Wiley.

Brendel, R. 2003. *Thin-film crystalline silicon solar cells.* New York: Wiley-VCH.

Catchpole, K. R. 2007. A conceptual model of the diffuse transmittance of lamellar diffraction gratings on solar cells. *J. Appl. Phys.* 102:013102.

Catchpole, K. R., and M. A. Green 2007. A conceptual model of light-coupling by pillar diffraction gratings. *J. Appl. Phys.* 101:063105.

Cotter, J. E. 1998. Optical intensity of light in layers of silicon with rear diffuse reflectors. *J. Appl. Phys.* 84:618.

Eisele, C., C. E. Nebel, et al. 2001. Periodic light coupler gratings in amorphous thin film solar cells. *J. Appl. Phys.* 89:7722–26.

Fahr, S., C. Rockstuhl, et al. 2008. Engineering the randomness for enhanced absorption in solar cells. *Appl. Phys. Lett.* 92:171114.

Grann, E. B., M. G. Moharam, et al. 1995. Optimal design for antireflective tapered two-dimensional sub-wavelength grating structures. *J. Optical Soc. Am.* 12A:333.

Green, M. A. 1995. *Silicon solar cells: Advanced principles and practice.* Sydney, Australia: University of South Wales.

Guo, L. J. 2007. Nanoimprint lithography: Methods and material requirements. *Adv. Mater.* 19:495.

Hecht, E. 2001. *Optics.* Reading, MA: Addison-Wesley.

Heijna, M., J. Loffler, et al. 2008. Nanoimprint lithography of light trapping structures in sol-gel coatings for thin film silicon solar cells. Paper presented at Materials Research Society, spring meeting, San Francisco.

Heine, C., and R. H. Morf. 1995. Submicrometer gratings for solar-energy applications. *Appl. Optics* 34:2476–82.

Inns, D., P. Campbell, et al. 2007. Wafer surface charge reversal as a method of simplifying nanosphere lithography for RIE texturing of solar cells. *Adv. Optoelectronics* Vol. 2007: 32707.

Jang, J.-H., C. K. Ullal, et al. 2007. 3D micro- and nanostructures via interference lithography. *Adv. Funct. Mater.* 17:3027–41.

Joannopoulos, J. D., R. D. Meade, et al. 1995. *Photonic crystals.* Princeton, NJ: Princeton University Press.

Kanamori, Y., M. Sasaki, et al. 1999. Broadband antireflection gratings fabricated on silicon substrates. *Optics Lett.* 24:1422.

Keevers, M. J., T. L. Young, et al. 2007. 10% efficient CSG minimodules. Paper presented at 22nd European Photovoltaic Solar Energy Conference, Milan, September 16.

Kikuta, H., H. Yoshida, et al. 1995. Ability and limitation of effective medium theory for subwavelength gratings. *Optical Rev.* 2:92.

Krc, J., F. Smole, et al. 2003. Potential of light trapping in microcrystalline silicon solar cells with textured substrates. *Progr. Photovolt.* 11:429–36.

Krc, J., M. Zeman, et al. 2004. Optical modelling of thin-film silicon solar cells deposited on textured substrates. *Thin Solid Films* 451–52:298–302.

Lalanne, P., J.-P. Hugonin, et al. 2006. Optical properties of deep lamellar gratings: A coupled bloch-mode insight. *J. Lightwave Technol.* 24:2442.

Lalanne, P., and M. Hutley. 2003. Artificial optical media properties—Subwavelength scale. In *Encyclopedia of optical engineering.* Boca Raton, FL: Taylor and Francis, pp. 62–71.

Lalanne, P., and D. Lemercier. 1996. On the effective medium theory of subwavelength periodic structures. *J. Modern Optics* 43:2063–85.

Mie, G. 1908. Beiträge zur optik trüber medien, speziell kolloidaler metallösungen. *Ann. Phys.* Vol. 330, pp. 377–445..

MiePlot. www.philiplaven.com/mieplot.htm.

Mischenko, M. I., J. W. Hovenier, et al. 2000. *Light scattering by non-spherical particles.* New York: Academic Press.

Mischenko, M. I., G. Videen, et al. 2004. T-matrix theory of electromagnetic scattering by particles and its applications: A comprehensive reference database. *J. Quant. Spect. Rad. Transfer* 88:357.

Moharam, M. G., E. B. Grann, et al. 1995. Formulation for stable and efficient implementation of the rigorous coupled-wave analysis of binary gratings. *J. Optical Soc. Am.* 12A:1068–76.

Morf, R. H., H. Kiess, et al. 1997. Diffractive optics for solar cells. In *Diffractive optics for industrial and commercial applications,* ed. J. Turunen and F. Wyrowski, 361 Berlin: Akademie Verlag.

Muskens, O. L., J. G. O. Rivas, et al. 2008. Design of light scattering in nanowire materials for photovoltaic applications. *Nano Lett.* 8:2638–42.

Nishimura, S., N. Abrams, et al. 2003. Standing wave enhancement of red absorbance and photocurrent in dye-sensitized titanium dioxide photoelectrodes coupled to photonic crystals. *J. Am. Chem. Soc.* 125:6306–10.

Pankove, J. I. 1971. *Optical processes in semiconductors*. Toronto: Dover.

Smestad, G., and H. Ries. 1992. Luminescence and current-voltage characteristics of solar cells and optoelectronic devices. *Solar Energy Mater. Solar Cells* 25:51.

Springer, J. 2005. Light-trapping and optical losses in microcrystalline silicon pin solar cells deposited on surface-textured glass/ZnO substrates. *Solar Energy Mater. Solar Cells* 85:1.

Stenzel, O. 2005. *The physics of thin film optical spectra*. Berlin: Springer.

Stiebig, H., N. Senoussaoui, et al. 2006. Silicon thin-film solar cells with rectangular-shaped grating couplers. *Progr. Photovolt.* 14:13.

Stuart, H. R., and D. G. Hall 1997. Thermodynamic limit to light trapping in thin planar structures. *J. Optical Soc. Am.* 14A:3001.

Taflove, A. 2000. *Computational electrodynamics: The finite-difference time-domain method.* Norwood, MA: Artech House.

Tsakalakos, L., J. E. Balch, et al. 2007. Strong broadband optical absorption in silicon nanowire films. *J. Nanophotonics* 1:013552.

Volakis, J. L., A. Chatterjee, et al. 1998. *Finite element method for electromagnetics: Antennas, microwave circuits and scattering applications.* New York: Wiley-IEEE.

Vuckovic, J., M. Loncar, et al. 2000. Surface plasmon enhanced light-emitting diode. *IEEE J. Quantum Electronics* 36:1131.

Wurfel, P. 1982. The chemical-potential of radiation. *J. Phys. C Solid State Phys.* 15:3967.

Yablonovitch, E. 1982. Statistical ray optics. *J. Optical Soc. Am.* 72:899.

Zeng, L., Y. Yi, et al. 2006. Efficiency enhancement in Si solar cells by textured photonic crystal back reflector. *Appl. Phys. Lett.* 89:111111.

3

Photovoltaic Device Physics on the Nanoscale

D. König

3.1 INTRODUCTION

Photovoltaic (PV) device physics on the mesoscopic scale combines quantum effects and ultrafast phenomena over several "units" of the device—quantum structures (QSs) or absorber-contact combinations—to merge into a classical macroscopic device behavior like a conventional p-i-n solar cell. Much about the energy band picture of QSs can be learned from their inverse counterpart as light-emitting and laser diodes based on superlattices (SLs). Other concepts like the intermediate band (IB) or hot carrier (HC) solar cells are in principle macroscopic devices, though they can be arguably realized best by QSs. Although hot carrier solar cells are based on ultrafast carrier dynamics combining suppressed cooling of generated excitons with very fast energy-selective carrier extraction, their device physics can be described in large part by classical semiconductor physics.

After a brief historical overview, we will introduce the device structures of third-generation PV energy conversion principles with respective material candidates. For brevity, we shall focus on the most important ones, as complete coverage is beyond the scope of this chapter. Section 3.2 evaluates the electronic and optical properties of PV QSs by investigating quantum dot (QD) SLs and their interdependence as a function of spatial and energetic parameters. Their tunable band gap enables them to be used in multiple-junction tandem PV cells. Section 3.3 deals with IB solar cells and possible implementations via localized and de-localized states of the IB. Hot carrier absorbers and solar cells are treated in Section 3.4 as a somewhat different concept to devices based on quantum confinement (QC), though core-shell nanocrystals (NCs) in a matrix are useful as hot carrier absorbers for phonon confinement. Multiple exciton generation (MEG) is an alternative process for harnessing the excess energy of free carriers above their respective band edges. Solar cell contacts are the most important interface to energy extraction, and thus one of the cornerstones for achieving a maximum conversion efficiency. Section 3.5 evaluates energy-selective contacts (ESCs) required for extracting high-energy carriers from hot carrier absorbers, low-loss contacts to QSs, and tandem cell interconnects.

Below, we give a brief historical summary of TG-PV, which is still a young and fast developing field of research. For single-junction, single-band-gap solar cells under nonconcentrated AM 1.5 conditions, the detailed balance limit of conversion efficiency is 30% (Shockley and Queisser 1961). In 1980, De Vos (1980) calculated the conversion efficiency of an infinite stack of tandem PV cells within the detailed balance limit to be 68% under 1 sun AM 1.5 conditions. This was closely matched in 1982 by the visionary work of Ross and Nozik (1982), who calculated efficiency limits of solar cells with the utilization of hot carrier populations, i.e., carrier extraction at energies prior to energy loss by phonon emission (thermalization). Their efficiency limit of 65% triggered some interest in nonconventional solar energy conversion concepts and paved the way to the concept of the hot carrier solar cell, which was introduced by Würfel (1997). In 1990, Barnham and Duggan (1990) introduced the concept of QWs into the i-layer of a AIGaAs p-i-n solar cell. Another concept, the intermediate band solar cell, was developed in the mid-1990s by Luque and Martí (1997). Both concepts converge if quantum confinement exists only for one carrier type, leading to one miniband within the band gap (Green 2000). Martin Green coined the term *third-generation PV* (TG-PV) for all nonconventional PV cell concepts (Green et al. 1999) and defined its goal to reach substantially higher conversion efficiencies at low manufacturing costs using abundant and environmentally benign materials (Green 2003).

Interest emerged for QD-based absorbers in the late 1990s by preliminary works of Lee and Tsakalakos (1997) on CdS-based systems and Gal et al. (1998), who introduced CdS quantum wires into a $CuInS_2$ i-layer and obtained higher short-circuit current densities, fill factors (FFs), and conversion efficiencies. Barnham et al. (2000) introduced the concept of a spectral QD concentrator based on a luminescent solar collector with InP QDs being used as optical downconverters via radiative transitions determined by the QD sizes and a quantum efficiency of radiation significantly exceeding 100%. The reemitted photons in a narrow spectral range are fed into a conventional solar cell having a band gap just below the photon energy, preventing the generation of hot charge carriers by high-energy photons that would dissipate their excess energy by phonon emission, heating up the solar cell. While this concept applies the tunable fundamental transition in QDs to an optical downconverter, it triggered research in Si-based QD-SLs due to prospective higher variability of the effective band gap, as opposed to QWs (Green et al. 2003). In addition to maximum quantum confinement given by structural constraints and leading to QDs as 0d systems, materials that have larger conduction and valence band offsets further increase the range of effective band gaps to higher energies. Combinations of these SLs culminate in the concept of the all-Si tandem solar cell (Green et al. 2005), with wide-band-gap matrix materials like Si_3N_4 or SiO_2.

3.1.1 Device Structures

Before we have a look at different device structures of TG-PV cells, we briefly evaluate a conventional tandem solar cell as a reference system.

The absorber layer in a conventional solar cell is primarily the so-called i-region, which is usually a medium-density doped semiconductor exposed to a drift field of a doped or field effect p/n junction. This drift field ensures carrier separation. Electrons diffuse to the n^+ doped region in order to replace electrons that have been collected. Holes diffuse to the p^+ doped region to recombine with electrons being fed back into the solar cell. The tunneling interconnection of the tandem cell will be discussed in Section 3.5. The drift field establishes a potential difference between the contacts of the solar cell. At maximum majority carrier concentration at the respective contacts, the majority diffusion current to the contacts is compensated by the drift of generated carriers back into the i-region and recombination losses. The solar cell has a maximum output voltage V_{OC} and no external current flow, with a maximum potential difference between the majority quasi-Fermi levels $E_{F,n}$ in the n^+- and $E_{F,p}$ in the p^+-region. Gradients of the majority quasi-Fermi level in the respective regions $\partial E_{F,n}/\partial x|_{i,n}$, $\partial E_{F,p}/\partial x|_{p,i}$ converge to zero and a minimum bending of the band edges E_C, E_V exists. With carriers being extracted from the contacts, a diffusion field is generated that drives carrier diffusion in order to refill the respective states emptied before by carrier extraction. This diffusion field is the cause of net carrier flux and works *against* the drift field implemented by the p/n junction. This is straightforward to see as the current in an illuminated solar cell flows *against* its potential difference between the contacts, which manifests the transition of the p/n junction from a passive dipole (resistor) to an active dipole (battery). The diffusion field increases with the flux rate (current) density of extracted carriers, thus decreasing the drift field—we get a higher net current density at a lower potential difference between the contacts. Eventually, the diffusion field converges against the drift field, resulting in the maximum net carrier flux density at a very small potential difference between the contacts of the solar cell, which is the short-circuit case (cf. Figure 3.1). The product of net current density and potential difference between the contacts of the solar cell yields the power that is extracted from the device. It has a maximum that is called the maximum power point (MPP), the optimum operating condition for a solar cell. The major modifications in TG-PV are made to the i-layer and to the contacts, the latter being a partial consequence of using modified absorbers.

It is instructive to look at a simple model of the current density-voltage (JV) characteristics of a p/n junction solar cell in order to explore the macroscopic options we have in increasing conversion efficiency. A very detailed discussion can be found in Böer (1992) and Green (1982). The JV characteristics of a one-diode solar cell model under illumination are described by the dark current density of the p/n junction shifted into the fourth JV quadrant by the short-circuit current density $\vec{j}_{SC}$:

$$\vec{j}(V_{bias}) = \vec{j}_0 \left[\exp\left(\frac{q V_{bias}}{A k_B T} \right) - 1 \right] - \vec{j}_{SC} \quad \text{for } V_{bias} \geq 0$$

or, with active parameters:

$$\cong \vec{j}_{SC} \left[\exp\left(\frac{q[V_{bias} - V_{OC}]}{k_B T} \right) - 1 \right]$$

(3.1)

V_{bias}, k_B, and T stand for the bias voltage, the Boltzmann constant, and the temperature, respectively. The open-circuit voltage, V_{OC}, is given by the V_{bias} of j_{SC} exactly compensating the forward drift current of the p/n junction. A is the ideality factor, which describes the recombination behavior over the p/n junction and should be close to 1 for minimizing carrier losses. For negligible recombination within the depletion layer of the i-region, A converges to 1. The saturation current density $\vec{j}_0$ is given by

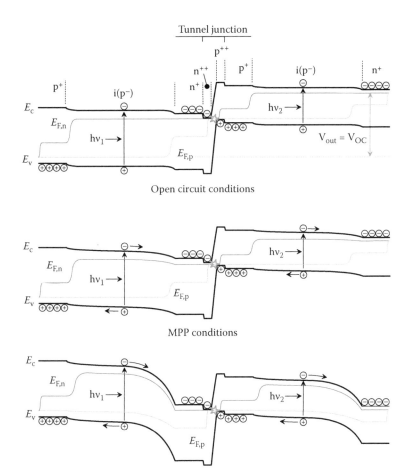

FIGURE 3.1

Band diagram of a conventional double-junction tandem solar cell with quasi-Fermi levels under illumination at open-circuit conditions (top), at maximum power point (MPP, middle), and at short-circuit conditions (bottom). Bending of band edges and quasi-Fermi levels increase with decreasing drift field strength, with the increasing diffusion field driving electrons/holes to the n^+/p^+-region. Extracted carriers perturb the equilibrium, with carriers from the i-region diffusing to the contacts for reestablishing the equilibrium situation, which is reached only at open-circuit conditions (zero net current).

$$\vec{j}_0 = q\left(n\frac{L_{\text{diff}}(n)}{\tau_n}\Big|_{\text{p-region}} + p\frac{L_{\text{diff}}(p)}{\tau_p}\Big|_{\text{n-region}}\right) \qquad (3.2)$$

where n, p, τ_n, τ_p, and L_{diff} stand for the electron and hole densities, the electron and hole lifetimes, and the diffusion length, respectively. Equation 3.2 considers the behavior of minorities—electrons in the p-region, holes in the n-region—and thereby constitutes an intrinsic recombination current density of minority carriers that results in carrier loss. We obtain a minimum of $\vec{j}_0$ for possibly small minority carrier densities (high doping levels), long minority carrier lifetimes (low defect density), and a short diffusion length of minority carriers. The latter requires that one highly doped region is very thin. Then, L_{diff} of the minority carrier type in the i-region (in Figure 3.1, these are electrons in the moderately p-doped i-region) is big enough to allow for minority carrier diffusion out of the depletion layer of the i-region in order to minimize recombination.

The short-circuit current density is given by the product of inpinging photon flux, $\vec{\Phi}$, and internal quantum efficiency, η_{IQ}, as a function of photon energy, $h\nu$, diminished by recombination losses in the bulk and at the interfaces of the semiconductor:

$$\vec{j}_{\text{SC}} = -q(\vec{\Phi}(h\nu)\eta_{\text{IQ}}(h\nu) - [\mathcal{R}\vec{d}_{\text{bulk}} + \mathcal{R}_{\|}\vec{d}_{\|}]) \qquad (3.3)$$

The recombination rates for the bulk $\mathcal{R}$ and the interfaces $\mathcal{R}_{\|}$ have to be integrated along the recombination path. As an approximation, we assume $\mathcal{R}$ and $\mathcal{R}_{\|}$ to be constant within the respective area, whereby the integration can be replaced by the product of the recombination rate over the corresponding layer thickness.

What can be derived from the above considerations for a maximum conversion efficiency? There are two major approaches. The obvious one is to minimize recombination losses. This requires a minimum density of recombination centers with their energies near the band edges, as evident from the Hall-Shockley-Read (HSR) formula (Hall 1952; Shockley and Read 1952) (cf. Equation 3.16). The saturation current density is a function of the carrier product $np = n_i^2$, whereby we have to rearrange this equation for the respective minority carrier type in the doped regions (cf. Equation 3.2). The intrinsic carrier density depends on the band gap energy, E_{gap}, as $n_i = \sqrt{N_C N_V}\exp(-E_{\text{gap}}/2k_BT)$, where N_C and N_V stand for the effective conduction and valence band density of states (DOS), respectively. For a wider band gap, the saturation current density, $\vec{j}_0$, diminishes, resulting in an increased V_{OC}. The diode characteristic is enhanced for a decreasing ideality factor A, requiring less recombination over the p/n junction. For a constant $\vec{j}_0$, V_{OC} decreases with A but is outweighed by an increasing fill factor, $FF = (V_{MPP} \vec{j}_{MPP})/(V_{OC} \vec{j}_{SC})$, so that the conversion efficiency still increases. Under illumination, the free carrier product is much bigger than the intrinsic carrier density, $np \gg n_i^2$. As a consequence, the split in quasi-Fermi levels, $E_{F,n}$-$E_{F,p}$, is large (cf. Figure 3.1). Recombination increases especially by deep recombination centers, which are located between $E_{F,n}$ and $E_{F,p}$.

Carrier recombination at surfaces and contacts is described by a parasitic resistor in parallel (R_P) to the ideal solar cell. The ideality factor contains another part of R_P given by recombination over the p/n junction and at interfaces under dark conditions. A loss in carrier energies occurs at the contacts, where current densities reach their maximum and an interface has to be penetrated. The latter results in carrier scattering and subsequent loss in carrier mobility. This can be compensated by a loss of carrier energy (potential) for acceleration into the contact metal. A similar loss is incurred by the finite conductivity of the solar cell, given by a carrier diffusion gradient and resulting diffusion potential. It increases with the diffusion carrier current density and occurs in regions with high diffusion current densities such as highly doped layers between contacts. These losses are given by a parasitic resistor in series (R_S) to the ideal solar cell. The right graph of Figure 3.2 shows the circuit model of a solar cell.

The other option for increasing the conversion efficiency is improving the internal quantum efficiency as a function of photon energy, $\eta_{IQ}(h\nu)$, for increasing the product $\vec{\Phi}(h\nu)\eta_{IQ}(h\nu)$. This is one basic ansatz of TG-PV. If $\eta_{IQ} \to 1$ over a wider range of $h\nu$, more free carriers are generated. Optical up- and downconversion by external cell structures go the opposite way by increasing $\vec{\Phi}(h\nu)$ while keeping $h\nu$ in the range of the band gap, without the need to change $\eta_{IQ}(h\nu)$ of the actual solar cell. In order to extend the range of $\eta_{IQ} \to 1$ to photon energies below the band gap of the nominal semiconductor, quantum structures like in Figure 3.3 or an intermediate band (IB) within the band gap is introduced for harvesting photons that otherwise would not generate free carriers.

The left graph in Figure 3.2 shows some dark and illuminated JV curves for $\mathcal{A} = 1.2$, with $\vec{j}_0$, R_S, and R_P as parameters.

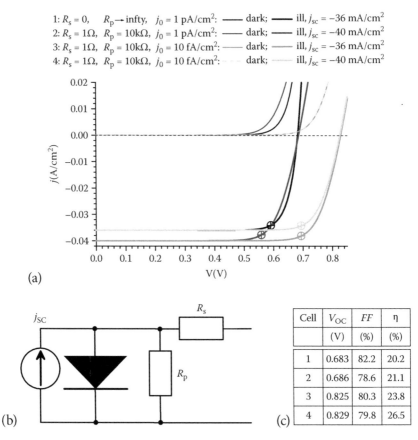

1: $R_s = 0$, $R_p \to$ infty, $j_0 = 1$ pA/cm²: ——— dark; ——— ill, $j_{sc} = -36$ mA/cm²
2: $R_s = 1\Omega$, $R_p = 10k\Omega$, $j_0 = 1$ pA/cm²: ——— dark; ——— ill, $j_{sc} = -40$ mA/cm²
3: $R_s = 1\Omega$, $R_p = 10k\Omega$, $j_0 = 10$ fA/cm²: ——— dark; ——— ill, $j_{sc} = -36$ mA/cm²
4: $R_s = 1\Omega$, $R_p = 10k\Omega$, $j_0 = 10$ fA/cm²: - - - dark; ········· ill, $j_{sc} = -40$ mA/cm²

(a)

(b)

Cell	V_{OC}	FF	η
	(V)	(%)	(%)
1	0.683	82.2	20.2
2	0.686	78.6	21.1
3	0.825	80.3	23.8
4	0.829	79.8	26.5

(c)

FIGURE 3.2

Dark and active JV curves of a single-junction solar cell with different $\vec{j}_0$, R_S, and R_P (left). Ideality factor is $\mathcal{A} = 1.2$ for all PV cells. Symbols show respective MPPs. The table lists the remaining parameters. The solar cell circuit is shown on the right.

First, we compare the solar cells with $\vec{j}_0 = 1$ pA/cm² (assuming same band gap host material), where one has $\vec{j}_{SC} = 36$ mA/cm² and negligible parasitic resistances, and the other has $R_S = 1\ \Omega$ and $R_p = 10$ kΩ, but a higher $\vec{j}_{SC}$ of 40 mA/cm². This corresponds to a comparison between a mass production high-efficiency solar cell and a solar cell that harvests more photons at the low-energy end, like the one depicted in Figure 3.3. The latter has a number of extra interfaces due to the QW-SL and was therefore deteriorated with a significant parasitic resistance, which can be seen in the lower fill factor. Still, the QS-SL cell arrives at a higher efficiency. Figure 3.2 shows that scattering losses can be overcompensated by exciton generation a few $k_B T$ below the band gap of the host semiconductor with subsequent thermionic

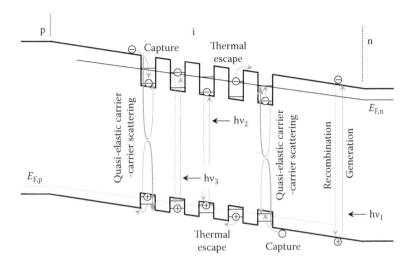

FIGURE 3.3

Band diagram of an illuminated solar cell with a QW-SL inserted into its i-region. (Based on Barnham, K. W. J., and Duggan, D., "A New Approach to High-Efficiency Multi-Band-Gap Solar Cells," *J. Appl. Phys.* 67 [1990]: 3490–93.) Apart from free carriers generated within the barrier material ($h\nu_1$), excitons confined to QWs can be generated at different excitation levels at lower photon energies ($h\nu_2$, $h\nu_3$). Quasi-elastic carrier-carrier scattering can excite carriers to an elevated energy level or even lead to carrier escape from the QW. At elevated QW levels, phonon absorption leads to thermionic emission from the QW.

carrier emission, resulting in an overall increase conversion efficiency. With an IB-SC, a bigger band gap can be used, with the IB providing the base for converting an appreciable part of photons with energies significantly below the band gap; see top graph of Figure 3.8. We consider two cases with a very low dark saturation current density of $\vec{j}_0 = 1$ fA/cm², accounting for the wider band gap like discussed above, and significant parasitic resistances, accounting for increased interface and defect scattering. One of them has $\vec{j}_{SC} = 36$ mA/cm² as the high-efficiency solar cell, and the other has $\vec{j}_{SC} = 40$ mA/cm², as shown for cells 3 and 4 in the left graph of Figure 3.2. Owing to the low value for $\vec{j}_0$ due to a bigger E_{gap}, efficiencies are much higher, while the extra bit of short-circuit current density can be delivered by the upconversion process via the IB. A larger band gap increases V_{OC} if $\vec{j}_0$ decreases. With the above considerations about the free carrier product $np = n_i^2$ under dark conditions, the ratio of minority carrier diffusion length to minority carrier lifetime emerges as the other main parameter. This is important in the context of a possible introduction of scattering and recombination centers by multiple interfaces or IB states. While these are likely to increase $\vec{j}_0$

and thereby diminish V_{OC} in addition to the parasitic resistances, a noncontinuous introduction of IB states as discussed in Section 3.3.3 may relieve the situation at the cost of loss in short-circuit current density, as given by solar cell 3 in Figure 3.2.

QS-SLs are very attractive due to their band gap varying with their spatial dimensions. Alternatively, if the barrier height is rather low, as in III-V SLs, low-energy photons or even phonons can be exploited for reexciting carriers (dissociating exitons) generated within the QSs onto the band edges of the barrier material. The former process bears close resemblance to an IB solar cell, though in a type I SL (cf. Figure 3.3) there would be two IBs given by the confinement of electrons *and* holes. The thermionic dissociation of confined excitons by phonon absorption is very attractive for thermophotovoltaics (TPV), where a massive phonon population is provided by a heat source. Phonon absorption by confined excitons results in a cooling of the absorber, which then works as a heat sink, and additional free carriers on the band edges of the wider-band-gap material (barrier) due to thermal dissociation of confined excitons in the quantum structures. While these effects are most pronounced for QD-SLs, the processing of QW-SLs is less difficult. Therefore, QW-SLs were of practical scientific interest initially. A first modification of the i-layer was discussed by Paxman et al. (1993), who developed the concept of inserting a AlGaAs/GaAs QW-SL as shown in Figure 3.3.

The original idea was to increase the short-circuit current density by additional carriers due to sub-band-gap exiton generation within the QWs. These excitons would dissociate by thermionic emission or by optical reexcitation due to low-energy photons, thereby providing additional electrons and holes to the carrier diffusion currents. Solar cells with strain-balanced growth of AlGaAs/GaAs QW-SLs were investigated experimentally by Barnham et al. (1996), who reported on a significant increase in open-circuit voltage, compared to control samples without a QW-SL, exceeding even the value of the best GaAs single-junction device at that time. This is a good example of a nominally higher $\vec{j}_{SC}$ combined with a significant R_S, leading to a lower fill factor, but still maintaining a high value of V_{OC}. The JV curve shift between cells 3 and 4 in Figure 3.2 depicts this situation to some extent. The issue of thermionic carrier emission from QW-SLs and its application to TPV devices was investigated in detail by Ekins-Daukes et al. (2003), who showed that an increase of the short-circuit current density is to be expected with an increasing number of InGaAs QWs of a SL included in the GaAs i-region. Recently, Oshima et al. (2008) obtained a 7% increase in the short-circuit current density of GaAs solar cells by inserting 20 layers of InAs QDs into N-diluted GaAs, thereby achieving strain-compensated growth with virtually no defects in the absorber. The increase in short-circuit current density occurred exlusively by the generation of sub-band-gap excitons in InAs QDs and subsequent thermionic emission onto the GaNAs band edges.

Depending on the magnitude of the band offsets, QSs can also be interpreted as an intermediate band solar cell. In this device, an asymmetric QS-SL creates a miniband at a favorable energy for a two-photon excitation from the valence band of the wide-band-gap material (matrix) to the miniband of the QS-SL, and from there to the conduction band of the matrix. An additional current is supplied by sub-band-gap photons and thereby increases the conversion efficiency with an additional carrier flow. This approach practically crunches three active PV layers into one and is the PV analogue to an optical upconverter where two lower-energy photons generate a localized exciton, which subsequently recombines radiatively at the band gap of the host matrix, thereby emitting a photon that can be reabsorbed by a wide-band-gap active PV layer. The inverse approach is optical downconversion, where a high-energy photon generates an exciton that then recombines radiatively via a mid-gap defect, emitting two photons of roughly half the energy of the initial excitation.

3.1.2 Material Considerations

There is an extensive range of materials applied to TG-PV devices in research. We classify the materials according to the device structures they are used for. Large-scale production of TG-PV devices has more restrictions, of which the abundance and toxicity of the chemical elements involved play a major role. Abundant and nontoxic materials were declared to be major boundary conditions for an environmentally benign large-scale production of TG-PV devices (Green 2003). Table 3.1 lists the relevant chemical elements with these properties.

If environmental concerns are ignored, Se, In, Te, and Hg are so scarce that a large-scale production is not feasible. These scarce materials are the most toxic as well, apart from In. The only element that is relatively abundant and very toxic is As. The restrictions to large-scale production will thus affect compounds containing these elements.

In this chapter, we will concentrate on Si as a prominent group IV material and on III-V compounds. Generally, the results presented here

TABLE 3.1

Abundance and Toxicity of Chemical Elements Used in Materials for TG-PV Devices

Chem. Element	Atomic Number	Abundance Weight % of Earth Crust	Toxicity	Used In	For
B	5	0.001	–	III-V	QS, HCA
C	6	0.02	–	IV	QS
Al	13	7.7	–	III-V	QS, HCA
Si	14	26.3	–	IV	QS, HCA
P	15	0.1	++	III-V	QS, HCA
S	16	0.03	–	I-VI, I-III-VI	QS
Cu	29	0.005	+	I-VI, I-III-VI	QS
Zn	30	0.007	+	II-VI	QS
Ga	31	0.0016	–	III-V	QS, HCA
Ge	32	1.4×10^{-4}	–	IV	QS
As	33	1.7×10^{-4}	+++	III-V	QS
Se	34	$\mathbf{5 \times 10^{-6}}$	+++	I-VI, I-III-VI	QS
Cd	48	2×10^{-5}	++	II-VI, II-III-VI	QS
In	49	$\mathbf{1 \times 10^{-5}}$	++	III-V, I-III-VI	QS, HCA
Sn	50	2×10^{-4}	–	IV	QS
Sb	51	2×10^{-5}	+	III-V	QS, HCA
Te	52	$\mathbf{1 \times 10^{-6}}$	+++	II-VI, I-III-VI	QS
Hg	80	$\mathbf{8 \times 10^{-6}}$	+++	II-VI	QS
Bi	83	2×10^{-5}	–	III-V, V	HCA

Source: Wiberg, N., *Lehrbuch der Anorganischen Chemie 101* [in German] (Berlin: De Guyter, 1995).

Note: N_2 and O_2 as nontoxic abundant gases left out for brevity. Numbers and symbols in boldface type show where chemical elements are likely to cause problems.

are transferable to other materials, though specific material properties may shift the significance of effects. This will be indicated in the text where applicable.

3.2 THE QUEST FOR OPTIMUM QUANTUM STRUCTURES: OPTICAL VS. ELECTRONIC PROPERTIES

One advantage of a QS-SL introduced into a solar cell is its tunable band gap as a function of quantum confinement. While it adds flexibility in spectral matching for QS-SL absorbers, compared to bulk-type absorbers, it is a mandatory property for tandem cells based on similar or even identical materials, like the all-Si tandem cell (Green et al. 2005). Another advantage is the tunable miniband width, which determines the optical and thermionic transition rates and carrier transport between QSs. Unfortunately, the optimum parameters for both phenomena are at the different ends of SL parameter space. This section attempts to show the options we have and sets out on a quest to find the mutual optimum eventually being the best solution for a highly efficient third-generation solar cell (TG-SC).

3.2.1 Parameters of Quantum Confinement

We briefly focus on quantum confinement of electrons, include the role of the barrier for quantum confinement, and investigate the DOS of quantum structures as a function of confinement. Holes can be treated in analogy, with respective boundary conditions like their effective mass within the quantum structure and in the barrier material.

The eigen-energies, $E_{tot,i}$, of confined carriers in a rectangular potential well of width a with finite potential barriers V_B can be solved iteratively with the following equations (Schiff 1968):

$$ka\tan(ka) = \sqrt{\gamma^2 - (ka)^2} \quad \text{for } E_{tot,i} \text{ with } i \in 2n+1$$

$$ka\cot(ka) = \sqrt{\gamma^2 - (ka)^2} \quad \text{for } E_{tot,i} \text{ with } i \in 2n, \text{whereby} \qquad (3.4)$$

$$\gamma^2 = \frac{2m_{\text{eff}}^W V_B a^2}{\hbar^2}$$

k is the 1d momentum, and m_{eff}^W is the effective mass of the confined electron within the potential well. With the $ka|_i$ known, we can determine the eigen-energies:

$$E_{tot,i} = \frac{\hbar^2 (ka|_i)^2}{2m_{\text{eff}}^W a^2} \qquad (3.5)$$

The eigen-energies are obtained analytically for $V_B \rightarrow \infty$ by

$$E_{tot}(n_x) = \frac{\hbar^2}{2m_{eff}^W}\left(\frac{2\pi}{a}\right)^2 n_x^2 \tag{3.6}$$

because the electron wave function does not leak into the potential barriers. For finite barrier heights, the electron wave function penetrates into the potential barriers and thus gets delocalized. Evaluating Equation 3.9 for small values of the transmission probability of the electron wave function, $T_T \leq 10^{-3}$, into the barrier and rearranging for the relevant parameters, we arrive at

$$a^2 m_{eff}^B (V_B - E_{tot}(n_x)) \geq \left[-\frac{\hbar \ln(T_T)}{2\sqrt{2q m_0}}\right]^2 \tag{3.7}$$

where the units for $V_B, E_{tot}(n_x)$, and m_{eff}^B are given in [eV] and [m_0], respectively. Equation 3.7 gives a lower limit for the parameter combination of QW extension, relative energy difference to the barrier height, and effective tunneling mass of the electron for which the case of $V_B \to \infty$ delivers sufficient accuracy. For the transmission probability of the electron wave function $T_T \leq 5 \times 10^{-4}$ into each barrier, the wave function is $\approx 99.9\%$ confined within the potential well, with an error in confinement energy of $\varepsilon(E) \approx 10^{-3} E_{tot}(n_x)$. With these initial conditions, we use the Si/SiO$_2$ 1d confinement system as an example, obtaining $V_B - E_{tot}(n_x) \geq 0.458$ eV for $a = 20$ Å and $V_B - E_{tot}(n_x) \geq 0.114$ eV for $a = 40$ Å, with $m_{eff}^B = 0.3m_0$ (König et al. 2007) as the lower limit above which we can use Equation 3.6. Other systems like GaAs/AlGaAs require much bigger values for a because they possess a small effective tunneling mass. The effect of quantum confinement decreases $\propto [m_{eff}^B]^{-1}$ as the barrier gets increasingly transparent to the electron wave function, counteracting electron confinement.

Equations 3.4 and 3.5 can be applied to 2d and 3d confinement cases by simple superposition due to the orthogonality of the electron wave functions, delivering more solutions per energy interval as confinement dimensionality increases. For the case of $V_B \to \infty$, superposition is expressed the same way. The general case of 3d confinement can then be described by

$$E_{\text{tot},\langle xyz\rangle} = \frac{\hbar^2}{2} \sum_i^{x,y,z} \left[\frac{2\pi}{a_i \sqrt{m_{\text{eff},i}^{\text{W}}}} n_i \right]^2 \qquad (3.8)$$

with Equation 3.6 obtained for 1d confinement. Further symmetry breaking of eigen-energies occurs for an anisotropic effective mass, $m_{\text{eff},x}^{\text{W}} \neq m_{\text{eff},y}^{\text{W}} \neq m_{\text{eff},z}^{\text{W}}$, and noncubic QDs, $a_x \neq a_y \neq a_z$. The estimate for the $\varepsilon(E)$ limit for finding out whether the case of $V_{\text{B}} \to \infty$ delivers results of required accuracy can also be applied to 2d and 3d confinement; only the contributions of the different spatial dimensions have to be added in analogy to Equation 3.8.

The simple 1d confinement case for infinitely high potential walls (Equation 3.6) shall be used now for deriving some general parameter relations of quantum confinement. The change in eigen-energies referring to the band edge of the bulk material is related to the extension of the potential well as $E_{\text{tot},n} \propto a^{-2}$, and to the effective carrier mass as $E_{\text{tot},n} \propto (m_{\text{eff}}^{\text{W}})^{-1}$. Some III-V materials like GaAs have very low carrier effective masses that shift the onset of quantum confinement to larger nanostructures. Together with its direct band gap, GaAs is a very good material for quantum structures, as it has a very high optical activity with a considerable active volume while still maintaining quantum confinement.

Another important magnitude for the energetics of quantum confinement is the DOS. Due to reduced dimensionality, the DOS of quantum structures experiences considerable changes. Figure 3.4 shows the DOS of a free electron in a bulk material together with the DOS of 1d (QW), 2d (QW), and 3d (QD) quantum confinement.

Quantum confinement increases from QWs via quantum wires to QDs, as evident from Equation 3.8. Hence, tuning of the effective band gap is accomplished best by QDs. The lateral DOS decreases with an increasing number of confinement directions. This increasingly limits the number of allowed optical transitions, and consequently the spectral range and intensity of optical absorption. Hence, the lowest carrier generation rate by photon absorption occurs at QDs. We can get *either* good control of the effective band gap *or* a high optical activity over a broad spectrum, with the two extrema drifting apart as we go from QW via quantum wires to QDs. Coupling of quantum structures through the barriers results in a further loss of effective band gap control (cf. top graph in

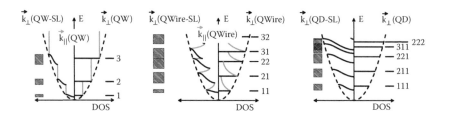

FIGURE 3.4

Electronic DOS over energy for 1d, 2d, and 3d isotropic confinement. Parabolas describe the bulk phase electronic DOS. Right half of the graphs describe DOS for isolated quantum structures with corresponding sets of quantum numbers; left half shows impact of inter-quantum structure coupling as in a SL, with minibands on the left. The DOS lateral to quantum confinement is shown in grey; the DOS in direction of confinement is shown in black.

Figure 3.6). However, there are two ways to tackle the somewhat small overlap in favorable electronic and optical properties. For III-V SLs, ternary compounds can be prepared such as $Al_xGa_{1-x}As$. Their electronic band gap varies with the ratio of cations (Al/Ga) and thereby can be tailored to the desired effective band gap. Very recently, it was shown by D'Costa et al. (2009) that the ternary group IV compound $Ge_{1-x-y}Si_xSn_y$ can be deposited by chemical vapor deposition (CVD) epitaxy as a single crystalline layer, whereby the electronic band gap can be tuned between 0.8 and 1.4 eV. This seminal result adds valuable flexibility to group IV compound SLs as only known from III-Vs to date. The other way is the broadening of the discrete levels into minibands with high optical activity, which is most important for QDs, and to some degree also for quantum wires. It can be tuned by the width and height of the barriers in a SL (see Section 3.2.2).

While the simple quantum mechanical model introduced here delivers valuable insight into the properties of confined structures, it should be mentioned that there are additional effects that cannot be treated within this formalism, but require density functional theory (DFT) methods instead. This refers to interface energetics and local defects like dangling bonds or atoms on interstitial or substitutional sites, vacancies, local stress, and modification of atomic bonds (König et al. 2008c; Ramos et al. 2004). The energetics at the interface and the associated charge transfer may result in considerable energy shifts of the electron and hole confined levels, with quantum confinement being only a perturbation effect below a certain size of the quantum structure. This refers in particular to highly polar interfaces, such as Si/SiO_2 or Si/Si_3N_4 (König et al. 2008a). Interface energetics has the biggest influence on QDs, as these have the biggest ratio of surface to interior atoms and can be exploited for QS-SL interconnects in tandem solar cells (see Section 3.5.2). III-V-based SLs are affected only to a minor degree, as their interface bonds have a polar nature like that of the bonds within the binary compounds. In addition, they can be grown epitaxially with strain compensation, and hence have very few defects (Oshima et al. 2008).

3.2.2 Super Lattice (SL) Properties as a Function of Quantum Confinement

Electron delocalization is achieved by the electron wave function penetrating throughout the barrier between adjacent potential wells. This is the essential step from isolated quantum structures to a SL. Therefore, a look at the mechanism of carrier tunneling through barriers is mandatory. For brevity, we consider the 1d case at a rectangular barrier as for a QW-SL (see also Figure 3.4).

For QWire- and QW-SLs, the 1d case refers to carrier confinement in the direction of carrier transport only (normal to the barriers, $\vec{k}_\perp$). Carriers can take on any energy above the confined ground level in the lateral direction (parallel to the barriers, $\vec{k}_\parallel$). This has major consequences for these SLs, as opposed to QD-SLs. In a QD-SL, there exists a *complete* separation between the minibands as long as they do not overlap by easing confinement in any of the directions of confinement, effectively changing the QD-SL toward a QWire- or QW-SL. The complete miniband separation allows for a prolonged lifetime of electron-hole pairs (excitons), provided the energy of miniband separation is significantly bigger than $k_B T$. Then, a interminiband transition requires the *simultaneous* emission of one or even more phonons of both, electron and hole, as otherwise the spin selection rules of a transition are not fulfilled (Kuzmany 1998), with the transition being forbidden.[*] Because of complete miniband separation, QD-SLs are the only solution for IB-SCs (see Section 3.3).

Equation 3.9 describes the probability of an electron tunneling through a rectangular barrier. Equation 3.10 describes electron tunneling over the

---

[*] In small nanostructures, there is a slight relaxation of selection rules, resulting not in a complete blocking, but a suppression of these transitions on the order of 10^{-3} relative to the intensity of the allowed transitions.

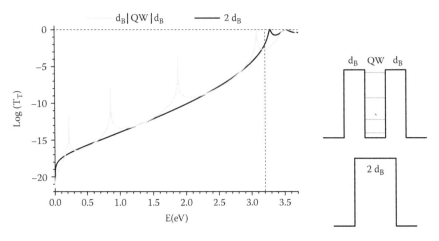

FIGURE 3.5

Logarithm of tunneling probability for QW with symmetric barriers and barrier of same thickness without QW, as shown in the right graph. Eigen-energies shown in grey. Barrier height is 3.2 eV, $d_{QW} = 60$ Å, $d_B = 20$ Å.

rectangular barrier. The latter case is also important for SL transport properties due to free carriers being scattered back with a probability $1 - T_T$ (cf. Figures 3.5 and 3.6).

$$T_T = \frac{4q^2 E(V_B - E)}{4q^2 E(V_B - E) + (qV_B \sinh[\alpha d_B])^2} \tag{3.9}$$

$$\alpha = \sqrt{\frac{2qm_{eff}^B}{\hbar^2}(V_B - E)} \text{ for } E < V_B$$

$$T_T = \frac{4q^2 E(E - V_B)}{4q^2 E(E - V_B) + (qV_B \sin[\alpha d_B])^2} \tag{3.10}$$

$$\alpha = \sqrt{-\frac{2qm_{eff}^B}{\hbar^2}(V_B - E)} \text{ for } E > V_B$$

In a detailed 2d and 3d treatment of tunneling, referring to quantum wires and QWs, there is a lateral electron momentum $\vec{k}_{\parallel}$ that modifies

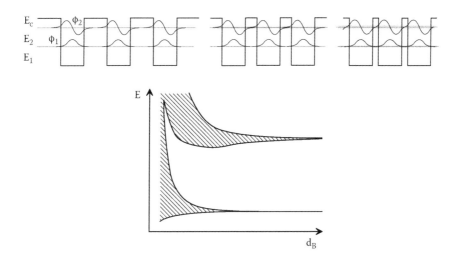

FIGURE 3.6
Electronic structure of a QW-SL with two confinement levels (top). With shrinking barrier width, originally separated electron wave functions increasingly overlap, forming minibands (grey). The second electron eigenfunctions, ϕ_2, show how delocalized wave functions (Bloch states) emerge from increasing overlap of individual wave functions. Broadening of discrete electron eigen-energies into minibands with decreasing barrier thickness (bottom). Effective band gap lowered by increasing spread of lowest miniband. For vanishing barriers, minibands merge into bulk conduction band.

T_T (Gray 1965). The simple 1d case used here is a good approximation for QW-SLs. In principle, the results are also applicable to quantum wire– and QD-SLs, though the number of eigen-energies inceases by the increasing number of permutations in quantum numbers, as shown in Figure 3.4. Tunneling current densities are modified by the product of DOS and occupation probabilities on both sides of the barriers, but depend in the first place on $T_T(E)$ (Schenk and Heiser 1997). Therefore, we evaluate T_T instead of the tunneling current density. For the latter one, knowledge of the DOS and Fermi–Dirac occupation probabilities on both sides of the barrier is mandatory, requiring considerations of different materials and bias ranges, which is beyond the scope of this chapter.

For confined electrons, Equation 3.9 shows that T_T is a function of the barrier thickness, $T_T \propto \exp(-2d_B)$; of the barrier height experienced by the confined electron (effective barrier height), $T_T \propto \exp(-[8V_B - E]^{-1/2})$; and the electron effective tunneling mass, $T_T \propto \exp(-[8m_{eff}^B]^{-1/2})$. The dominant parameter of T_T is the barrier thickness, followed by the barrier height and electron effective tunneling mass. The dependence of T_T on m_{eff}^B is important for barriers with a moderate to high defect density, allowing for trap-assisted tunneling. This process increases T_T, which can be pictured by tunneling of a carrier from an allowed state to a trap and a carrier tunneling from this trap into an allowed state on the other side of the barrier at the same time. Consequently, two carriers tunnel through barriers less than d_B. The total tunneling process per carrier is thereby split into individual processes as a function of the number of traps within the tunneling path, whereby the reciprocal total tunneling probability is given by the sum of the reciprocal individual events,

$$(T_T)^{-1} = \sum_{l=1}^{m} (T_{T,l})^{-1}$$

(Böer 1990), reflecting the simultaneous occurrence of individual tunneling processes. The maximum of T_T is reached for equal partitioning of the barrier. This effect can be described to some extent by varying m_{eff}^B. If the trap DOS over energy is known, it can be used to weight m_{eff}^B as a function of $V_B - E$ for describing the impact of traps within the barrier onto tunneling.

The eigen-energies concominant with quantum confinement lead to resonant transmission characteristics of the carrier flux if the barriers have a finite thickness. On a qualitative base, this is straightforward to see if we consider the potential well as a conductor with its discrete eigen-energies as preferential transmission states. These constitute the energies with a maximum transition probability as they effectively decrease the barrier thickness given by the potential well. Figure 3.5 depicts the situation.

In the discussion above it emerged that the penetration of the electron wave function into the barrier lowers the eigen-energies of a quantum structure by delocalizing the electron. SLs are formed by the periodic arrangement of many identical quantum structures. These behave like isolated quantum structures only if the barriers are thick enough to prevent even a small overlap of the electron wave functions confined in two adjacent potential wells. With decreasing barrier thickness, T_T increases, so that resonant tunneling due to the presence of eigen-energies becomes appreciable. For thin barriers, eigen-energies spread into minibands as shown in the top graph of Figure 3.6.

In a real SL, a deviation in barrier thickness, in the extension of the confining potential well and in $m_{\text{eff}}^{\text{B}}$, exists. The latter is a consequence of the trap density within the barrier changing over energy and spatial position. Local stress is another parameter that modifies the dispersion of the electronic bands in reciprocal space, $E(\vec{k})$, that modifies the electron effective tunneling mass via $\left|m_{\text{eff}}\right| = \left|\hbar^2/(\partial^2 E/\partial \vec{k}^2)\right|$. Processing tolerances induce deviations in spatial dimensions, modifying both tunneling probability and quantum confinement. The latter dominates the electronic behavior until the thickness ratio of barriers to confining potential well is $\leq 1/5$ to $1/10$.

The deviations discussed above shall be illustrated by an example consisting of Si QWs in SiO_2. We first consider a single QW, then go to ten QWs, accounting for a QW-SL. The results of Figure 3.7 can serve as a qualitative guide for other combinations of QW and barrier materials. A normal distribution was assumed around a nominal value, with a limit of $\pm 6\sigma$, where σ stands for the standard deviation. Any value outside these limits was assigned to the nominal value. From the top graphs of Figure 3.7 we see that the change in barrier thickness d_{B} does not modify T_{T} as heavily as discussed above. This is due to its rather large value, given the barrier height of $V_{\text{B}} = 3.2$ eV. The QW thickness has a much bigger influence on the broadening of the minibands as it sets the energy values for resonant transport. The electron effective tunneling mass also has a much bigger influence on T_{T} than d_{B}. A closer look reveals that the tunneling events with minimum values of both d_{B} and $m_{\text{eff}}^{\text{B}}$ are responsible for further broadening of the resonance. In a practical QW structure, these regions with minimum barrier thickness and lower $m_{\text{eff}}^{\text{B}}$ are the dominant paths for tunneling currents, inducing an appreciable lateral carrier flux.

In quantum wire and especially in QD-SLs, miniband broadening would not be as strong as for a QW-SL. On the other hand, there are more minibands with the dimensions of quantum confinement increasing as the single quantum number of a QW is replaced by doublets for quantum wires and triplets for QDs. This may lead to a miniband overlap, although electron localization is stronger than in a QW (cf. Figure 3.4).

Tunneling is the dominant but not the only transport process in QS-SLs. In particular for QS-SLs with an appreciable density of traps in the barrier material, thermal hopping exists. The current density for thermally activated hopping around T = 300 K can be described by (Sze 1967)

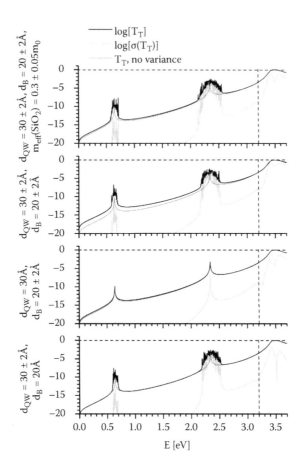

FIGURE 3.7

(a) T_T for one Si QW in SiO$_2$ with normal distributed uncertainty margin Δ. Average values and standard deviation $\sigma(T_T)$ obtained by one thousand runs and shown together with exact (zero deviation) solution for comparison. Top graph includes a Δ on $m^*_{T,n}$ (SiO$_2$) accounting for local deviation in defect state density. Single QW with same parameters, but bigger Δ (b, top left). Normal distribution ensemble for Δ limited by assigning all values beyond $\Delta = 6\sigma(T_T)$ to zero, corresponding to nominal value (b, bottom left). Ten QWs forming a SL with double QW size and Δ referring to single QWs. While single QWs show a more energy-selective behavior, these SLs yield information about optical and electronic transport properties as a function of deviation in spatial sizes and m^B_{eff} .

Continued

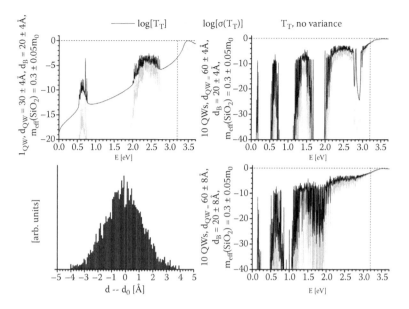

FIGURE 3.7
Continued.

$$\vec{j}_{tah} = \frac{G_0 \vec{F}_{A-B}}{d_{A-B}} \exp\left(-\frac{E_{act}}{k_B T}\right) \tag{3.11}$$

where G_0 is the nominal conductance between traps A and B without hopping barrier, with an intertrap distance d_{A-B} and concominant electrostatic field $\vec{F}_{A-B}$. The trap activation energy E_{act} presents the effective hopping barrier. The field can be interpreted as a local diffusion pressure subject to the charge state of the traps considered. With a net current throughout the device, a global diffusion field exists that results in directed thermally assisted hopping via the local field vector aligning to the diffusion field, $\vec{F}_{A-B} \uparrow\uparrow \vec{F}_{diff}$. For barrier heights less than VB $\approx 15\,k_B T$, thermionic emission begins to modify transport properties. The thermionic emission current is given by Sze (1981) and Böer (1992)

$$\vec{j}_{te} = A_R T^2 \exp\left(-\frac{E_{act}}{k_B T}\right) \left[\exp\left(\frac{q\sqrt{(q\vec{F})/(4\pi\varepsilon_0\varepsilon_w)}}{k_B T}\right) - 1\right] \tag{3.12}$$

with the effective Richardson constant

$$A_R = \frac{q m_{eff}^B k_B^2}{2\pi^2 \hbar^3}$$

The second exponential function describes the local field $\vec{F}$, which acts upon the emitted electron, whereby it joins the local flow of free carriers. The relative dielectric constant of the potential well is given by ε_W, and ε_0 stands for the dielectric constant in vacuum.

At thick barriers, thermionic emission current densities become more dominant, as they are not a function of d_B in contrast to T_T. Thermionic emission is exploited in QW-SLs as a means to excite confined free carriers onto the band edges of the barrier material (cf. Figure 3.3), where they join the carrier diffusion current of the solar cell. The coefficients in Equations 3.11 and 3.12 show that thermally assisted hopping is proportional to the electrostatic field divided by intertrap distance, while the thermionic emission depends quadratically on the temperature. For high carrier diffusion currents, the ratio of $\vec{j}_{tah}$ to $\vec{j}_{te}$ will increase by the increased diffusion field. For a hot carrier population at a barrier, as in a NC-HC solar cell (cf. Section 3.4.5), the elevated carrier temperature results in a strong increase of $\vec{j}_{te}$.

3.3 INTERMEDIATE BAND (IB) ABSORBERS

Solar cells can increase the energy range in which free carriers can be generated by electronic states within the actual band gap (Wolf 1960). The resulting impurity photovoltaic (IPV) effect was discussed by Güttler and Queisser (1970), who found it to be detrimental to a Si absorber as recombination outweighs the gain in photocurrent, though they did not rule out a gain in conversion efficiency for wider-band-gap materials such as CdS. The topic was revived in the 1990s (Keevers and Green 1994), triggering a debate about the benefit of the IPV effect implemented by deep impurities for solar cell efficiency (Schmeits and Mani 1999; Karazhanov 2001).

The intermediate band (IB) solar cell supplies a miniband within the band gap of the host material by quantum structure (QS)-SLs and was introduced by Luque and Martí (1997), who showed that the theoretical efficiency limit of an IB-SC is 63.1% under ideal conditions. Experimental confirmation of electron-hole generation by a two-photon absorption process via an IB was reported once more by Martí et al. (2006), although at a low rate.

3.3.1 Working Principle

The IB replaces the impurity level and works as an electronic upconverter. Sub-band-gap photons can excite electrons from the valence band to the IB and from there to the conduction band. This process competes with recombination, which proceeds the opposite way. The top graph of Figure 3.8 shows the band diagram of an IB-SC with the IB implemented by a QD-SL, as proposed by Green (2000).

A review of IB-SC devices based on GaAs/InAs QD-SLs has been published by Martí et al. (2008). We focus on the device operation of IB-SCs, building upon Martí's work and adding complementary aspects.

As in a conventional solar cell, there is the optical generation with its rate $\mathcal{G}_V^C$ and the carrier recombination as the reverse process with its rate $\mathcal{R}_V^C$ between the band edges of the wide-band-gap absorber. The circuit model of the IB-SC in Figure 3.8 shows additional parasitic resistances that account for carrier recombination [R_P (V,C)] and potential loss for diffusion transport to and carrier collection at the contacts (R_S[(V,C)]) (cf. Section 3.1.1). The electronic transitions involving the IB can be split into two PV subdevices, with their parasitic shunt conductances per unit area, $\Box G$, and current density sources. The latter can be described by

$$\vec{j}_\mathcal{G}(IB,C) = -q \int\limits_0^{d_{IB}} \int\limits_{h\nu_{IB}^C}^{h\nu_V^{IB}} \mathcal{G}(h\nu)\,d\nu\,dz \cong -q\bar{\mathcal{G}}_{IB}^C d_{IB} \quad \text{for transition IB} \rightarrow \text{C} \quad (3.13)$$

and

$$\vec{j}_\mathcal{G}(V,IB) = -q \int\limits_0^{d_{IB}} \int\limits_{h\nu_V^{IB}}^{h\nu_V^C} \mathcal{G}(h\nu)\,d\nu\,dz \cong -q\bar{\mathcal{G}}_V^{IB} d_{IB} \quad \text{for transition VB} \rightarrow \text{IB}$$

$\mathcal{G}(h\nu) = -\nabla_z\vec{\Phi}(h\nu)\eta_{IQ}(h\nu)$ stands for the optical generation rate, with $\nabla_z\vec{\Phi}(h\nu)$ as the divergence of the photon flux throughout the IB region

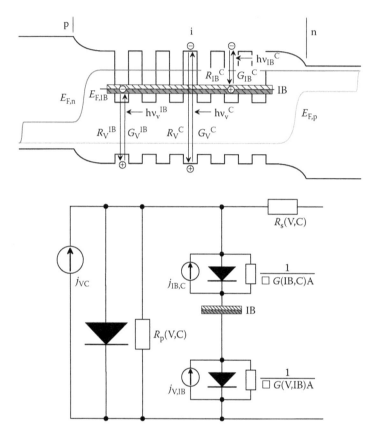

FIGURE 3.8

Band diagram of an IB-SC under illumination and V_{OC} conditions. (Top: Based on Luque, A., and Martí, A., "Increasing the Efficiency of Ideal Solar Cells by Photon Induced Transitions at Intermediate Levels," *Phys. Rev. Lett.* 78 [1997]: 5014–17, and Green, M. A., "Potential for Low Dimensional Structures in Photovoltaics." *Mater. Sci. Eng. B* 74 [2000]: 118–24.) The three different optical transitions, $h\nu_V^{IB}$, $h\nu_{IB}^C$, $h\nu_V^C$, are shown with the respective emission (generation) rates $\mathcal{G}$ and capture (recombination) rates $\mathcal{R}$. Circuit model of the IB-SC with parasitic conductances presenting carrier loss mechanisms under illumination. (Bottom: Based on Martí, A., Antolín, E., Cánovas, E., López, N., Linares, P. G., Luque, A., et al., "Elements of Design and Analysis of Quantum-Dot Intermediate Band Solar Cells," *Thin Solid Films* 516 [2008]: 6716–22.)

and $\eta_{IQ}(h\nu)$ as the internal quantum efficiency (probability for generating an electron-hole pair per photon); $h\nu$ is the photon energy; and q is the elementary charge. The negative signs in Equation 3.13 indicate that electrons flow *against* the recombination current, as is the case in a bias (drift) field of the p/n junction under dark conditions (cf. Section 3.1.1). The integration range of the photon energies is given by the absorption onset of the respective transition and the absorption onset of the transition with the next higher energy. In our case (cf. band diagram in Figure 3.8), the transition from the IB to the conduction band has the lowest energy and is integrated from its own minimum transition energy to the onset of the transition from the valence band to the IB. The latter, in return, is integrated from its own onset to the onset of the transition from the valence band to the conduction band. The absorption extends into the valence and conduction band away from the band edges and also over the occupied and unoccupied DOS of the IB until the transition energy is reached for the higher-energy process. The dispersion of the IB over energy is rather small, so that this effect occurs to a major degree in the main bands of the absorber. A small energetic dispersion of the IB prevents significant energy loss of carriers excited to the IB, as the energy range for carrier relaxation within the IB is then very limited. In our considerations, we neglect energy losses by carrier relaxation within the IB. The integration over the region where the IB is implemented, d_{IB}, in transport direction z accounts for the spatial variation of $\mathcal{G}$ for the respective transition. Assuming a constant (average) optical generation rate for the respective transition throughout the IB region, the integral over d_{IB} can be replaced by the product of average optical generation rate and length of the IB region as shown in the rightmost terms in Equation 3.13.

The shunt conductivities per square, $\square G$, present the inverse process to the excitations and thereby describe trapping (C $\rightarrow$ IB) and recombination (IB $\rightarrow$ V) losses. Their description requires the carrier capture/recombination rates, $\mathcal{R}_{IB}^{C}$ and $\mathcal{R}_{V}^{IB}$, to be known. The shunt conductivities per square can be described by

$$\square G_{IB}^{C} = \frac{q^2}{E_{IB}^{C}} \int_0^{d_{IB}} \mathcal{R}_{IB}^{C} \, dz \cong \frac{q^2}{E_{IB}^{C}} \bar{\mathcal{R}}_{IB}^{C} d_{IB} \qquad (3.14)$$

and

$$\Box G_V^{IB} = \frac{q^2}{E_V^{IB}} \int_0^{d_{IB}} \mathcal{R}_V^{IB} \, dz \cong \frac{q^2}{E_V^{IB}} \overline{\mathcal{R}}_V^{IB} d_{IB}$$

The transition energies between the IB and the conduction band and between the valence band and the IB are given by E_{IB}^C and E_V^{IB}, respectively. Again, the integration over the IB region can be replaced by multiplication with d_{IB} if we assume an average rate for the carrier capture/recombination rates, $\overline{\mathcal{R}}_{IB}^C$, $\overline{\mathcal{R}}_V^{IB}$. Integration of trapping/recombination rates over energy, as for the optical generation rates (cf. Equation 3.13), is not carried out as complete electron cooling to the respective band edge is assumed prior to recombination. The respective shunts are given by the inverse product of the shunt conductivities and the area A considered, $R = 1/(\Box G \, A)$.

The losses over the shunts per square, $\Box G_{IB}^C$ and $\Box G_V^{IB}$, can be different so that $\vec{j}_G(IB,C) = \vec{j}_G(V,IB)$ does not necessarily hold as suggested by the series connection of the current sources (cf. Figure 3.8). However, the two-step generation current density via the IB is determined by the smaller current density as a consequence of the series connection. There is an interdependence of $\vec{j}_G(IB,C)$ and $\vec{j}_G(V,IB)$ as a function of the IB occupation probability. In steady state, the differences of the respective current density and its recombination current density $\vec{j}_\mathcal{R}(a,b) = \Box G_a^b E_a^b / q$ over the entire IB are equal, arriving at the balance equation for the valence band–IB–conduction band current densities:

$$\vec{j}_G(IB, C) + \vec{j}_\mathcal{R}(IB, C) = \vec{j}_G(V, IB) + \vec{j}_\mathcal{R}(V, IB)$$

or, in detail

$$-q \int_0^{d_{IB}} \left[\int_{h\nu_{IB}^C}^{h\nu_V^{IB}} \mathcal{G}(h\nu) d\nu - \mathcal{R}_{IB}^C \right] dz = -q \int_0^{d_{IB}} \left[\int_{h\nu_V^{IB}}^{h\nu_V^C} \mathcal{G}(h\nu) d\nu - \mathcal{R}_V^{IB} \right] dz \quad (3.15)$$

The implementation of an IB rules out quantum wire– and QW-SLs as they possess one and two dimensions, respectively, where no confinement exists. Consequently, the DOSs in the directions without confinement are continuous (cf. Figure 3.4), what connects the IB with one of the two main bands. Therefore, QD-SLs are the only way to introduce an IB with QS-SLs. Indeed, all IB-SCs that have been prepared so far employed QD-SLs (Martí et al. 2008).

3.3.2 Energetic Position of the Intermediate Band

Obviously, the energetic position of the IB has a great influence on the two-step carrier generation process. We will have a brief look into recombination before we look at spectral matching of the solar photon flux, $\vec{\Phi}(h\nu)$, to the optical generation rates, $G(h\nu)$. Then, we investigate the optimum energetic position of the IB.

In nondegenerate semiconductors like in the i-region of an absorber with an IB, the main contribution to carrier recombination occurs via defect-assisted transitions, which can be described by the Hall–Shockley–Read recombination rate (Hall 1952; Shockley and Read 1952):

$$\mathcal{R}_{\text{HSR}} = \frac{n\,p - n_i^2}{\tau_p\left(n + n_i \exp\left[\frac{E_t - E_{\text{F,i}}}{k_{\text{B}} T}\right]\right) + \tau_n\left(p + n_i \exp\left[\frac{E_{\text{F,i}} - E_t}{k_{\text{B}} T}\right]\right)} \tag{3.16}$$

whereby E_t and $E_{\text{F,i}}$ stand for the energy of the trap level and the Fermi energy of the undoped (intrinsic) wide-band-gap semiconductor, and τ_p and τ_n are the lifetimes of holes and electrons. For the IB-SC, E_t has to be replaced by the energy of the IB E_{IB}. Assuming constant carrier densities and lifetimes, $\mathcal{R}_{\text{HSR}}$ increases for E_{IB} approaching the intrinsic Fermi level, with its maximum at $E_{\text{IB}} = E_{\text{F,i}}$. Therefore, the IB should be possibly near to one of the band edges.

A midgap position of the IB is also not beneficial for optical carrier generation. It results in no optical carrier generation for photons with energies $h\nu < E_{\text{gap}}/2$ and a split of $\vec{\Phi}(h\nu \geq E_{\text{gap}}/2)$ into both transitions V → IB and IB → C. The latter decreases the internal quantum efficiency by 50%, as *one* photon with $h\nu \geq E_{\text{gap}}/2$ can excite an electron *either* from the valence band to the IB *or* from the IB to the conduction band. The other extreme is having the IB possibly near to one of the band edges, as is ideal for a minimum $\mathcal{R}_{\text{HSR}}$. This would correspond to a conventional solar cell with a shallow dopant density so high that the majority carrier Fermi level is equal to the dopant energy. At such a shallow IB, the high-energy transition is only a few $k_{\text{B}}T$ below E_{gap}, whereby this small energy margin is also the spectral energy range absorbed exclusively by the higher-energy transition of the IB.

In order to find the optimum position of the IB within the band gap, we investigate the interdependence of the respective trapping/recombination rates. From Figure 3.8 we see that the recombination process via the IB is a serial process, with the shunt resistance $R_V^C = R_V^{IB} + R_{IB}^C$. In the above discussion we stated that the shunt is related to the shunt conductivity per square, R = $1/(\square G\ A)$. With Equation 3.14, $\mathcal{R}_V^C = \mathcal{R}_{HSR}$, and $E_V^C = E_{gap}$, we can derive a relation of the recombination rates,

$$\frac{1}{\square G_V^C\ A} = \frac{1}{\square G_V^{IB}\ A} + \frac{1}{\square G_{IB}^C\ A} \rightarrow \frac{E_{gap}}{Aq^2 \int\limits_0^{d_{IB}} \mathcal{R}_{HSR}\, dz} = \frac{E_V^{IB}}{Aq^2 \int\limits_0^{d_{IB}} \mathcal{R}_V^{IB}\, dz} + \frac{E_{IB}^C}{Aq^2 \int\limits_0^{d_{IB}} \mathcal{R}_{IB}^C\, dz}$$

from which we eventually obtain

$$\int\limits_0^{d_{IB}} \mathcal{R}_{HSR}\, dz = \frac{\int\limits_0^{d_{IB}} \mathcal{R}_V^{IB}\, dz \int\limits_0^{d_{IB}} \mathcal{R}_{IB}^C\, dz}{\dfrac{E_{IB}^C}{E_{gap}} \int\limits_0^{d_{IB}} \mathcal{R}_V^{IB}\, dz + \dfrac{E_V^{IB}}{E_{gap}} \int\limits_0^{d_{IB}} \mathcal{R}_{IB}^C\, dz}$$

or, with average rates

$$\bar{\mathcal{R}}_{HSR} = \frac{\bar{\mathcal{R}}_V^{IB}\, \bar{\mathcal{R}}_{IB}^C}{\dfrac{E_{IB}^C}{E_{gap}}\, \bar{\mathcal{R}}_V^{IB} + \dfrac{E_V^{IB}}{E_{gap}}\, \bar{\mathcal{R}}_{IB}^C} \tag{3.17}$$

Equation 3.17 does not provide us with a closed solution as long as we do not know either $\bar{\mathcal{R}}_V^{IB}$ or $\bar{\mathcal{R}}_{IB}^C$. However, the weighting with the energy ratio shows that the total recombination rate, $\bar{\mathcal{R}}_{HSR}$, is more sensitive to the recombination rate over the IB transition with the higher energy. This recombination rate—in our case $\bar{\mathcal{R}}_V^{IB}$—is weighted in the denominator with the smaller ratio of transition energies. Thus, the minimization of the recombination rate with the higher energy is the most effective way to suppress recombination losses over the entire band gap via the IB.

With the assumption of a high carrier mobility within the IB, we can find the optimum position of the IB by matching the photon flux as a function of photon energy to the optical generation rate for the respective transition. The high mobility is assumed to provide a nearly constant carrier density within the IB that yields nearly identical electronic conditions over the IB prior to any transition. We further assume that the generation rates and trapping/recombination rates over the IB are constant, though this is a rather crude approximation due to $\vec{\Phi}(h\nu)$ diminishing with increasing penetration depth. The average values $\mathcal{R}$ and $\mathcal{G}$ can be replaced by the integration over the IB region for more accurate results. With $\bar{\mathcal{G}}(h\nu) = [\vec{\Phi}(h\nu,0) - \vec{\Phi}(h\nu,d_{IB})]\eta_{IQ}(h\nu)/d_{IB}$ and the assumptions mentioned above, Equation 3.15 takes the form

$$\int_{h\nu_{IB}^C}^{h\nu_V^{IB}} [\vec{\Phi}(h\nu,0) - \vec{\Phi}(h\nu,d_{IB})]\frac{\eta_{IQ}(h\nu)}{d_{IB}}d\nu - \bar{\mathcal{R}}_{IB}^C =$$

$$\int_{h\nu_V^{IB}}^{h\nu_V^C} [\vec{\Phi}(h\nu,0) - \vec{\Phi}(h\nu,d_{IB})]\frac{\eta_{IQ}(h\nu)}{d_{IB}}d\nu - \bar{\mathcal{R}}_V^{IB}$$

(3.18)

Further evaluation requires the explicit use of material parameters and is therefore not treated in more detail. The only publication about a working IB-SC device did not show an appreciable increase in conversion efficiency over its reference sample (Martí et al. 2006). Cuadra et al. (2004) investigated the ratio of transition energies under the constraints of the solar photon flux within the framework of a thermodynamic model. All trapping/recombination processes were assumed to be radiative, with complete reabsorption of the resulting photons and consequential regeneration of carriers. This assumption appears to be somewhat far from reality. Nonradiative dominates over radiative recombination and is a significant loss mechanism in high-efficiency solar cells. Radiation efficiencies of these cells have been reported to be around 1% (Green et al. 2004). Cuadra and coworkers (2004) included the energetic overlap of absorption for the different transitions that were assumed herein to be zero, as in the earlier work of the group (Luque and Martí 1997). They obtained $E_{IB}^C = 0.77$ eV, $E_V^{IB} = 1.32$ eV, and $E_{gap} = 1.99$ eV as the optimum transition energies, assuming the sun as an ideal blackbody radiator with $T = 6{,}000$ K and a photon flux equivalent to 1,000 suns. Conversion efficiencies were found to be around 55% for a spectral overlap of the absorption coefficients of 0.5 eV, a ratio of the absorption coefficients of $\alpha_{IB}^C : \alpha_V^{IB} : \alpha_V^C = 1:10:100$, an IB-SC with an optimum thickness of $d_{IB} = 40$ μm and a perfect back reflector. The absorption coefficients can be related to the photon flux via $\alpha = 1/d_{IB} \ln[\vec{\Phi}(0)/\vec{\Phi}(d_{IB})]$. For the same conditions of photon spectrum, photon flux, and spectral overlap, but with $\alpha_{IB}^C = \alpha_V^{IB} = \alpha_V^C$ and a resulting optimum thickness of $d_{IB} = 0.7$ μm, the conversion efficiency drops to ca. 45%. The concominant increase in energy flux for this drop in efficiency by increased absorption involving the IB shows a typical thermodynamic phenomenon. Ultimate efficiencies are only possible at vanishing energy flux. Therefore, the latter case is better for solar energy conversion, although the conversion efficiency is lower.

3.3.3 Localized vs. Delocalized Intermediate Band Density of States (DOS)

In the last section it became clear that recombination losses via the IB are an obstacle for high conversion efficiencies of IB-SCs. The task of the IB is to harvest sub-band-gap photons for generating additional free carriers over the band gap. Carrier transport within the IB is not necessary and can increase carrier recombination. This section introduces a different concept that minimizes recombination losses: the localized IB.

IBs are implemented by miniband formation in QD-SLs with inter-QD coupling in all three dimensions, extending throughout the part of the absorber where the QD-SL is located (Martí et al. 2006). A recombination center has a spherical cross-section for electrons and holes extending throughout the QD-SL. This case is shown in the left picture of Figure 3.9, which requires some explanation about the presentation. If we consider a 1d carrier current density along z within the x,z plane, the picture gives the right dimensionality of QD rows and recombination cross-sections as a function of energy. For the 3d case, we would have to unfold the y direction as a fourth dimension as the energy of the electronic structure was substituted for y in order to depict the situation. This means that one row of QDs along x stand for the entire QD array. For the same reason, the third spatial dimension of the recombination cross-section is not included. It would have the same extension as in the x direction, yielding a round sphere for the left picture, and getting increasingly flattened to an ellipsoid for decreasing inter-QD array coupling (see below).

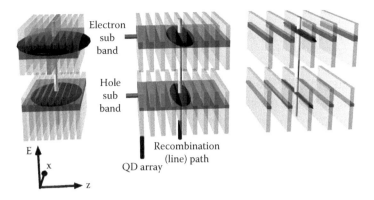

FIGURE 3.9
Transition from QD-SL band structure to band structure of periodic arrangement of QD arrays. Recombination center in central QD array (white) shown with its cross-section for electrons and holes (black) together with energy bands. The remaining two dimensions show spatial extension of QD-SL, with one QD row shown per QD array as $\bar{y}$ dropped for E (see text).

In principle, the 3d continuity of the IB can increase conversion efficiency by transporting carriers to locations of enhanced carrier generation. However, the second law of thermodynamics renders this effect to be outweighed by enhanced recombination as entropy is produced in any realistic (irreversible) thermodynamic process. This is even more the case with an appreciable energy flux. Therefore, it would be beneficial to split the 3d miniband into sections of 2d minibands in order to limit the impact of recombination centers from a spherical to a very flat elliptical recombination cross-section (cf. Figure 3.9). For decoupled QD arrays, the miniband would still be continuous in 2d, which is required for a high optical activity in the two-step carrier generation process. For optical carrier generation, the QD arrays must be normal to the vector of photon flux, $\vec{\Phi}$. Otherwise, slabs devoid of QDs exist where sub-band-gap photons are not absorbed. For the same reason, the logical continuation of the concept to rows of QDs is not attractive, as a large number of sub-band-gap photons do not have a QD in their path throughout the absorber. The thickness of the IB absorber d_{IB} is increased by the additional inter-QD array distance, with more wide-band-gap material between the same number of QD arrays for achieving the same absorption of sub-band-gap photons.

3.4 HOT CARRIER ABSORBERS AND SOLAR CELLS

Conventional solar cells absorb all photons with an energy exceeding their band gap. Carrier extraction occurs at energy levels within the band gap in the energetic proximity of the respective band edge (cf. Figure 3.1). There is a large difference in the average energy of absorbed photons and the average energy of extracted carriers, which is lost by carrier cooling. The concepts of hot carrier absorbers (HCAs) and hot carrier solar cells (HC-SCs) attempt to minimize these losses, thereby considerably increasing the conversion efficiency of solar irradiation.

We will elaborate a bit more on this TG-SC type as the working principle of a HC-SC is completely different from other TG approaches. Very little literature is available for HC-SCs compared to TG-SCs based on QD-SLs, owing to their novelty and rather unconventional material requirements.

A. J. Nozik (2001) proposed multiple exciton generation (MEG) as an alternative process for harnessing the excess energy of hot carriers, which was later confirmed by Schaller and Klimov (2004). If a hot exciton with its energy in the range of threefold the effective band gap of a QD is generated, it starts to interact with ground-state electrons, generating multiple excitons and thereby cooling down (Ellingson et al. 2005). So far, MEG has been shown experimentally only in colloidal solutions of NCs (Ellingson et al. 2005), which are rather unsuitable for environmentally robust absorbers. The QDs must be devoid of surface defects with possibly covalent bonds to attached ligand molecules (Guyot-Sionnest et al. 2005; Nozik 2008). Hence, MEG shall only be mentioned briefly, though it is very attractive, because energy loss by carrier cooling is not an issue. Qualitatively, MEG can be interpreted as a parallel generation of carriers out of one hot electron-hole pair. Hot carrier solar cells attempt to extract the carriers at their elevated energy, which resembles a serial process (increased V_{OC}) as the excess energy is not constituted in a multitude of cold free carriers (increased $\vec{j}_{SC}$).

3.4.1 Kinetics of Hot Carrier Cooling

Before we investigate the working principle of HC-SCs, we take a look at the basic kinetics of hot carrier cooling. Figure 3.10 shows the kinetics of carrier cooling along with the major processes involved.

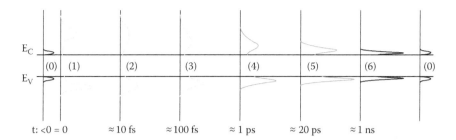

FIGURE 3.10
Kinetics of carrier cooling in a bulk semiconductor. (Based on Green, M. A., et al., *2002 Annual Report of the Photovoltaics Special Research Centre* [University of New South Wales, 2002], http://www.pv.unsw.edu.au/documents/Annual%20Report%202002/ HotCarrier.pdf.) Shading of the curves corresponds to average energy/temperature of the carrier ensemble, from high energy or hot populations (light grey) to completely thermalized carrier distributions (black). Thermal equilibrium at room temperature (0), instantaneous carrier distribution of noninteracting carriers (1), carrier-carrier scattering and impact ionization set in (2), renormalization of carrier energies and a common carrier temperature with Fermi–Dirac distribution (3), net energy loss by inelastic scattering of hot carriers (4) and decay of optical phonons into acoustic phonons and photon-polaritons result in considerable carrier cooling (5), further inelastic scattering decreases carrier temperature down to ambient conditions, with carrier recombination setting in (6).

The optical generation of charge carriers is an instantaneous process, with the generated carrier population being an energetic blueprint of the photon flux convoluted with the electronic DOS in the HCA. Carriers have individual energies, do not undergo any interaction yet apart from selection rules of excitation such as Pauli blocking (spin selectivty of states), and do not behave like an ensemble of carriers. After ca. 100 fs, carriers begin to interact with each other via elastic scattering. This occurs between excited carriers, leading to energy renormalization, or between excited and ground-state carriers, generating more excited carriers via impact ionization. Energy renormalization via elastic scattering and impact ionization results in a carrier population with one common temperature that can be described by a Fermi–Dirac distribution like a cold electron population, yet with a very high carrier temperature— hence the name *hot carrier*. In contrast to the electron temperature, the lattice temperature is in equilibrium with the environment ($T = 300$ K) as no additional acoustic phonons were generated yet. At this stage, about 1 ps has elapsed since carriers were generated.

Electron-optical phonon interaction sets in by local electrostatic lattice distortion around each hot carrier (polaron formation). This distortion is dynamic and leads to an oscillating lattice in the vicinity of each hot carrier, which can be described as electronic energy loss due to electrostatic vibrational excitation of the lattice. Second quantization (wave → particle) of these lattice vibrations yields optical phonons—the electron undergoes energy loss by the emission of optical phonons. The reverse process—optical phonon absorption by a hot electron—occurs with a much lower probability on the same timescale. For high electron densities, n, their collective oscillations (plasmons) have a frequency $\omega_{pls} \propto \sqrt{n}$ in the range of the optical phonon frequencies ω_{opt}. Electron-optical phonon coupling becomes much stronger, with optical phonons getting absorbed by hot electrons, as evident by the decreasing lifetime of optical phonon modes at high electron densities (Srivastava 2008). The electron absorbs the electromagnetic momentum of the optical phonon, leaving its mechanical momentum (low-frequency lattice vibration with very low charge elongation) behind, which is described in second quantization as an acoustic phonon. By absorbing the electromagnetic momentum, the electron gains energy and is thus reheated.

The optical phonons decay further into acoustic phonons or via generation of photon-polaritons (decoupling of the electromagnetic momentum from the optical phonon, leaving the mechanical momentum of the optical phonon behind) into an infrared (IR) photon and an acoustic phonon. The latter process can also be seen as quantum antennas radiating off their electromagnetic momentum. Optical phonon decay carries on until the carrier temperature converges against ambient conditions, a stage occurring ca. 100 ps after optical carrier generation. Now, the lattice may have been heated significantly by acoustic phonons, which will affect magnitudes like intrinsic carrier density increasing and carrier mobility decreasing.

While the carriers now lost all their excessive energy above the band edge, the free carrier product is still much bigger than the square of the intrinsic carrier density, $np \gg n_i^2$. Free carriers yield to direct (radiative) and indirect (Hall–Shockley–Read) recombination, whereby the latter is subject to capture time constants and capture cross-sections of the defects providing the recombination path. The time constants of carrier recombination are in the range of a few 100 ps to few 100 μs, and thus beyond the scope of hot carrier kinetics.

The timescale of hot carrier cooling clearly displays the challenge of HC-SCs—how can we minimize the cooling process and gain enough time in order to allow for hot carrier extraction? For answering this question, we have a brief look into the different decay mechanisms and their working principles before we work on the details of how to slow down carrier cooling. We choose the way from the slowest to the fastest decay mechanism, as we need a description of optical phonons for electron polarons and photon-polaritons as well.

We evaluate phononics by investigating a 1d diatomic chain presenting the case of a 1d binary compound. For a detailed introduction to phononics, refer to Elliott (1998). The optical (ω_+) and acoustic (ω_-) phonon dispersion of a 1d diatomic chain in reciprocal space is given within the harmonic approximation by Elliott (1998)

$$\omega_\pm = \sqrt{\gamma\left[\frac{1}{\mu_m} \pm \sqrt{\left(\frac{1}{\mu_m}\right)^2 - \frac{4}{m_b m_s}\sin^2(\vec{a}\vec{k}/2)}\,\right]} \quad \text{with} \quad \mu_m = \frac{m_b m_s}{m_b + m_s} \quad (3.19)$$

where γ (Nm^{-1}) is the force constant of the vibrating bond, m_b, m_s (kg) are the masses of the heavy and light atoms, respectively, μ_m (kg) is the reduced mass of the oscillating system, $\vec{k}$ (m^{-1}) is the wave vector in reciprocal space, and $\vec{a}$ (m) is the length of the 1d unit cell—here two bond lengths for a diatomic chain. The first derivative of the phonon frequency $\partial\omega/\partial\vec{k}$ of the 1d diatomic chain is analytical (König 2008),

$$\frac{\partial\omega_\pm}{\partial\vec{k}} =$$

$$\mp \frac{\sqrt{\gamma}\sin(\vec{a}\vec{k})}{(m_b + m_s)\sqrt{\left[\frac{1}{\mu_m} - \frac{4\sin^2(\vec{a}\vec{k}/2)}{m_b + m_s}\right] \pm \left[\frac{1}{(\mu_m)^{2/3}} - \frac{4(\mu_m)^{1/3}\sin^2(\vec{a}\vec{k}/2)}{m_b + m_s}\right]^{3/2}}} \quad (3.20)$$

and yields the propagation speed of the vibrational mode, which is equivalent to the group velocity, $\vec{v}_{gr}(\vec{k})$, of the corresponding phonon. The left graph of Figure 3.11 shows the phonon dispersion and its wave vector derivative in $\vec{k}$ space together with the corresponding atomic vibrations in real space.

From Figure 3.11 we see that for optical phonon modes, the group velocity is zero at the edge and the center of the Brillouin zone. Consequently, optical phonon modes at these $\vec{k}$ values are *standing* waves. For all other $\vec{k}$ values, optical phonon modes are *moving* waves. The atomic elongation of an optical phonon mode is in antiphase. This yields a period of two atoms for the phonon frequency that is identical to the vibration constituting an optical phonon, provided the phonon mode is a standing wave, i.e., $\vec{v}_{gr} = 0$. The diatomic vibration at $\vec{k} \neq \vec{0}$ has a periodic increase and decrease in amplitude moving with $\vec{v}_{gr}$ through the solid, whereby the atomic elongations are generally smaller than the standing waves. The vibration unit constituting an optical phonon is thus not given by two atoms as for the standing waves, but by a bigger number of atoms vibrating with a lower amplitude. An analogue to this envelope periodicity comprising several diatomic vibrations is frequency modulation in signal theory.[*] The wave packet composed of several diatomic vibrations moves through the solid at the speed of the group velocity and becomes a propagating optical phonon in second quantization. The integral of vibrational amplitude over the atoms involved in a collective vibration mode constituting one phonon is therefore constant, as its square modulus presents the probability of one phonon to exist. This is reflected in the amplitude of atomic vibration in Figure 3.11, with the extension of the wave packet presenting a phonon shown underneath.

The absolute value of the $\partial\omega / \partial\vec{k}$ maxima of the acoustic branch is the low-frequency limit of the group velocity and presents the speed of sound in the solid and is typically on the order of 5 to 10×10^4 ms^{-1} (Adachi 2004a, 2004b) for compounds with the heavy core having at least the atomic mass of Si. An important feature apparent from Figure 3.11 is the lower group velocity of optical modes for all $\vec{k}$ except for the Brillouin zone edge, where $\vec{v}_{gr}(\vec{k} = \pi/\vec{a}) = 0$. For optical phonons, the same static situation occurs at the center of the Brillouin zone.

An optical phonon can decay into two acoustic phonons out of a standing optical phonon mode, as discovered by Klemens (1966). This so-called Klemens decay is the main loss mechanism of optical phonons. It requires two acoustic phonon states with opposite momentum vector, $\vec{k}$, for momentum conservation and the energy of the optical phonon matching the sum of the acoustic phonon energies for energy conservation, $\hbar\omega_{opt}(\vec{k}_0) = \hbar\omega_{ac,1}(\vec{k}_0 + \vec{k}) + \hbar\omega_{ac,2}(\vec{k}_0 - \vec{k})$. There are $\vec{k}$ selection rules that allow for either a longitudinal optical (LO) phonon to decay into two

---

[*] Transversal optical modes may be more illustrative as they relate to amplitude modulation. However, strictly speaking, there are no transversal modes in a 1d system.

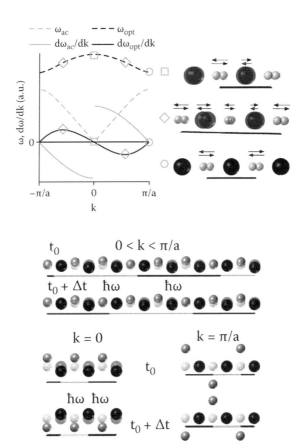

FIGURE 3.11

Phonon dispersion of 1d diatomic chain, $\omega(\bar{k})$, in reciprocal space with their first momentum derivatives, $\partial\omega/\partial\bar{k}$ (top). The grey symbols show the minima and maxima positions of $\partial\omega/\partial\bar{k}$ for the optical branch, with their oscillation patterns (top-right). The length of vibrational wave packet equivalent to a phonon is shown as a black strip. Dynamics involved in diatomic vibration, shown on transversal optical (TO) modes for better visibility; see also page 110 (bottom). The propagating wave packet for $\bar{k}\neq0$ is shown together with phonon modes at the center and the edge of the Brillouin zone.

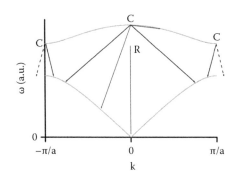

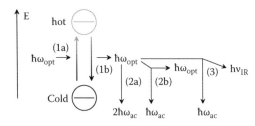

FIGURE 3.12

Phonon dispersion of 1d diatomic chain in reciprocal space: Principle of Klemens (C) and Ridley (R) phonon decay mechanisms shown on the phonon branches (top). The Klemens mechanism at the Brillouin zone edge is due to an *Umklapp* process. Diagram showing the inelastic processes affecting the energy of a hot electron (bottom): Reheating of electrons by optical phonon absorption (1a), cooling of electrons by optical phonon emission (1b), Klemens (2a) and Ridley decay (2b), decay of optical phonon into an IR photon and an acoustic phonon (3) (photon-polariton).

transversal acoustical (TA) phonons or a transversal optical (TO) phonon to decay into two longitudinal acoustic (LA) phonons (König 2008). For a 1d diatomic chain, the Klemens mechanism does not exist as we only have longitudinal modes. With a quasi-2d expansion by allowing for amplitudes normal to the diatomic chain, we get a rather simple situation due to the symmetric dipersion curves, which allow only for the Klemens decay at the local extrema of the optical phonon dispersion with LO → 2 TA as the only possible 1d transition, as shown in the top graph of Figure 3.12. The *Umklapp* process from the ajacent Brillouin zone in the extended $\vec{k}$ space scheme (shown as dashed arrows) is a result of periodicity and straightforward to understand if the adjacent Brillouin zone in the extended scheme is projected onto the reduced scheme. The Klemens decay can be prevented by using solids with a phononic band gap that is bigger than the maximum energy of acoustic phonons, $E_{gap}(\hbar\omega) > Max. (\hbar\omega_{ac})$. In principle, this is achieved when $m_b > 4m_s$ (see Equation 3.19). In real 3d solids, the space group symmetry is another parameter, as it determines the degeneracy of the phononic branches. For a high-symmetry (face-centered cubic [fcc]) solid, this decreases the dispersion of the phonon branches and thereby keeps the phonon gap open.

Ridley (1989) described an optical phonon decay into an acoustic phonon and another optical phonon with lower energy. The Ridley decay obeys the same $\vec{k}$ conservation and selection rules as well as energy conservation shown for the Klemens decay. Hence, the Ridley decay proceeds like $\hbar\omega_{TO}(\vec{k}_0) \rightarrow \hbar\omega_{LO}(\vec{k}_0 + \vec{k}) + \hbar\omega_{TA}(\vec{k}_0 - \vec{k})$ or $\hbar\omega_{LO}(\vec{k}_0) \rightarrow \hbar\omega_{TO}(\vec{k}_0 + \vec{k}) + \hbar\omega_{LA}(\vec{k}_0 - \vec{k})$ (König 2008). Again, there is no Ridley decay for a 1d diatomic chain, but a quasi-2d extension allowing for atomic elongation normal to the chain extension yields one possible mechanism for Ridley decay, LO → TO + LA. This decay mechanism is shown in the top graph of Figure 3.12. Unlike Klemens decay, a large phononic gap does not prevent the Ridley decay due to the large energetic difference in the energies of the decayed optical and acoustic phonon. Still, one of the decay products, the optical phonon, possesses a comparatively high energy and may contribute to the optical phonon population, which eventually can be reabsorbed by free electrons.

An electron loses energy by forming a polaron with the lattice atoms within its vicinity defined by the polaron radius (Böer 1990), which is in the range of 20 to 90 Å for most solids. The dynamic dislocation of the lattice atoms with their dipole moment of their atomic bonds then constitutes the emission of an optical phonon. The coupling strength of an electron to optical phonons is described by the Fröhlich interaction. The material-specific Fröhlich coupling constant $\alpha_{Frö}$ is proportional to $1/\varepsilon_{eff}\sqrt{m_{pol}/T}$ (Böer 1990), which is approximately $1/\varepsilon_{eff}\sqrt{m_{eff}^n/T}$ for small values of $\alpha_{Frö}$, which are the ones of interest for suppressing optical phonon emission. The Fröhlich interaction is small for solids with a large dielectric constant, ε_{eff} (small band gap), and small effective electron mass, m_{eff}^n. The latter is also advantageous for generating a hot carrier population as the free electron energy is $E(e^-) = (\hbar\vec{k})^2/2m_{eff}^n$. Di Ciolo et al. (2009) recently approximated the electron-phonon interaction in a solid with a high electron density (strong electron correlation). Their results indicate that the electron-phonon interaction decreases with increasing electron correlation, provided the plasmon frequency is not in the range of ω_{opt}. An increasing electron correlation decreases charge fluctuations within the electron ensemble. Smaller charge fluctuations provide a more homogeneous electrostatic field, which in return provides less dynamic coupling to the atomic charges as required for vibrational excitation of the atomic lattice by optical phonon emission of an electron.

Beneath the Fröhlich interaction, the energy quantum emitted per optical phonon is another value that describes the kinetics of energy loss. Although the emission of an optical phonon is an ultrafast process, it occurs in a finite amount of time. Hence, the lower the energy of the optical phonon, $\hbar\omega_{opt}$, the more time is needed in order to radiate off the same amount of energy. The electron can also undergo multiphonon scattering, emitting a few optical phonons in a single scattering event. However, the transition probabilities of these multiphonon emissions are somewhat low due to high restrictions of the involved phononic and electronic DOS. In addition to cumulative energy and momentum conservations, all transitions have to occur within a few femtoseconds, as given by the Heisenberg uncertainty relation, with the phonon energies as parameters. The minimum of ω_{opt} can be reached by binary compounds with a possibly big core mass of the lighter chemical element (see also Equation 3.19).

Another loss mechanism also scales with the energy of the optical phonons $\hbar\omega_{opt}$ and is given by the photon-polariton, the decoupling of the electromagnetic field of an optical phonon constituting a photon of the energy $h\nu = \hbar\omega_{opt} - \hbar\omega_{ac}$, which is radiated off. The acoustic phonon describes the mechanical low-frequency, high-momentum vibration remaining within the lattice. The probability of an optical phonon to undergo photon-polariton decay is given within Fermi's golden rule by the oscillator strength f_{OS} of the transition. For an optical transition, the oscillator strength depends quadratically on the dipole moment of the intial state ($\hbar\omega_{opt}$), $f_{os} \propto P_D^2$ (Ludwig et al. 2003). The probability of photon-polariton decay is thus minimized for a minimum dipole moment that occurs when the atomic bonds are covalent.

Phonons belong to the class of particles with integer spin (bosons); therefore, the occupation probability of phonon states is governed by Bose–Einstein statistics, with occupation numbers reaching infinity for an infinitely large solid (Elliott 1998). In practice, and in particular for nanostructures, this occupation probability has a thermal limit due to the number of phonons being proportional to the amplitude of lattice vibration. If the phonon occupation number gets very large, the solid will disintegrate, which is described macroscopically as melting at elevated temperature. This sets a practical limit to phonon confinement, subject to material properties, photon-polariton decay rates, phonon reabsorption by hot electrons, and the photon flux density absorbed within the volume of the HCA.

3.4.2 Working Principle of a Hot Carrier Solar Cell (HC-SC)

An HC-SC consists of a hot carrier absorber (HCA) with energy-selective contacts (ESCs) for electrons and holes and a contact material of appropriate electron work function for matching the extraction energy of the ESC. Figure 3.13 shows the schematic of the electronic band structure of a HC-SC. Properties and design of HCA materials and ESCs will be covered in detail in Sections 3.4.3 and 3.5.3, respectively. For now, we focus on device operation.

HC-SCs do not require a p/n junction or any doping. In fact, doping would accelerate carrier cooling by introducing additional scattering centers and by introducing an additional phononic DOS in the phononic band gap. Carriers are generated and diffuse to the respective ESCs. The effective carrier velocity of electrons/holes, $\vec{v}_{\mathrm{rms},n/p}$, can be approximated as (Böer 1990)

$$\vec{v}_{\mathrm{rms},i} \approx \sqrt{\frac{3kT}{m^i_{\mathrm{eff}}}}; \quad i = n, p \tag{3.21}$$

under the assumption that *no scattering occurs*. For hot carrier extraction, this ballistic transport is essential in order to avoid energy loss by inelastic scattering. With carrier cooling kinetics as observed in a GaAs bulk phase and in AlGaAs/GaAs QW-SLs (cf. Figure 3.15) (Nozik 2001), the estimate of the hot electron mean free path assuming no electron-phonon scattering is $\lambda_{\mathrm{mfp}} \approx 1$ to 3 µm in bulk GaAs and $\lambda_{\mathrm{mfp}} \approx 50$ to 200 µm in AlGaAs/GaAs QW-SLs for an electron temperature of $T = 1,200$ to 600 K. Carrier-carrier

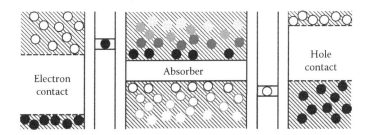

FIGURE 3.13
Energy band structure of a hot carrier solar cell. (Based on Würfel, P., "Solar Energy Conversion with Hot Electrons from Impact Ionisation, *Solar Energy Mater. Solar Cells* 46 [1997]: 43–52.) Within the HCA, carriers are increasingly hot at elevated energies, as shown by their shades of gray (see Figure 3.10).

and inelastic scattering decrease these estimates by at least two orders of magnitude. In a crystalline lattice, the mean free path for electron ballistic transport at $T = 300$ K is in the range of $\lambda_{mfp}(e^-) = 80$ to 200 Å for Si (Banoo et al. 2001; Kojima and Koshida 2005), 170 Å for InGaAs/InP SLs (Hieke et al. 2000), 200 Å for CdSe (Beard et al. 2002), 180 Å in InP (Teissier et al. 1998), and 200 Å in GaAs/AlGaAs SLs (Schneider et al. 2000). HCAs have a very high phonon density, which increases the scattering probability of hot carriers with the lattice and thereby decreases the carrier mobility $\mu_{n,p}$ and thus λ_{mfp}. The major contribution to carrier scattering at $T \geq 300$ K originates from acoustic phonons in covalent solids and from optical phonons in polar solids (Yu and Cardona 2003). This is a description for the electron-optical phonon (Fröhlich) interaction as in Section 3.4.1, but in terms of scattering rates. As a consequence, acoustic phonon modes should not be confined to a HCA. Their potential upconversion to an optical phonon mode by a phase-coherent superposition is grossly outweighed by the massive increase in the electron scattering rate.

At the mean free path, 50% of the electrons underwent an inelastic scattering process so that for quasi-ballistic transport of the hot electron population, the diameter of the HCA is in the range of 100 to 150 Å. This ensures that the majority of hot electrons reach the ESC without inelastic scattering. The mean free path of ballistic transport undergoes a stronger decrease due to numerous interfaces normal to the current path in a SL. These increase the area density of inelastic scattering centers as a function of their area density of defects per energy interval, and also invoke electron back-scattering by changes in the conduction band energy. In this point, epitaxially grown III-V SLs have a clear advantage over group IV–based SLs, where a higher defect density exists.

Massive photon flux may result in bleaching of electronic states near the band edges. However, free carrier absorption—see (5) in Figure 3.17—will absorb a considerable part of the photon flux around the band gap of the HCA, so that bleaching is not a primary issue. Nozik (2008) showed experimentally that the absorption cross-section of free carriers at elevated energies is considerably larger than for carriers undergoing a fundamental excitation to the band edges.

Takeda et al. (2009) investigated the achievable thermodynamic efficiency of HCAs under realistic conditions for both material parameters and energy dissipation. They obtained an optimum density of hot electrons in the range of 10^{12} to 10^{15} cm^{-3} for photon fluxes of 1 and 1,000 suns, respectively. These rather low carrier densities are a consequence of ultra fast carrier extraction and show that bleaching does not occur in an HC-SC. A conversion efficiency in the range of 50 to 55% was determined under 1,000 suns concentrated solar irradiation, assuming a carrier thermalization time constant of τ_{th} = 1 ns. Carrier thermalization can also be expressed in a relative way by the ratio of carrier retention time, τ_{re}, within the HCA to τ_{th}. Remarkably, the maximum achievable efficiency under different boundary conditions is reached for a ratio of τ_{re}/τ_{th} = 0.1. The optimum band gap of the HCA illuminated with 1,000 suns was estimated to be ca. 0.8 eV.

Carrier transport through the ESCs occurs by tunneling, with negligible time constants, compared to hot carrier diffusion. The ESCs work as a semipermeable barrier to hot carriers: only one type can penetrate the respective ESC, while the other one is reflected. In practice, the latter will penetrate the barrier with a certain probability, resulting in an energy loss (see Section 3.5.3). This carrier selection process is done by the p/n junction in conventional solar cells, requiring a change of majority carrier type over the device. The width and position of the energy-selective level (ESL) are important parameters for the energy flux and the conversion efficiency of the device. The position of the ESL depends on the average carrier temperature within the HCA. While the carrier current decreases with increasing energetic position of the ESL, the energetic difference of extracted electrons and holes increases, with an optimum located between these cases, in analogy to the MPP case located between the short-circuit and open-circuit cases. In an HC-SC, there are additional processes that have an influence on conversion efficiency. Quasi-elastic electron-electron scattering not only generates a carrier ensemble with one temperature, but also provides ultrafast refilling of the electronic states in the energy range of the ESL. Hot carrier extraction proceeds by ballistic transport within the same timeframe. A large width or high position (lower density of hot carriers) of the ESL may lead to electronic bleaching of the HCA as the number of extracted carriers exceeds the numbers of hot carriers with an energy in the range of the ESL.

A very small ESL width yields a very high conversion efficiency, as a high energy selectivity results in minimum cooling of charge carriers during their extraction from the HCA (O'Dwyer et al. 2005). On the other hand, the energy flux given by the current of extracted hot carriers decreases for a narrowing width of the ESL. An ideal HC-SC approaches the Carnot efficiency at *zero* energy flux, as opposed to a wide ESL with considerable energy flux but a very low device efficiency of the HC-SC due to massive cooling of hot carriers. Obviously, the optimum for HC-SC operation is located at an appreciable energy flux through the ESC while still maintaining a rather high conversion efficiency given by a low carrier cooling rate. This optimum depends on the applied materials, involved scattering processes, and HC distribution within the HCA as a function of the density and the spectral distribution of the absorbed photon flux. A brief discussion of this issue is presented in Section 3.5.3, though exact numbers require a working HC-SC, which is still a field of intense research.

Large reservoirs of carriers at ambient temperature are required at the outer side of ESC as macroscopic contacts. These reservoirs should have an empty/full (electron/hole) population, which allows for sufficient carrier relaxation in order to prevent carriers to tunnel back into the HCA via the ESL. The carrier relaxation should be limited as it presents a loss in carrier energy. A consideration of a resonant tunneling contact yields 6 kT, with ambient temperature T, as the potential difference between ESL and the Fermi energy, E_F, of the carrier reservoir. This results in a good compromise between energy loss by carriers tunneling back into the HCA and by cooling due to carrier relaxation. Carrier reservoirs prevent further losses if no allowed states exist below their relaxation level (see also Figure 3.13). This prevents carriers of the opposite type—hot holes at the ESC for hot electrons—to recombine with the carriers in the reservoir—cold electrons at the ESC for hot electrons—by leaking through the tunneling barrier.

3.4.3 Materials for Hot Carrier Absorbers

Section 3.4.1 delivered the material properties for slowing down carrier cooling. These properties are summarized here and complemented with optical, electronic, and structural parameters of solids in order to come up with real materials fulfilling all demands of an HCA. Structural parameters are relevant in particular for nanostructured HCAs such as SLs.

Klemens decay of optical phonons can be prevented by having a phononic band gap exceeding the maximum energy of the acoustic phonons, requiring the heavy core to possess more than a fourfold mass of the light one. We can directly relate the ratio of core masses to the ratio of phonon band gap to maximum energy of the acoustic phonon by $E_{gap}(\hbar\omega) / Max(\hbar\omega_{ac}) = \sqrt{m_b / m_s} - 1$ (see also Equation 3.19 and Table 3.2). A minimum energy of optical phonons $\hbar\omega_{opt}$ requires a hot electron to emit more optical phonons for losing a certain amount of energy. It also minimizes photon-polariton losses, because the IR photon radiated off has a smaller energy. The electron-optical phonon (Fröhlich) interaction

TABLE 3.2

Properties of Material Candidates for HCAs

Solid	$\hbar\omega$ Gap [$\hbar\omega_{ac}$ (π/a)]	m_s [m_H]	ε_{opt}	$\alpha_{Frö}$	m_{eff}^n [m_0]	m_{eff}^p [m_0]	e^--Gap [eV], Type	a [Å]	Space Group
BAs[1,3,10]	1.63	10.81	?	?	0.05	0.08	1.45, 1.2*, i	4.78*	zb*
BSb[4,10]	2.56	10.81	11.3	?	?	?	0.5*, i*	5.18*	zb*
BBi[6,8]	3.40	10.81	?	?	?	?	0.91*, i*	5.53*	zb*
AlSb[1]	1.12	26.98	10.2	0.02	0.14	0.13	2.24, i	6.14	zb
AlBi[4,6]	1.78	26.98	?	?	?	?	0.04*, d*	6.46*	zb*
GaN[1,12]	1.23	14.01	5.35	0.44	0.15	0.20	3.42, d	4.52	zb
InN[1]	1.86	14.01	6.7	0.24	0.11	0.25	0.7–0.9, d	3.54, 5.74	w
InP[1]	0.92	30.97	9.9	0.13	0.08	0.11	1.35, d	5.87	zb
BiN[5]	2.86	14.01	?	?	?	?	≈1.4	4.98*	zb*
Bi$_2$S$_3$[9,11]	1.55	32.06	10.9	?	0.68	0.02	1.3, 1.52, d	11, 11, 4	orh
AlAs[1]	0.67	26.98	8.18	0.13	0.12	0.16	3.01, i	5.66	zb
GaP[1]	0.50	30.97	8.8	0.20	0.09	0.17	2.76, i	5.45	zb
GaAs[1]	0.04	69.72	10.9	0.08	0.07	0.08	1.43, d	5.65	zb
SiC[2]	0.53	12.01	6.48	0.26	0.31	0.23	2.39, i	4.36	zb
SiGe[7]	0.61	28.09	14.5	?	0.20	≈0.3	0.99, i	5.54	zb

Note: For solids having different space groups, the data refer to the phase of highest symmetry. Values of band gap refer to T = 300 K, m_{eff}^p, given for light holes. Values were obtained experimentally unless noticed otherwise. The bottom section contains commonplace semiconductors for comparison. Space groups are zinc blende (zb), wurzite (w), and orthorhombic (orh).

Source: All atom masses are from (Sargent-Welch Scientific Company 1980).

1, Adachi (2004a); 2, Adachi (2004b); 3, Kalvoda et al. (1997); 4, Wang and Ye (2002); 5, Carrier and Wei (2004); 6, Ferhat and Zaoui (2006); 7, Kasper and Lyutovich (2000); 8, Deligoz et al. (2007); 9, Pejova and Grozdanov (2006); 10, Touat et al. (2006); 11, Black et al. (1957); 12, Barker and Ilgems (1973).

is small for solids with a high dielectric constant, ε_{eff}, and small electron effective mass, m_{eff}^n. A possibly covalent atomic bond has a minimum dipole moment, P_D, which further decreases photon-polariton losses. The optimum phononic material should therefore have $m_b > 4m_s$, with m_s as big as possible without violating the preceding statement and being a possibly covalent solid with a low direct band gap, with the latter improving photon absorption.

Small values for m_{eff}^n, m_{eff}^p are also desirable for electronic properties. The generation of a hot carrier population is enhanced as $E \propto 1/m_{eff}$,

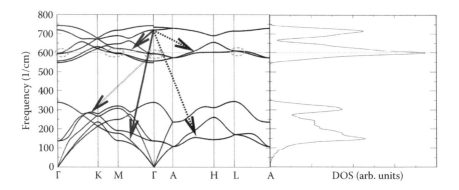

FIGURE 3.14

Phonon dispersion and DOS of GaN (Srivastava 2008). Dark full grey and dashed dark grey arrows show Ridley decays, and dotted grey arrow shows the Barman–Srivastava decay. Anticrossings of optical phonon branches around the phononic DOS maximum at 600 cm⁻¹ are encircled by light grey dashed lines. (Reproduced with kind permission of the American Physics Society.)

and carrier mobilities increase. The latter provides a longer mean free path to carriers before they undergo (inelastic) scattering, keeping them hot when diffusing to the respective ESC. As an optical property, a high absorption coefficient is desirable for a massive carrier generation rate. This refers mainly to the carriers near the band edges, as states farther away from the band edges have a larger excitation cross-section and free carriers reabsorb photons at a high rate. The former is also due to higher-lying band minima, which start to participate in carrier generation at higher photon energies.

Structural properties refer to HCAs consisting of two or even three solids, as in a SL. The primary material parameters are lattice constant and space group symmetry for an epitaxial growth with a minimum number of defects. There is some tolerance to stress built up by small mismatches in lattice constants, which is exploited for strain-induced epitaxial growth of III-V QD- and QW-SLs.

We are now in a position to look at the properties of some real materials listed in Table 3.2. Older literature states the band gap of InN to be around 1.7 eV. This is likely to occur due to InN quickly oxidizing in air to InO_xN_y (Shrestha and Clady 2008).

The requirement of a heavy core mass with the other core mass being less than one-quarter of the former points to binary solids that have not been subject to detailed investigation yet. Some compounds appear to be rather exotic as opposed to commonplace semiconductor compounds. This is evident in particular for Bi compounds, which have a very high figure of merit (*ZT*) in thermoelectric devices due to high electrical conductivity and low thermal conductivity (Goldsmid 2006)—a good indicator that Bi compounds are promising candidates for HCAs. Compounds with a high core mass have a low melting temperature, which is advantageous for device processing. An example is Bi_2S_3, which can be deposited in acidic solution at 60°C and shows a low-defect nanocrystalline structure after a heat treatment for 3 h at 250°C (Pejova and Grozdanov 2006).

It should be noted that the phononic band gap given in Table 3.2 is a maximum value, as it was derived from the 1d model and should be used as a guide only. As an example, BSb and BAs have a phononic band gap derived from density functional theory (DFT)—generalized gradient approximation (GCA) calculations of $E_{gap}(\hbar\omega)/Max(\hbar\omega_{ac}) = 1.44$ and 0.963, respectively (Touat et al. 2006), while the 1d diatomic chain model yields 2.56 for BSb and 1.63 for BAs.

3.4.4 Super Lattice (SL) Approaches for Phononic Confinement

Employing phononic nanostructures is a means to delay carrier cooling (cf. Figure 3.15), with several effects contributing. All of these can be

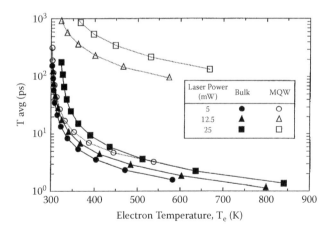

FIGURE 3.15

Evolution of the average electron temperature in GaAs bulk and in $GaAs/Al_{0.32}Ga_{0.68}As$ QW-SLs over time, excited with different photon flux densities (LASER powers) at 600 nm. (Reproduced from Nozik, A. J., "Spectroscopy and Hot Electron Relaxation Dynamics in Semiconductor Quantum Wells and Quantum Dots," *Annu. Rev. Phys. Chem.* 52 [2001]: 193–231. With kind permission of Annual Reviews.)

tackled to some extent by phononic engineering. This section attempts to explore the different effects in phononic nanostructures and to evaluate their contribution to the delay of carrier cooling.

The length of the vibrational wave packet corresponding to an optical phonon is very small. This oscillation length can also be considered as the characteristic extension of the phonon (cf. Figure 3.11). For standing modes at the center or edge of the Brillouin zone of binary compounds, the characteristic phonon extension is defined by twice the interatomic distance, which is roughly the distance between two adjacent lattice planes. This value increases to a distance over several lattice planes for traveling phonon modes, but is still small compared to the effective Bohr radius of a carrier confined to a QS. Although optical phonons have a considerable electromagnetic momentum, the change in the electromagnetic momentum (change in the electromagnetic field) can only travel at the speed of the *mechanical* elongation of the polarized atoms, which presents the group velocity of the optical phonon, $\vec{v}_{gr}(\hbar\omega_{opt})$. From Figure 3.11 we see that the maximum of $\vec{v}_{gr}(\hbar\omega_{opt})$ is smaller than $\vec{v}_{gr}(\hbar\omega_{ac}[\vec{k} \to 0])$, and hence smaller than the speed of sound. This also holds in real 3d solids (cf. Figure 3.14). In an optical phonon mode, the change of the electromagnetic field cannot be detached from the elongation of the polarized atoms. If the optical phonon radiates off its electromagnetic momentum as an infrared photon (photon-polariton) or excites an electron, the optical phonon decays into an acoustic phonon plus infrared photon or gets absorbed by an electron.

Another aspect we have to consider is the phononic DOS. We refer to a simple 1d picture, and then unfold the results in 3d for showing the implications. Starting from the 1d phononic dispersion of a diatomic chain (cf. Figure 3.11) in reciprocal ($\vec{k}$) space, the first derivative of the angular frequency yields the velocity of the proparating wave. With second quantization (wave → particle) this is equivalent to the phonon group velocity, thus $\partial\omega/\partial\vec{k} = \vec{v}_{gr}(\hbar\omega)$. Next, we find that $\partial\omega/\partial\vec{k} \propto [\text{DOS}(\hbar\omega)]^{-1}$. This is straightforward to see as the minimum characteristic phonon extension of a phonon mode is given by a standing wave ($\vec{v}_{gr} = \vec{0}$), leading to a maximum number of phonon modes per unit length. The characteristic phonon extension increases with the group velocity, what decreases the number of phonon modes per unit length, and thereby DOS($\hbar\omega$) at the corresponding $\vec{k}$. As a result, the DOS($\hbar\omega$) has maxima at the minima of the characteristic phonon extension, which makes phononic confinement most effective for structures consisting of two lattice planes—a binary compound. For *2n* lattice planes, we get *n* minigaps in the phonon dispersion, whereby the sum of these energy gaps stays constant, leading to a decreasing energy per minigap for an increasing number of lattice planes. In 3d spatial space, these minigaps also get increasingly localized in $\vec{k}$ space with the number of lattice planes, allowing for a continuous phonon DOS over all $\vec{k}$. The situation deteriorates for the 3d case due to the orthogonality of phonon momenta ($\vec{k}_x \perp \vec{k}_y \perp \vec{k}_z$). This orthogonality condition delivers additional degrees of freedom for a phonon of a certain energy $\hbar\omega$ if

$$\omega = \sum_{v}^{x,y,z} \omega(\vec{k}_v) = \text{constant}$$

In other words, a multitude of wave vector combinations exist at a constant phonon angular frequency ω. Minigaps still slow down phonon decay, in particular if they block $\omega_{opt}(\vec{k})$ regions where there would be a high degeneracy of optical phonon modes in a bulk phase of the same material. Figure 3.16(a) illustrates the effect of a QD-SL on the phononic DOS and gives an impression on the structure size required.

The current lower limit in SL engineering is ca. 10 MLs (ca. 24 Å for GaAs, similar for Si), which is far off from the required size for phonon confinement. More importantly, the required dimensions of phononic confinement are about one order of magnitude smaller than the range of quantum confinement defined by the exciton radius, which is typically in the range of 40 to 100 Å for the materials of interest. Phonon confinement and quantum confinement of charge carriers occur at different structure sizes, with no or very little overlap. For generating an HC population by photon absorption, at least a quasi-continuous carrier DOS is required. Otherwise, many energetic transitions are not allowed due to discrete energies and momenta, leading to a selective absorption of photons and suppressed carrier-carrier energy renormalization, impact ionization, and free carrier reabsorption. In fact, hot carriers with a Fermi-Dirac distribution cannot exist in a discrete carrier DOS. For these reasons, the NC-HCA has to be significantly bigger than the onset of quantum confinement. Klein et al. (1990) found out that the electron-phonon (Fröhlich) coupling in NCs is inversely proportional to the square root of the NC diameter. Therefore, the Fröhlich interaction is enhanced in small NCs, resulting in accelerated electron cooling by increased phonon emission. The phononic mini-Brillouin zone of an SL in the size range of NCs can have several minigaps in the range of up to a few meV, if wide phononic band gap

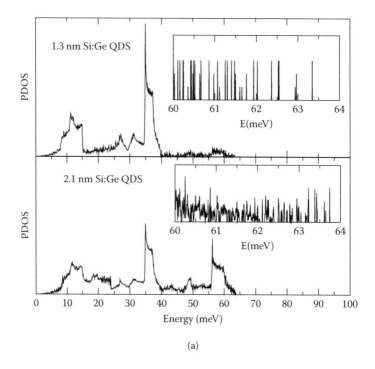

(a)

FIGURE 3.16

Effect of an SL on phononics of zero phononic band gap solids Si (NC) and Ge (matrix) shown on the phononic DOS (PDOS). (a) The impact of *size effect* shows that phononic minigaps exist, but are extremely small in the range of the energy of optical phonons and already convert to quasi-continuum for a Si QD size of 2.1 nm. (From Huang, L. M., Zafran, J., Le Bris, A., Olsson, P., Domain, C., and Guillemoles, J. F., "Phonon Modes in Si-Ge Nano-structures for Hot Carrier Solar Cells," in *Proceedings of the 23rd European PV Science & Engineering Conference*, Valencia, Spain, September 1–5, 2008, pp. 689–91.) (b) The *interface effect* shows that optical phonon modes of Si cannot propagate into Ge as there are no phononic states available at this energy (frequency). (From Conibeer, G., Patterson, R., Huang, L. M., Guillemoles, J. F., König, D., Shrestha, S., et al., "Hot Carrier Solar Cell Absorbers," in *Proceedings of the 23rd European PV Science & Engineering Conference*, Valencia, Spain, September 1–5, 2008, pp. 156–62 [1BO.6.2].) Si as NC and Ge as matrix only used for investigation of phononic behavior of high-space group symmetry solids. *Continued*

materials are used for the NC and the matrix. While these do slow down carrier cooling in principle, their benefit depends strongly on where they are located in the mini-Brillouin zone. A local phononic band gap in a $\vec{k}$ space region where many branches of the optical phonon dispersion overlap (high degeneracy of phononic DOS) would have the biggest effect on carrier cooling. These optical modes would be forbidden along with the phonon-polariton anticrossings, which allow for energy transfer between

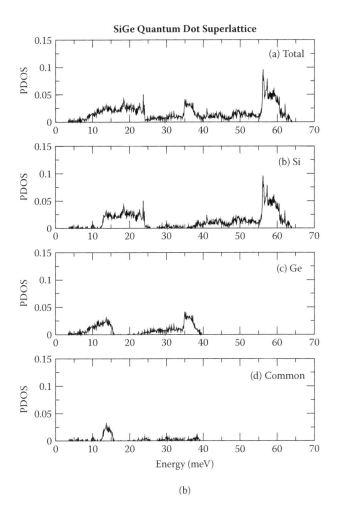

(b)

FIGURE 3.16
Continued.

different optical modes. Such anticrossings are encircled by light grey dotted lines in Figure 3.14 for the maximum DOS of optical phonons around the wave number of 600 cm^{-1} ($\hbar\omega_{opt}$ = 74.4 meV).

The analogy with engineered 3d photonic SLs showing a complete gap (supressed photon propagation) in all spatial directions together with the bosonic nature of photons and phonons is often used as an argument for a phononic SL approach. However, the distance between two nodes of a photon wave function is within the range of a few hundred nanometers to a few micrometer, which exceeds the phononic oscillation length by three to four orders of magnitude. It is relatively easy to engineer an SL with a periodicity of a few hundred nanometers as opposed to a few lattice planes. While it is clear that nanoscopic SLs are a useful means of slowing down carrier cooling (see also Figure 3.15), it remains to be shown to what extent phonon confinement *by size effects*, i.e., in analogy to electronic quantum confinement in QS-SLs, is beneficial for HCAs.

The other cause for slowed carrier cooling in QS-SLs is phonon confinement *by material effects*, where phononic modes are reflected back into the NC, either due to interface engineering or by a vibrationally mismatched matrix material. The analogy in the wave mechanics picture is an impedance mismatch of atomic oscillations over the interface by either very loose or very stiff coupling, corresponding to a very soft or very hard interface, respectively. The impact of such interfaces provides a powerful means of phonon confinement, with no strong dependence on the NC size and inter-NC distance. If the optical phonon frequencies of an NC are significantly above the optical phonon frequencies of the matrix, the matrix atoms are unable to follow the vibrational modes of the NC, leading to phonon confinement by reflecting the optical phonon modes back into the NC. Similar to the interface effect described above, this impedance mismatch does not show a strong dependence on NC size. Both effects lift the constraint of an exact periodicity and allow for bigger tolerances in NC size and location. As a result, a superstructure (superlattice with relaxed periodicity constraints) would be sufficient. Figure 3.16(b) shows the phononic total and partial DOS of Si NCs in a Ge matrix. There is no phononic DOS in Ge that would allow Si optical modes to propagate (escape) from the Si NC.

The electronic structure of a QD can also be used to slow down carrier cooling. Discrete eigen-energies of the QD result in relaxation transitions limited by selection rules, as energy conservation only allows for the emission of phonons with energy matching the transition between two eigen-energies, $\hbar\omega = E_{n+i} - E_n$. However, a QD with a discrete energy spectrum only absorbs photons at these discrete states. While this can be exploited for IB solar cells or QS-SLs with combined low-energy photon and thermionic excitation, the photons with energies not matching the allowed optical transitions cannot be used to heat up the carrier population in a QD. A compromise can be found by implementing QD-SLs with a deliberate deviation in QD size such that in a thick QD-SL, every photon will undergo absorption eventually by hitting a QD with a matching energy transition. However, this is not an HC-SC device, as no carrier population with a common temperature exists.

3.4.5 Concept of a Nanodot Hot Carrier Solar Cell

From the preceding sections it has become clear that HC extraction from an HCA is a real challenge. Phononic material properties can be modified in a narrow range only, e.g., by introducing strain. Below, we have a look into a HC-SC concept that utilizes structural modifications and device design to a hybrid between NC-HC-SCs embedded in a wide-band-gap conventional solar cell. The concept is shown in Figure 3.17.

For hot carrier extraction, ballistic transport is essential in order to avoid inelastic scattering, yielding a spatial extension of the HCA in the range of 150 Å (cf. Section 3.4.2). NCs with such a diameter have a quasi-continuous electronic DOS mandatory for hot carrier generation. On the other hand, an HCA consisting of one layer with 200 Å thickness absorbs only a small fraction of the solar photon flux. Even metals like Al or Cu are semitransparent to sunlight despite their high density of free electrons and zero band gap.

An obvious design is the embedding of the NC-HCAs into a shell that acts as an ESC. This shell can simultaneously work as an interfacial layer reflecting phonon modes back into the HCA. Covalent materials in which acoustic phonons of the NC can propagate, while optical phonon modes of the NC are reflected, are a good choice. The impedance mismatch for optical phonons can be increased further by a shell material that does not have a phononic DOS in the range of the NC (see also Figure 3.16[b]). Another attractive material group for the shell around the NC is porous solids. Their porosity does not allow for bulk phonon transport and their softness attenuates phonon modes of the HCA so that they cannot propagate beyond the NC. While the concomitant confinement of acoustic phonon modes is a disadvantage for ballistic carrier transport (cf. Section 3.4.2), a porous intermediate layer allows for combining NC and matrix solids that would not be compatible in crystalline structure (lattice constant, space group). The ESL within the shell material can be introduced by iso-valent impurities like B-doped GaN, which yields a state at 0.6 eV below the conduction band (Jenkins and Dow 1989). Iso-valent impurities have the advantage over QDs that their electronic properties may be modified only by their first and second next neighbor atoms; size effects can be ruled out as they are point defects. Intrinsic Coulomb blockade effects do not exist as the iso-valent impurity level is neutral. The wide electronic and phononic band gap of GaN:B, with B being a lighter atom than N, make it a good candidate for an ESC shell with good phononic properties. A combination

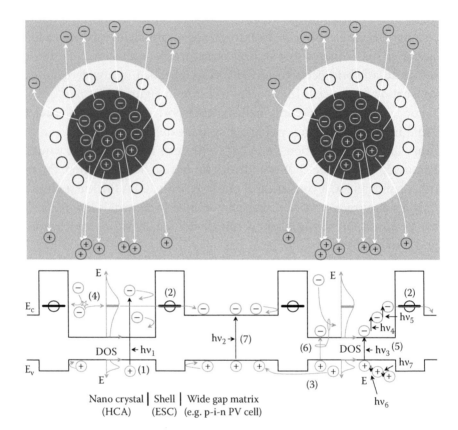

FIGURE 3.17

Physical picture of the ND-HC solar cell (top) with a carrier diffusion field (core-shell thickness ratio not to scale). The shade of the materials represents the band gap and absorption strength, increasing toward lower energies (higher absorption). In the same way, faint carriers are hot, darker carriers are cooled down. Iso-valent impurities providing an energy-selective level for electrons are shown schematically as black circles within the barrier. Diffusion paths are shown by white arrows. Band diagram of the ND-HC Solar Cell matched to the physical structure (bottom). Excitations involving photons are shown in black. Processes shown: Direct generation of hot exciton (1), energy-selective extraction of hot electron from NC-HCA (2), thermionic emission/diffusion of holes over the barrier (3), electron-electron scattering with renormalization of carrier energies and possible energy loss by phonon emission (4), free carrier reabsorption (5), impact ionization (6), and carrier generation in the wide-band-gap matrix (7).

with a porous inner shell like silica aerogel as used for insulating metal interconnects in very large-scale integration would increase the phonon confinement effect. On the downside, the introduction of point defects as ESLs would allow only for one carrier type to be extracted at an elevated energy. Introducing yet another impurity level at the right energy for the second carrier type appears to be not feasible, as both defects would have to exist in the same location. Table 3.2 shows that for most compounds, there is a great difference in the effective mass for electrons and holes. The carrier type with the higher effective mass does not get heated up as much, whereby an extraction by thermionic emission over the barrier becomes attractive as in a thermoelectric device (Takeda 2009). Free carrier reabsorption supports the emission process, which is shown for holes in Figure 3.17.

The core-shell NCs must be embedded into a matrix that receives the carriers at the elevated energy level from which they were extracted, subject to a small decrease of carrier energy. An i-region of a conventional wide-band-gap solar cell would provide the transport of extracted carriers along its band edges. Because the electronic band gap scales as the inverse of the dielectric constant, light trapping into the NCs is enhanced. A massive flux of photons with energies below the wide-band-gap solar cell can heat up the carrier population within the NCs due to free carrier reabsorption and impact ionization.

The concept bears some resemblance to the IB solar cell, but differs in a few important aspects. The NC-HCAs have a continuous electronic DOS and thereby absorb all photons with an energy bigger than their narrow electronic band gap. In principle, photons with an energy exceeding the wide electronic band gap of the matrix can still contribute to the generation of hot carriers in the NC-HCAs if they are not absorbed within the matrix. This may be of importance for matrix materials with an indirect band gap, where the absorption coefficient around the band gap energy is rather low.

3.5 CONTACTS

Contacts to TG-PV conversion layers like QS-SL IB or HC absorbers have some additional requirements owing to the different working principle of the active absorber layers compared to conventional p-i-n devices. We will start with contacts to IB-SCs as the most conventional case, then move on to tandem cell interconnects, and eventually investigate contacts to QS-SLs and HCAs as devices with a tunneling barrier through which one carrier type has to be extracted. Special emphasis is given in this last section to the energy selectivity of the ESC for HCAs.

3.5.1 Contacts to Intermediate Band (IB) Solar Cells

Contacts to IB-SCs have in principle the same requirements as contacts to conventional solar cells, though with a wider range of electron work functions required in order to extract carriers at the respective edges of a wider band gap. Tunneling of minority (nonselected) carriers through the n- or p-region is not a major issue, though at conventional high-efficiency SCs, these losses show up as minority carrier recombination around the contacts and present one of the tackling points by which conversion efficiency can be increased. The choice of contact materials to IB-SCs is thus less restrictive than the choice of materials to QS-SLs and HCAs.

Metal-semiconductor (Schottky) contacts can be used under the proposition that contact is established to highly doped regions (n^{++}, p^{++}). The extension of the Schottky space charge region is (Böer 1992)

$$d_{Sch}^{SCR} = \sqrt{\frac{2\varepsilon_0\varepsilon_{mat}\Psi_{Sch}}{qN_{dope}}} \tag{3.22}$$

with ε_{mat} as the relative dielectric constant of the n^{++}-, p^{++}-regions, N_{dope} the respective doping density, and Ψ_{Sch} the work function difference between the highly doped region and the contact metal. Assuming typical values of $\varepsilon_{mat} = 11$, $N_{dope} = 10^{20}$ cm^{-3}, and $\Psi_{Sch} = 0.8$ eV, we obtain $d_{Sch}^{SCR} = 31$ Å. Majority carriers can easily tunnel through such an ultrathin Schottky barrier. This is routinely used at Al-diffused p^{++}-regions contacted with Al when forming a back surface field on p-type base Si solar cells (Rohatgi et al. 1996). Highly doped contact regions may not be desirable for ultrahigh efficiency devices due to band gap narrowing (Zerga et al. 2003; Kerr et al. 2001). The loss in carrier potential for a doping density of $N_{dope} = 10^{20}$ cm^{-3} in Si amounts to 0.1 eV, or 9% of E_{gap} (Si) (Schenk 1998), where losses are doubled by the second highly doped region at the other contact. Though band gap narrowing can become a significant loss mechanism, the relative contribution to band gap narrowing decreases with increasing band gap for a fixed value of carrier density. This is due to the exchange-correlation energy of the carrier-carrier interaction depending only on the majority carrier density, but not on the energetic position of the dopant. Hence, a very shallow dopant in a wide-band-gap material decreases band gap narrowing relative to the band gap of the highly doped semiconductor.

As an alternative, metals of appropriate electron work function can be used on doped regions where band gap narrowing does not become significant. This is the case for N_{dope} not exceeding 10% of the DOS of the respective band edge, yielding $N_{\text{dope}} \approx 5 \times 10^{16}$ to 3×10^{18}, with the lower and upper limits set by GaAs and Si, respectively. For electron extraction, Mg with $E_{\text{WF}} = 3.77$ eV or Hf with $E_{\text{WF}} = 3.89$ eV (Halas and Durakiewicz 1998) are abundant and nontoxic candidates. A layer thickness of ca. 100 nm, establishing a metallic phase, will provide the electron work function required. Therefore, Mg or Hf layers can be capped with a metal less prone to oxidation so that the contact can withstand deterioration by aging. The capping of Mg should be carried out under vacuum *in situ* for preventing oxidation of the Mg layer directly at the IB absorber (evaporation/sputtering of capping metal within the same deposition system). Hf is less prone to oxidation and should not require such precautions. At the upper end, Ni, Ir, and Pt with the respective work functions of $E_{\text{WF}} = 5.15$, 5.31, and 5.55 eV (Halas and Durakiewicz 1998) are good candidates for hole extraction, with Ni being a good compromise between cost and required energetics. Again, a layer of ca. 100 nm will provide the metallic phase, which can be reinforced by other, less expensive metals. The difference in electron work function of $\Delta E_{\text{WF}} = 1.26$ to 1.78 eV gives us an idea about the maximum band gap, which must exceed this work function difference, as the carriers require some drift velocity to be pushed from the absorber into the contact metal. With $\Delta E_{\text{WF}} = 1.78$ eV, the estimate for the total band gap of the IB absorber $E_{\text{gap}}^{\text{tot}}$ is ca. 2 eV, which covers $E_{\text{gap}}^{\text{tot}} = 1.93$ eV found for the maximum efficiency of an IB-SC (Luque and Martí 1997). Metal contacts on semiconductors introduce dipole states at the interface (interface dipoles), which may alter ΔE_{WF} between the contact metal and the IB absorber by up to a few 100 meV (Ibach 2006). For a moderately doped semiconductor, the resulting Schottky space charge region is too thick for being penetrated by tunneling carriers. For this reason, the interface properties of metal-semiconductor contacts have to be included in the experimental investigation. The reverse effect—a decrease in ΔE_{WF} between the contact metal and the IB absorber—is also possible. The impact of interface dipoles on contact properties requires a detailed evaluation of concrete material combinations and crystal planes of the semiconductor surface and is beyond the scope of this chapter.

3.5.2 Tandem Cell Interconnects

The task of tandem cell interconnects is a possibly smooth transition between the energy levels of electrons and holes of adjacent single-junction solar cells (cf. Figure 3.1). In conventional tandem cells based on III-Vs, this is achieved by a n^{++}/p^{++} tunnel junction, usually deposited monolithically in a molecular beam epitaxy (MBE) (Sugiura et al. 1988) or a metal-organic chemical vapor deposition (MO-CVD) process (Seidel et al. 2008). Holes from the p^{++} region of one solar cell recombine with electrons from the n^{++} of the adjacent solar cell, thereby passing on the carrier flux—the single-junction solar cells are connected in series. A potential difference in the involved band edges does not hamper carrier recombination as long as carriers gain energy by moving toward the antipolar doped region for recombination, and the potential difference is small enough for promoting carrier tunneling. However, a potential step promoting carriers to recombine by tunneling is a loss of the potential energy for the carriers. In device performance, this decreases the open-circuit voltage V_{OC}, but promotes diffusion currents to the recombination at the tunnel junction. Another loss mechanism due to the required massive doping is band gap narrowing (see also section 3.5.1).

For III-V-based QS-SC interconnects, a few monolayers (MLs) with massive doping can be grown in the epitaxial process of SL formation. This so-called δ (delta) doping can be done by MBE along with the deposition of the SL (Cheng and Ploog 1985). For SLs based on group IV compounds, namely Si, the situation is different. QS and matrix materials are not compatible in lattice constants and space group symmetry. In fact, the relevant matrix materials SiO_2 and Si_3N_4 are amorphous. Only SiC annealed at very high temperatures or deposited at very high plasma powers around 1,200 to 1,300°C develops a nanocrystalline structure (Wagner et al. 2003; Seo et al. 2002). Thin layers with high doping densities cannot be used for several reasons. Group IV semiconductors give much less variability in the band gap than binary, ternary, or even quarternary III-V compounds for aligning the electron/hole transport level to the respective miniband or discrete energy level of the SL. As an example, the band edges of $Al_xGa_{1-x}As$ can be shifted over a wide range as a function of composition (Barnham and Vredensky 2001). Si-based SL precursor layers have to undergo a segregation anneal in order to form Si NCs within the SiC, Si_3N_4, or SiO_2 matrix, typically carried out in the range of 1,000 to 1,200°C over 15 to 240 min. Thin bulk layers with a degenerate doping density result in a massive out-diffusion of the dopants for the temperature-time integrals mentioned above. A n^{++}/p^{++} tunnel junction cannot be prepared with the required steep junction profile.

Si-based QSs have a polar interface to the surrounding matrix as a function of the anion of the matrix compound (C, N, O). This effect is more significant than in III-V compounds, as the latter have an intrinsic polarity included by the very existence as a binary compound. Extensive density functional theory (DFT) calculations showed that interface polarity results in a charge shift from the Si NC to the interface anion as a function of the ionicity of the interface bond (IOB) (König et al. 2008a). As a consequence, the electronic structure of the Si NC experiences a large shift, with the highest occupied molecular orbital (HOMO) and lowest unoccupied molecular orbital (LUMO). This energetic shift as a function of the anion in the matrix compound was estimated to govern the electronic structure to a NC diameter of up to ca. 40 Å (König et al. 2008a). Figure 3.18 illustrates the effect for Si core approximants completely terminated with functional groups presenting the dielectric.

A complementary method to the DFT bottom-up aproach is the semiclassical theory of interface dipoles (Ibach 2006). Louie and Cohen (1976) found out that a dominant impact of an Al layer on the Si DOS exists for a few MLs by electron wave functions decaying into the adjacent material. Evaluation of Si/SiO_2 interfaces within the Airy function formalism of decaying electron wave functions from Si into the SiO_2 band gap yields similar characteristic lengths of the interface dipole. Moreover, its impact is increased by a factor of 6 when going from a bulk interface to a QD, arriving at a QD diameter of ca. 45 Å (König 2007), which is in close agreement to above DFT results. Experimental verification is currently being pursued, though not easy to accomplish (Zimina et al. 2006).

The interface energetics has potential to be employed as a tunnel junction between p- and n-type QS-SLs if appropriate doping achieves a further increase of the energetic shift. In this case, Si QDs in SiO_2 require electrons and Si QDs in Si_3N_4 require holes as majorities.

3.5.3 Energy-Selective Contacts and Contacts to QS-SLs

Contacts to QS-SLs and HCAs have in common that the carrier reservoir of the opposite type to be extracted—unoccupied states for taking

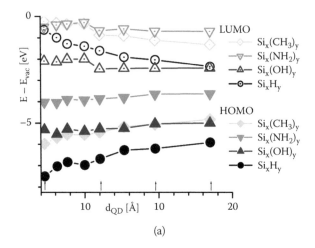

(a)

FIGURE 3.18

(a) HOMO and LUMO levels of Si core approximants completely terminated with functional groups. Computations were carried out with the B3LYP hybride DF and an all-electron Gaussian 6-31G(d) molecular orbital basis set. (From König, D., Rudd, J., Green, M. A., and Conibeer, G., "Role of the Interface for the Electronic Structure of Si Quantum Dots," *Phys. Rev. B* 78 [2008a]: 035339 1–9, and König, D., Rudd, J., Green, M. A., and Conibeer, G., "Critical Discussion of Computation Accuracy," EPAPS article to [57], 2008b, ftp://ftp.aip.org/epaps/phys_rev_b/E-PRBMDO-78-101827/EPAPS_Rev3. pdf) OH-, NH_2-, and CH_3-groups emulate SiO_2, Si_3N_4, and SiC as a dielectric matrix. (b) Example of optimized $Si_{165}(OH)_{100}$ approximant, corresponding to a Si NC diameter of 18.5 Å. O, Si, and H atoms are shown in dark grey, grey, and light grey, respectively. (See color insert following page 206.) *Continued*

up electrons, occupied states for injecting electrons into holes—is needed adjacent to a tunneling barrier. Simultaneously, a suppression of tunneling for the carrier type not to be extracted has to be accomplished, providing the carrier selectivity realized by p^+- and n^+-regions in p-i-n solar cells. This has several consequences that apply to both types of contacts. We will evaluate these and subsequently focus on the particular requirements of the energy-selective level (ESL) of the ESC.

For QS-SLs and HCAs, the thin tunneling barriers require wide-band-gap semiconductors with low electron affinities or high ionization energies for electron or hole extraction, respectively. This also refers to QS-SLs and NC-HCAs in the i-region of a conventional p-i-n solar cell; only there the contact material is constituted by the i-region as the embedding matrix. IB-SCs have this embedding by employing p- and n-regions of a wide-band-gap material, which take over the roll of carrier selection. The band gap of the contact material must be big enough

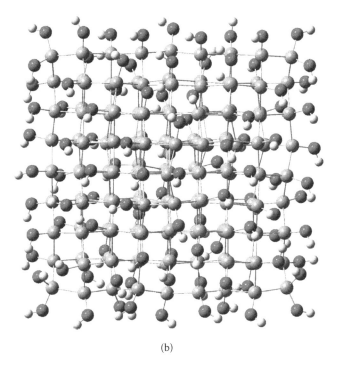

(b)

FIGURE 3.18
Continued.

to extend beyond the energy level of the carrier type not to be extracted (cf. Figure 3.19).

Selective carrier extraction proceeds around the miniband ionization energies and electron affinities found in common QS-SLs, which are in the range of 3 to 6 eV below the vacuum level. For electron extraction at elevated energies, AlAs are a good candidate as a wide-band-gap semiconductor, with the energetic difference of the band edges to the vacuum level, E_{vac}, being $E_{vac} - E_C = 3.5$ eV (Milnes and Feucht 1972) and $E_{vac} - E_V = 6.5$ eV, the latter value obtained by $E_V = E_C + E_{gap}$ from Adachi (2004a). Another good candidate is AlSb, with an electron affinity of $E_{vac} - E_C = 3.65$ eV (Milnes and Feucht 1972). With a band gap of 3.01 eV (Adachi 2004a), we obtain $E_{vac} - E_V = 5.87$ eV. GaN with 3.3 eV and 6.72 eV, respectively, is also a suitable compound for electron extraction at elevated energies. A wide-band-gap semiconductor that has a low binding energy of the valence band states and an electron affinity close to the vacuum level is needed for extracting confined holes, reflecting confined electrons back into the QS-SL. Nitrides with a possibly covalent bond to

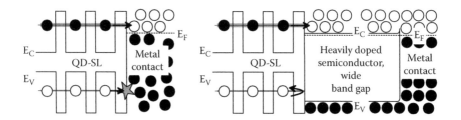

FIGURE 3.19

Contact to a QD-SL established by a metal (left). Carrier selectivity is destroyed by holes being able to recombine with electrons of the metal. Contact to a QS-SL established by a wide-band-gap semiconductor, with the hole miniband being located within the band gap of the semiconductor contact (right). Carrier selectivity given by a high ratio of transmission probabilities due to holes having no occupied counterpart to undergo recombinative tunneling. Principle shown for electron extraction applies also to holes and for the ESC to an HCA—only there, carriers are allowed to escape from HCA within a narrow energy range (cf. Figure 3.13).

the cation are a favorite group of materials for hole extraction.[*] As an example, Ge_3N_4 has $E_{vac} - E_C = 1.78$ eV and $E_{vac} - E_V = 5.77$ eV (Wang et al. 2006), as derived by relating the band offsets to the band edges of Ge (Böer 1990).

ESCs have an additional energy selectivity on top of the carrier selectivity. In addition, they must reflect the phonon modes of the HCA. The ESL can be accomplished by QDs or by doping. Arrays of small QDs have one discrete energy level, which can be used for energy-selective carrier extraction. The ESL should have a certain width in order to allow for a high carrier (energy) flux through the ESC with a small decrease of conversion efficiency. The latter is due to carriers thermalizing to the minimum energy of the ESL during extraction. For the spread of the originally discrete energy level per QD, a miniband formation by close inter-QD spacing is not required. Instead, the unavoidable deviation of QD sizes delivers a finite width of the ESL in analogy to a low dispersion miniband (see Figure 3.7(a)). This requires a precise engineering of the QD size, though the deviation from a nominal inter-QD distance is more forgiving.

Energy-selective extraction of hot carriers has been accomplished at room temperature by optically assisted IV (König et al. 2008d). A 40-nm thick nanocrystalline n^+ Si layer was excited with a photon flux of ca. 100

---

[*] Unfortunately, an evaluation of the quantum chemical nature of the bond N atom–cation is beyond the scope of this book.

suns in the spectral range of 950 to 500 nm (1.3 to 2.5 eV). Samples were not heated above 20°C, referring to ambient (room) temperature. Electrons were extracted through a QD array embedded into SiO_2. An n^+ Si wafer served as counterelectrode, deliberately aligning the polarity and the doping density to the nanocrystalline Si top layer for eliminating the influence of space charge region effects on the IV characteristics. The faint resonance under dark conditions could be increased by a factor of 14 under intense illumination, with the resonance feature shifted down to 60% of the bias voltage, and its bias range decreased to 50% of its original value under dark conditions (König et al. 2006). This experiment showed that a hot carrier population can exist at room temperature within the range of ballistic transport even in a semiconductor with very unfavorable phononic properties such as Si, and that an energy-selective extraction is feasible with an ESC based on a Si QD array consisting of ca. 2×10^{10} QDs.

An easier and more elegant way of introducing an ESL is the use of dopants, which form a defect level within the band gap of the ESC barrier (Conibeer et al. 2003). Further examination of this concept shows that iso-valent dopants are the optimum type of defect. They are neutral and thus do not show any Coulomb or exchange-interaction effects that would hamper carrier transport. Their iso-valent configuration does not create dangling bonds presenting scattering centers. Both effects ensure optimum ballistic transport over the ESC. An example is given by GaN:B in Section 3.4.5. The phononic properties can be realized either by a porous solid or by a material with a wide phononic band gap and strong mismatch to the optical phonon modes of the HCA (see Section 3.4.5). A modification of the latter approach can be realized with a free-standing QD array sandwiched between SiO_2 barriers, as shown by Takeda et al. (2006). The sparse mechanical contact to the HCA results in a high suppression of phonon propagation.

The energy spread ΔE of the ESL has been discussed in detail by O'Dwyer et al. (2005). Their work confirms the statement that systems with 3d confinement such as QDs or atomic impurities show the best energy selectivity compared to quantum wires or QWs where one, $\vec{k}_{\|x}$, or even two, $\vec{k}_{\|x}$ and $\vec{k}_{\|y}$, momenta in lateral direction to the ESC allow for carrier cooling. For the same reason, QD-based ESCs are more sensitive to carrier cooling if ΔE increases, once more showing that iso-valent impurities are superior.

Tunneling barriers without an ESL achieve carrier extraction with a lower limit and are thus only half selective. This results in an energy loss by cooling of the carriers with energies above the lower limit. Thermoelectric devices work with such semiselective energy contacts. For HC-SCs an ESL integrated into an otherwise inpenetrable barrier is mandatory for reflecting hot carriers with energies other than the range of the ESL back into the HCA. A contact as to a thermoelectric device would result in a considerable energy loss, with a concominant serious degradation of conversion efficiency. For HCAs with very different effective carrier masses, an exception is given by the thermionic escape of the carrier type with a high effective mass over the barrier, as discussed in Section 3.4.5. The energy of a free carrier is $E(i) = (\hbar \vec{k_i})^2 / 2m_{\text{eff}}^i$, whereby $i = n$, p, with the excitation energy passed on to the electron/hole in the inverse ratio of the effective masses, $E(n)/E(p) = m_{\text{eff}}^p / m_{\text{eff}}^n$. If this ratio is large, it leaves one of the generated carrier types with little excess energy. These "warm" carriers can then be extracted with little energy loss over a low barrier by thermionic emission or thermal diffusion, as shown for holes in Figure 3.17.

3.6 SUMMARY

In this chapter, we have analyzed the physics relevant for nanoscale solar cells. Starting from a conventional solar cell, we identified two approaches for increasing the conversion efficiency η_{tot}. The carrier recombination rate, $\mathcal{R}$, can be diminished by lowering the dark saturation current, $\vec{j}_0$, over the junction, increasing the open-circuit voltage, V_{OC}. This can be achieved with a wider-band-gap material and high-quality interfaces. The other approach is to maximize the optical generation rate $\mathcal{G}_{\text{opt}}$ by utilizing sub-band-gap photons, increasing the short-circuit current density, $\vec{j}_{\text{SC}}$. Implementations of TG-PV structures were done into i-regions of conventional solar cells with the attempt to increase both V_{OC} and $\vec{j}_{\text{SC}}$. Minimum carrier scattering/recombination rates and maximum carrier mobilities minimize efficiency losses by parasitic resistances.

GaAs/AlGaAs QW-SLs inserted into an i-region led to an increase of V_{OC} (Barnham et al. 1996). Inclusion of an InAs QD-SL into an N-diluted GaAs matrix of a GaAs solar cell led to an increase in $\vec{j}_{\text{SC}}$ of 7%, exclusively due to the generation of sub-band-gap excitons and subsequent thermionic emission (Oshima et al. 2008). The concept of carrier generation by two-photon absorption in intermediate band solar cells (IB-SCs) was proven experimentally by Martí et al. (2006), although the generation rate was rather low. An implementation of a neutral (half-filled) IB by quantum structures requires QD arrays or SLs, as these are the only quantum structures that have a full miniband separation. Deep dopants used for the realization of the IB in an IB-SC were found to be detrimental to solar cell performance due to carrier recombination outweighing the two-step generation process. The working principles of both QS-SLs for sub-band-gap carrier generation with thermionic emission and carrier generation by two-photon absorption via an IB are qualitatively the same (Green 2000). In terms of optical generation rates, the IB-SC should have a better efficiency regarding the utilization of sub-band-gap photons. Unfortunately, the carrier recombination rate increases with the energetic distance of the IB from the band edges, while thermionic carrier emission decreases. Both effects render the achieved conversion efficiency inferior to QW-SLs with minibands close to the conduction band. The scheme of a local IB implemented by isolated QD arrays was proposed for limiting the recombination cross-section of defects and thereby the carrier recombination rate.

For QS-SLs, a trade-off between the controllability of the effective band gap and the miniband dispersion for improved optical activity and transport exists, increasing with the dimensionality of quantum confinement (wells → wires → dots). It can be relieved by choosing ternary III-V or group IV compounds, with their band gap depending on their chemical composition. Another way to increase the miniband dispersion is given by the lowering of the barrier height, which also increases the carrier mobility. Stacks of electrically interconnected QW- and QD-SLs are attractive for TG tandem solar cells, especially if they consist of similar compounds like III-V or group IV solids. Appropriate cell interconnects require the blocking of the minority carrier type so that the majorities recombine with the (antipolar) majorities of the adjacent cell. The electronic structure of Si-QDs below $d_{QD} = 40$ Å depends to a major part on the polar bonds to the dielectric matrix, an effect that does not lead to appreciable changes for III-V QDs due to their intrinsic polar nature as a binary compound. This modification of Si-QDs can be exploited for QD array-based tandem cell interconnects. For III-V compounds, delta doping of contact layers is applied to tandem cell interconnects.

Hot carrier solar cells work after a different principle, namely, the extraction of free carriers at their initial (nonthermalized) energetic position. They require an absorber that delays carrier cooling by a minimum electron-phonon coupling and a low decay rate of optical phonons. Binary covalent solids with a maximum atomic mass of the heavy core and a mass ratio of $\geq 1/4$ between the light and heavy core are candidates for a low decay rate of optical phonons. For a minimum energy loss per optical phonon emitted by the electrons, the light core mass must be possibly heavy without violating the ratio given above. Bi compounds like Bi_2S_3, BiN, or BBi are promising candidates. Bi compounds have a high thermoelectric figure of merit and are applied very successfully at thermoelectric devices—another hint that Bi-based compounds are a good choice for HCAs. Super structure approaches can be applied for improving phonon confinement, whereby interface and material engineering in terms of a vibrational mismatch at the HCA surface are very promising. Phonon confinement by size effects in analogy to electron confinement in QS-SLs does not improve a HCA, because the structures required are too small to be engineered and the electronic properties of such small structures would preclude any existence of a hot carrier population. The timescale of carrier cooling requires the HCA to be operated within the quasi-ballistic transport regime, allowing only for carrier-carrier scattering and impact ionization. This limits the extension of the HCA to about 200 Å, with NC of this size as the only option, because carrier cooling must be suppressed also in lateral directions. Based on the observations above, a core-shell NC-HC-SC embedded into a wide-band-gap semiconductor is proposed, with the shell also working as an ESC. Multiple exciton generation (MEG) is another approach to harvest the excess energy of electron-hole pairs. MEG has been shown experimentally only in colloidal solutions of NCs, which are rather unsuitable for environmentally robust absorbers. Evidence of MEG was measured indirectly by photoluminescence, leaving carrier collection as another hurdle to be overcome. Nevertheless, MEG is a very attractive method as energy loss by carrier cooling is only a minor issue.

Contacts to TG structures must be carrier selective in order to prevent carrier recombination by minorities tunneling through a contact barrier and then recombine with the majorities. This can be achieved by employing heavily doped wide-band-gap semiconductors of appropriate conduction and valence band energies. These can extract majority carriers and have no states at the minority energy level. Minority carriers cannot tunnel into the contact, but are reflected back into the SL instead. For ESCs, an energy-selective level (ESL) is required for extracting carriers only within a narrow energy range. QD arrays are a means for introducing such an ESL. Experimental proof of a hot carrier population and energy-selective carrier extraction at room temperature has been obtained with a Si QD array embedded in SiO_2 and Si as a somewhat nonideal HCA (König et al. 2006). A more attractive implementation is the use of iso-valent dopants in a barrier. As a candidate we mentioned B-doped GaN for energy-selective extraction of hot electrons.

ACKNOWLEDGMENTS

The author acknowledges some fruitful discussions with his colleague I. Perez-Würfl regarding section 3.1.1. This work was financially supported by the Australian Research Council (ARC) Centre of Excellence funding scheme and by the Global Energy Climate Project (GECP).

REFERENCES

Adachi, S. 2004a. *Handbook on physical properties of semiconductors: III-V compound semiconductors.* Vol. II. Boston: Kluwer Academic Press.

Adachi, S. 2004b. *Handbook on physical properties of semiconductors: Group IV semiconductors.* Vol. I. Boston: Kluwer Academic Press.

Banoo, K., Rhew, J.-H., Lundstrom, M., Shu, C.-W., and Jerome, J. W. 2001. Simulating quasi-ballistic transport in Si nanotransistors. *VLSI Design* 13:5–13.

Barker Jr., A. S., and Ilgems, M. 1973. Infrared lattice vibrations and electron-free dispersion in GaN. *Phys. Rev. B* 7:743–50.

Barnham, K. W. J., and Duggan, D. 1990. A new approach to high-efficiency multi-bandgap solar cells. *J. Appl. Phys.* 67:3490–93.

Barnham, K., Collony, J., Griffin, P., Haarpaintner, G., Nelson, J., Tsui, E., et al. 1996. Voltage enhancement in quantum well solar cells. *J. Appl. Phys.* 80:1201–6.

Barnham, K., Marques, J. L., Hassard, J., and O'Brien, P. 2000. Quantum-dot concentrator and thermodynamic model for the global redshift. *Appl. Phys. Lett.* 76:1197–9.

Barnham, K., and Vredensky, V. V., eds. 2001. *Low-dimensional semiconductor structures: Fundamentals and device applications.* Cambridge: Cambridge University Press.

Beard, M. C., Turner, G. M., and Schmuttenmaer, C. A. 2002. Size-dependent photoconductivity in CdSe nanoparticles as measured by time-resolved terahertz spectroscopy. *Nano Lett.* 2:983–7.

Black, J., Conwell, E. M., Seigle, L., and Spencer, C. W. 1957. Electrical and optical properties of some $M-2^{V-B}N_3^{VI-B}$ semiconductors. *J. Phys. Chem. Solids* 2:240–51.

Böer, K. W. 1990. *Survey of semiconductor physics.* Vol. I. New York: Van Nostrand Reinhold.

Böer, K. W. 1992. *Survey of semiconductor physics.* Vol. II. New York: Van Nostrand Reinhold.

Carrier, P., and Wei, S.-H. 2004. Calculated spin-orbit splitting of all diamondlike and zincblende semiconductors: Effects of $P_{1/2}$ local orbitals and chemical trends. *Phys. Rev. B* 70:035212 1–9.

Cheng, L. L., and Ploog, K., eds. 1985. *Molecular beam epitaxy and heterostructures.* NATO ASI Series E, no. 87. Dordrecht: Martinius Nijhoff.

Conibeer, G., Jiang, C.-W., Green, M. A., Harder, N.-P., and Straub, A. 2003. Selective energy contacts for potential applications to hot carrier PV cells. In *Technical Digest of the 3rd World PV Science & Engineering Conference*, Osaka, Japan, May 11–15, pp. 2730–33 (S1P-A7-13).

Conibeer, G., Patterson, R., Huang, L. M., Guillemoles, J. F., König, D., Shrestha, S., et al. 2008. Hot carrier solar cell absorbers. In *Proceedings of the 23rd European PV Science & Engineering Conference*, Valencia, Spain, September 1–5, pp. 156–62 (1BO.6.2).

Cuadra, L., Martí, A., and Luque, A. 2004. Influence of the overlap between the absorption coefficients on the efficiency on the intermediate band solar cell. *IEEE Trans. ED* 51:1002–7.

D'Costa, V. R., Fang, Y.-Y., Tolle, J., Kouvetakis, J., and Menéndez. 2009. Tunable optical gap at a fixed lattice constant in group-IV semiconductor alloys. *Phys. Rev. Lett.* 102:107403 1–4.

Deligoz, E., Colakoglu, K., Ciftci, Y. O., and Ozisik, H. 2007. The first principles study on boron bismuth compound. *Comput. Mater. Sci.* 39:533–40.

De Vos, A. 1980. Detailed balance limit of the efficiency of tandem solar cells. *J. Phys. Appl. Phys.* 13D:839–46.

Di Ciolo, A., Lorenzana, J., Grilli, M., and Seibold, G. 2009. Charge instabilities and electron-phonon interaction in the Hubbard-Holstein model. *Phys. Rev. B* 79:085101 1–14.

Ekins-Daukes, N. J., Ballard, I., Calder, C. D. J., Barnham, K. W. J., Hill, G., and Roberts, J. S. 2003. Photovoltaic efficiency enhancement through thermal up-conversion. *Appl. Phys. Lett.* 82:1974–76.

Ellingson, R. J., Beard, M. C., Johnson, J. C., Pingrong, Y., Micic, O. I., Nozik, A. J., et al. 2005. Highly efficient multiple exciton generation in colloidal PbSe and PbS quantum dots. *Nano Lett.* 5:865–71.

Elliott, S. R. 1998. *The physics and chemistry of solids.* Chichester, UK: Wiley & Sons.

Ferhat, M., and Zaoui, A. 2006. Structural and electronic properties of III-V bismuth compounds. *Phys. Rev. B* 73:115107 1–7.

Gal, D., Hodes, G., Hariskos, D., Braunger, D., and Schock, H.-W. 1998. Size-quantized CdS films in thin film $CuInS_2$ solar cells. *Appl. Phys. Lett.* 73:3135–37.

Goldsmid, H. J. 2006. Bismuth—The thermoelectric material of the future? In *Proceedings of the 25th International Conference on Thermoelectrics*, Vienna, August 6–10, Session I, pp. 5–10.

Gray, P. 1965. Tunneling from metal to semiconductors. *Phys. Rev. A* 140:179–86.

Green, M. A. 1982. *Solar cells.* New York: Prentice-Hall.

Green, M. A. 2000. Potential for low dimensional structures in photovoltaics. *Mater. Sci. Eng. B* 74:118–24.

Green, M. A. 2003. *Third generation solar cells.* New York: Springer.

Green, M. A., et al. 1999. *1999 annual report of the Photovoltaics Special Research Centre.* University of New South Wales. http://www.pv.unsw.edu.au/documents/Annual%20 Report%201999/Photovoltaics%20Special%20Research%20Centre.pdf.

Green, M. A., et al. 2002. *2002 annual report of the Photovoltaics Special Research Centre.* University of New South Wales. http://www.pv.unsw.edu.au/documents/Annual%20 Report%202002/HotCarrier.pdf.

Green, M. A., et al. 2003. *2003 annual report of the Photovoltaics Special Research Centre.* University of New South Wales. http://www.pv.unsw.edu.au/documents/Annual%20 Report%202003/Research.pdf.

Green, M. A., Cho, E.-C., Cho, Y., Huang, Y., Pink, E., Trupke, T., et al. 2005. All-silicon tandem cells based on "artificial" semiconductor synthesised using silicon quantum dots in a dielectric matrix. In *Proceedings of the 20th European PV Science & Engineering Conference*, Barcelona, Spain, June 6–10, pp. 3–7 (1AP1.1).

Green, M. A., Zhao, J., Wang, A., Reece, P. J., and Gal, M. 2004. Efficient silicon light-emitting diodes. *Nature* 412:805–8.

Guo, Q., and Yoshida, A. 1994. Temperature dependence and band gap change in InN and AlN. *Jpn. J. Appl. Phys. Part I* 33:2453–56.

Güttler, G., and Queisser, H. J. 1970. Impurity photovoltaic effect in silicon. *Energy Conversion* 10:51–55.

Guyot-Sionnest, P., Wehrenberg, B., and Yu, D. 2005. Intraband relaxation in CdSe nanocrystals and the strong influence of the surface ligands. *J. Chem. Phys.* 123:74709 1–7.

Halas, S., and Durakiewicz, T. 1998. Work function of elements expressed in terms of the Fermi energy and the density of free electrons. *J. Phys. Condens. Matter* 10:10815–26.

Hall, R. N. 1952. Electron-hole recombination in germanium. *Phys. Rev.* 87:387.

Hieke, K., Wesström, J.-O., Forsberg, E., and Carlström, C.-F. 2000. Ballistic transport at room temperature in deeply etched cross-junctions. *Semicond. Sci. Technol.* 15:272–76.

Huang, L. M., Zafran, J., Le Bris, A., Olsson, P., Domain, C., and Guillemoles, J. F. 2008. Phonon modes in Si-Ge nano-structures for hot carrier solar cells. In *Proceedings of the 23rd European PV Science & Engineering Conference*, Valencia, Spain, September 1–5, pp. 689–91.

Ibach, H. 2006. *Physics of surfaces and interfaces*. Berlin: Springer.

Jenkins, D. W., and Dow, J. D. 1989. Electronic structures and doping of InN, $In_xGa_{1-x}N$, and $In_xAl_{1-x}N$. *Phys. Rev. B* 39:3317–29.

Kalvoda, S., Paulus, B., Fulde, P., and Stoll, H. 1997. Influence of electron correlations on ground-state properties of III-V semiconductors. *Phys. Rev. B* 55:4027–30.

Karazhanov, S. Zh. 2001. Impurity photovoltaic effect in indium-doped silicon solar cells. *J. Appl. Phys.* 89:4030–36.

Kasper, E., and Lyutovich, C., eds. 2000. *Properties of silicon germanium and SiGe:carbon*. London: INSPEC.

Keevers, M. J., and Green, M. A. 1994. Efficiency improvements of silicon solar cells by the impurity photovoltaic effect. *J. Appl. Phys.* 75:4022–31.

Kerr, M. J., Schmidt, J., Cuevas, A., and Bultman, J. H. 2001. Surface recombination velocity of phosphorous-diffused silicon solar cell emitters passivated with plasma enhanced chemical vapour deposited silicon nitride and thermal silicon oxide. *J. Appl. Phys.* 89:3821–26.

Klein, M. C., Hache, F., Ricard, D., and Flytzanis, C. 1990. Size dependence of electron-phonon coupling in semiconductor nanospheres: The case of CdSe. *Phys. Rev. B* 42:11123–32.

Klemens, P. G. 1966. Anharmonic decay of optical phonons. *Phys. Rev.* 148:845–48.

Kojima, A., and Koshida, N. 2005. Ballistic transport mode detected by picosecond time-of-flight measurements for nanocrystalline porous silicon layer. *Appl. Phys. Lett.* 86:022102 1–3.

König, D. 2007. The intrinsic QD p/n junction. Internal seminar. Sydney: ARC Photovoltaics Centre of Excellence, UNSW.

König, D. 2008. Bulk phase phononic properties of hot carrier absorbers. Internal seminar. Sydney: ARC Photovoltaics Centre of Excellence, UNSW.

König, D., Jiang, C. W., Conibeer, G., Takeda, Y., Ito, T., Motohiro, et al. 2006. Static distribution of hot electrons and energy selective extraction observed by optically assisted IV measurements. In *Proceedings of the 21st European PVSEC*, Dresden, Germany, September 4–8, pp. 366–69.

König, D., Rennau, M., and Henker, M. 2007. Direct tunneling effective mass of electrons determined by intrinsic charge-up process. *Solid State Electronics* 51:650–54.

König, D., Rudd, J., Green, M. A., and Conibeer, G. 2008a. Role of the interface for the electronic structure of Si quantum dots. *Phys. Rev. B* 78:035339 1–9.

König, D., Rudd, J., Green, M. A., and Conibeer, G. 2008b. Critical discussion of computation accuracy. EPAPS article to [57]. ftp://ftp.aip.org/epaps/phys_rev_b/E-PRB-MDO-78-101827/EPAPS_Rev3.pdf.

König, D., Rudd, J., Conibeer, G., and Green, M. A. 2008c. Impact of bridge- and double-bonded oxygen onto OH-terminated Si quantum dots: A density-functional–Hartree-Fock study. *Mater. Sci. Eng. B*, in press. doi:10.1016/j.mseb.2008.11.022.

König, D., Flynn, C., Conibeer, G., and Green, M. A. Sydney. 2008d. Investigation of static hot carrier populations and energy selective contacts by optically assisted IV. Paper presented at 17th International Conference on Photochemical Conversion and Storage of Solar Energy, July 27–August 1.

Kuzmany, H. 1998. *Solid-state spectroscopy*. Berlin: Springer.

Lee, J., and Tsakalakos, T. 1997. Influences of growth conditions on physical, optical properties, and quantum size effects of CdS nanocluster thin films. *NanoStructured Mater.* 8:381–98.

Louie, S. G., and Cohen, M. L. 1976. Electronic structure of a metal-semiconductor interface. *Phys. Rev. B* 13:2461–69.

Ludwig, H., Runge, E., and Zimmermann, R. 2003. Exact calculation distribution for excitonic oscillator strength and inverse participation ratio in disordered quantum wires. *Phys. Rev. B* 67:205302 1–10.

Luque, A., and Martí, A. 1997. Increasing the efficiency of ideal solar cells by photon induced transitions at intermediate levels. *Phys. Rev. Lett.* 78:5014–17.

Martí, A., Antolín, E., Cánovas, E., López, N., Linares, P. G., Luque, A., et al. 2008. Elements of design and analysis of quantum-dot intermediate band solar cells. *Thin Solid Films* 516:6716–22.

Martí, A., Antolín, E., Stanley, C. R., Farmer, C. D., López Díaz, P., et al. 2006. Production of photocurrent due to intermediate-to-conduction-band transitions: A demonstration of a key operating principle of the intermediate-band solar cell. *Phys. Rev. Lett.* 97:247701. 1–4.

Milnes, A. G., and Feucht, D. L. 1972. *Heterojunctions and metal-semiconductor junctions*. New York: Academic Press.

Nozik, A. J. 2001. Spectroscopy and hot electron relaxation dynamics in semiconductor quantum wells and quantum dots. *Annu. Rev. Phys. Chem.* 52:193–231.

Nozik, A. J. 2008. Multiple exciton generation in semiconductor quantum dots. *Chem. Phys. Lett.* 457:3–11.

O'Dwyer, M. F., Lewis, R. A., Zhang, C., and Humphrey, T. E. 2005. Electronic efficiency in nanostructured thermionic and thermoelectric devices. *Phys. Rev. B* 72:205330 1–10.

Oshima, R., Takata, A., and Okada, Y. 2008. Strain-compensated InAs/GaNAs quantum dots for use in high-efficiency solar cells. *Appl. Phys. Lett.* 93:083111 1–3.

Paxman, M., Nelson, J., Braun, B., et al. 1993. Modeling the spectral response of the quantum well solar cell. *J. Appl. Phys.* 74:614–21.

Pejova, B., and Grozdanov, I. 2006. Structural and optical properties of chemically deposited thin films of quantum-sized bismuth(III) sulfide. *Mater. Chem. Phys.* 99:39–49.

Ramos, L. E., Furthmüller, J., and Bechstedt, F. 2004. Effect of backbond oxidation of silicon nanocrystallites. *Phys. Rev. B* 70:033311 1–4.

Ridley, B. K. 1989. Electron scattering by confined LO polar phonons in a quantum well. *Phys. Rev. B* 39:5282–86.

Rohatgi, A., Narashima, S., Kamra, S., Doshi, P., Khattak, C. P., Emery, K., et al. 1996. Record high 18.6% efficiency solar cell on HEM multicrystalline material. In *Proceedings of the 25th IEEE Photovoltaics Specialists Conference*, Washinton, DC, May 13–17, pp. 741–44.

Roskovcová, L., and Pastrnák, J. 1980. The "Urbach" absorption edge in AlN. *Czech. J. Phys.* 30:586–91.

Ross, R. T., and Nozik, A. J. 1982. Efficiency of hot-carrier solar energy converters. *J. Appl. Phys.* 53:3813–18.

Sargent-Welch Scientific Company. 1980. *Table of periodic properties of the elements*. Skokie, IL: Sargent-Welch Scientific Company.

Schaller, R. D., and Klimov, V. I. 2004. High efficiency multiplication in PbSe nanocrystals: Implications for solar energy conversion. *Phys. Rev. Lett.* 92:186601 1–4.

Schenk, A. 1998. Finite-temperature full random-phase approximation model of band gap narrowing for silicon device simulation. *J. Appl. Phys.* 84:3684–95.

Schenk, A., and Heiser, G. 1997. Modeling and simulation of tunneling through ultra-thin gate dielectrics. *J. Appl. Phys.* 81:7900–8.

Schiff, L. I. 1968. *Quantum mechanics*. 3rd ed. Singapore: McGraw-Hill.

Schmeits, M., and Mani, A. A. 1999. Impurity photovoltaic effect in c-Si solar cells. A numerical study. *J. Appl. Phys.* 85:2207–12.

Schneider, H., Schönbein, C., Schwarz, K., and Walther, M. 1998. Ballistic effects and inter-subband excitations in multiple quantum well structures. *Physica E* 2:28–34.

Seidel, U., Sağol, B. E., Szabò, N., Schwarzburg, K., and Hannappel, T. 2008. InGaAs/GaAsSb-interface studies in a tunnel junction of a low band gap tandem solar cell. *Thin Solid Films* 516:6723–28.

Seo, J.-Y., Yoon, S.-Y., Niihara, K., and Kim, K. H. 2002. Growth and microhardness of SiC films by plasma-chemical vapor deposition. *Thin Solid Films* 406:138–44.

Shockley, W., and Queisser, H. J. 1961. Detailed balance limit of efficiency of p-n junction solar cells. *J. Appl. Phys.* 32:510–19.

Shockley, W., and Read, W. T. 1952. Statistics of the recombinations of holes and electrons. *Phys. Rev.* 87:835–42.

Shrestha, S., and Clady, R. 2008. Internal research report. Sydney: ARC Photovoltaics Centre of Excellence, UNSW.

Srivastava, G. P. 2008. Origin of the hot phonon effect in group-III nitrides. *Phys. Rev. B* 77:155205 1–6.

Sugiura, H., Amano, C., Yamamoto, A., and Yamaguchi, M. 1988. Double heterostructure GaAs tunnel junction for a AlGaAs/GaAs tandem solar cell. *Jpn. J. Appl. Phys.* 27:269–72.

Sze, S. M. 1967. Current transport and maximum dielectric strength of silicon nitride films. *J. Appl. Phys.* 38:2951–56.

Sze, S. M. 1981. *Physics of semiconductor devices*. 2nd ed. New York: Wiley & Sons.

Takeda, Y. 2009. Private communication. April 2009.

Takeda, Y., Ito, T., Motohiro, T., König, D., Shrestha, S., and Conibeer, G. 2009. Hot carrier solar cells operating under practical conditions. *J. Appl. Phys.* 105:1–10.

Takeda, Y., Ito, T., Motohiro, T., Nagashima, T., König, D., and Conibeer, G. 2006. A novel method based on oblique depositions to fabricate quantum dot arrays. In *Proceedings of the 4th World PV Science & Engineering Conference*, Waikoloa/Hawaii, May 7–12, pp. 75–78.

Teissier, R., Pelouard, J.-L., Mollot, F. 1998. Direct measurement of ballistic electron distribution and relaxation length in InP-based heterojunction bipolar transistors using electroluminescence spectroscopy. *Appl. Phys. Lett.* 72:2730–32.

Touat, D., Ferhat, M., and Zaoui, A. 2006. Dynamical behaviour in the boron III-V group: A first-principles study. *J. Phys. Condens. Matter* 18:3647–54.

Wagner, G., Schulz, D., and Siche, D. 2003. Vapour phase growth of epitaxial silicon carbide layers. *Process Crystal Growth Characterization Mater.* 47:139–65.

Wang, S. J., Chai, J. W., Pan, S. J., and Huan, A. C. H. 2006. Thermal stability and band alignments for ge_3N_4 dielectrics on Ge. *Appl. Phys. Lett.* 89:022105 1–3.

Wang, S. Q., and Ye, H. Q. 2002. Plane-wave pseudopotential study on mechanical and electronic properties for IV and III-V crystalline phases with zinc-blende structure. *Phys. Rev. B* 66:235111 1–7.

Wiberg, N. 1995. *Lehrbuch der Anorganischen Chemie 101* [in German]. Berlin: De Guyter.

Wolf, M. 1960. Limitations and possibilities for improvement of photovoltaic solar energy converters. *Proc. Institution Radio Eng.* 48:1246–63.

Wu, C. I., and Kahn, A. 1999. Electronic states at aluminum nitride (0001)-1×1 surfaces. *Appl. Phys. Lett.* 73:1346–8.

Wu, C. I., Kahn, A., Hellma, E. S., and Buchanan, D. N. E. 1998. Electron affinity at aluminum nitride surfaces. *Appl. Phys. Lett.* 73:1346–8.

Würfel, P. 1997. Solar energy conversion with hot electrons from impact ionisation. *Solar Energy Mater. Solar Cells* 46:43–52.

Yu, P. Y., and Cardona, M. 2003. *Fundamentals of semiconductors*. 3rd ed. Berlin: Springer

Zerga, A., Benosman, M., Dujardin, F., Benyoucef, B., and Charles, J.-P. 2003. New coefficients of the minority carrier lifetime and band gap narrowing models in the transparent emitter of thin film silicon solar cells. *Active Passive Elec. Comp.* 26:1–10.

Zimina, A., Eisebitt, S., Eberhard, W., Heitmann, J., and Zacharias, M. 2006. Electronic structure and chemical environment of silicon nanoclusters embedded in a silicon dioxide matrix. *Appl. Phys. Lett.* 88:163103 1–3.

4

Nanostructured Organic Solar Cells

James T. McLeskey Jr. and Qiquan Qiao

4.1 INTRODUCTION

Existing single-crystal silicon (c-Si) solar cells require sophisticated high-temperature processing, high-quality silicon, and complex engineering, and therefore are not cost-effective as an energy source for most applications (Sun and Sariciftci 2005). In addition, these cells have no or very limited mechanical flexibility. Organic polymer solar cells and dye-sensitized solar cells have become a low-cost alternative to silicon solar cells because they can be fabricated using solution-based processing such as inexpensive painting (K. Kim et al. 2007; Reyes-Reyes et al. 2005a). Other advantages of these cells are their significant flexibility and their ability to be directly fabricated onto most surfaces, including plastics. The most efficient of these devices use a polymer or dye as an electron donor and a second material as the electron acceptor. This chapter describes the application of bulk nanostructured materials to two closely related devices: organic polymer solar cells and dye-sensitized solar cells. The focus is on the successful device structures, important material electronic properties, and nanoscale morphology.

4.2 BACKGROUND

The first organic solar cells were fabricated in the late 1960s and early 1970s and often consisted of a single organic layer (e.g., tetracene) sandwiched between a low-work-function metal layer (aluminum) and a high-work-function metal (gold) (Figure 4.1) (Fang et al. 1970; Ghosh and Feng 1973;

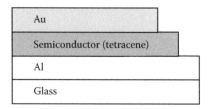

FIGURE 4.1
Early homojunction solar cells consisting of a single organic layer between metal electrodes with different work functions.

Lyons and Newman 1971; Mukherjee 1970; Reucroft et al. 1968, 1969). Thanks to the growth of a thin oxide layer on the low-work-function material, these Schottky-type devices formed metal-insulator-semiconductor (MIS) structures that acted as photodiodes (Chamberlain 1983) and demonstrated a photovoltaic effect. The efficiencies, however, were extremely low (~10^{-5}%) (Ghosh and Feng 1973). Over the next several years, a wide array of materials was tried by different research groups, but efficiencies for these homojunctions remained below 1% (Chamberlain 1983).

A major advance occurred in the mid-1980s when first Harima et al. (1984) and later Tang (1986) reported on a two-component organic photovoltaic cell (Figure 4.2). Harima's device utilized thin films of zinc phthalocyanine (ZnPc) and a porphyrin derivative (TPyP). Tang's device consisted of thin films of copper phthalocyanine (CuPc) and a perylene tetracarboxylic (PV) derivative and achieved power conversion efficiency of 1% under AM2 illumination. These devices showed that at the interface between two materials with different electron affinities, charge transfer is energetically favorable (Huynh et al. 1999). This discovery led to the

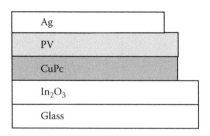

FIGURE 4.2
Early two-component organic solar cell demonstrating the importance of having an interface between materials with different electron affinities to enhance charge transfer. (After Tang, C. W., "Two-Layer Organic Photovoltaic Cell," *Appl. Phys. Lett.* 48 [1986]: 183–85.)

development of heterojunction polymer solar cells consisting of distinct electron donor and acceptor layers.

Since this breakthrough, successful organic solar cells have been based on the use of two different materials. This concept is much the same as that found in photosynthesis in plants, algae, and bacteria. In photosynthesis, light is first absorbed by chlorophyll and an electron is released. The chlorophyll donates that electron to an electron acceptor (pheophytin) (Klimov 2003). This process starts the flow of electrons down the electron transport chain. In photovoltaic cells, the organic semiconductor acts as the light-absorbing and electron-donating material.

Since 1986, continuous efforts have been made to raise the efficiency of organic solar cells while taking advantage of the potential for lower costs. In this chapter, we review the progress made since that time in the application of bulk nanostructured materials to organic solar cells. After reviewing the basic operating principles, we will look at the different device structures that have been fabricated with a focus on the materials, nano-scale morphology, and characteristics of each structure.

4.3 OPERATING PRINCIPLES OF ORGANIC SOLAR CELLS

As introduced above, the most successful organic solar cells utilize two different materials, where at least one of these materials is an organic semiconductor. In order to understand the challenges facing organic solar cells and some of the methods being envisioned to overcome these challenges, the typical operation is outlined in Figure 4.3 and consists of the steps described below.

4.3.1 Light Absorption and Charge Generation

In organic solar cell structures, absorption of light occurs primarily in the organic material. Unlike traditional semiconductors (and solar cells) where light absorption results in the generation of a free electron, light absorption in organic semiconductors results in the formation of a mobile excited state consisting of a tightly bound electron-hole pair known as a Frenkel exciton (Gregg 2005).

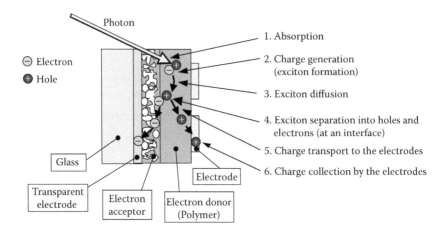

FIGURE 4.3
Operational steps of an organic solar cell.

4.3.2 Exciton Diffusion and Separation

Once the exciton is formed by photon absorption, it must be separated into free electrons and holes. This occurs primarily at the interface between two materials with different electron affinities (Hoppe et al. 2004). Therefore, the exciton must move (by diffusion) to an interface.

4.3.3 Charge Carrier Transport and Collection at the Electrodes

Once the excitons have been separated into distinct electrons and holes, these charge carriers must move to the electrodes. In polymer devices, the polymer itself typically acts as the electron-donating layer and the hole transport layer while the nanostructured material is generally an electron-accepting and transport layer. In dye-sensitized solar cells (DSSCs), a separate electrolyte serves as the hole transport material.

In photovoltaic applications under the short-circuit conditions, the electric field in Schottky devices (often referred to as homojunctions) is caused by the difference between the work functions of the electrodes when no external voltage is applied. The excitons need to diffuse to the organic-metal interface to be dissociated by charge transfer. The electric field then causes the separated photogenerated charges to be transported to their corresponding contacts (Figure 4.4b). In two-component devices (or donor/acceptor heterojunctions), the differences in potential energy between two materials, if larger than the exciton binding energy, causes the exciton dissociation and charge transfer (Figure 4.4d).

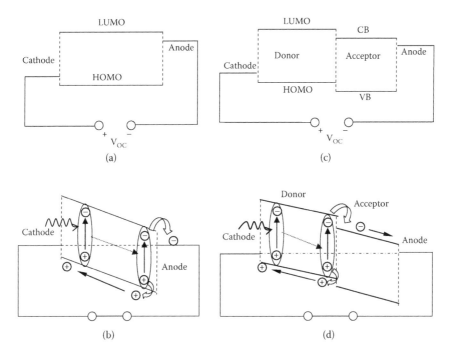

FIGURE 4.4

Schematic illustrations of the band diagrams for homojunctions in (a) open-circuit mode and (b) short-circuit mode, and for planar donor/acceptor heterojunctions in (c) open-circuit mode and (d) short-circuit mode.

4.4 CHALLENGES

Based on the previous section, we can better understand the challenges to achieving higher efficiencies in organic solar cells.

4.4.1 Poor Light Harvesting

Although organic semiconductors have strong absorption coefficients ($>10^5$ cm^{-1}), most have relatively large band gaps (>2 eV) (Hoppe and Sariciftci 2004) (Figure 4.5). Therefore, light harvesting is efficient for blue photons but poor for red. In order to improve absorption in the red and increase the overall efficiency of the devices, the band gap of these materials must be reduced to approach the optimal value of 1.4 eV (Shaheen et al. 2005). In theory, it should be possible to tune the absorption spectrum of organic photovoltaic devices by modifying the chemistry of the polymers or the absorbing dyes. There have been some successes in introducing

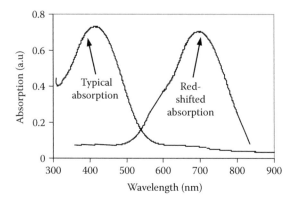

FIGURE 4.5
Schematic showing the absorption spectra of typical organic semiconductors and the red-shifted organic semiconductors currently under investigation.

red-shifted dyes (Nazeeruddin et al. 2001) but relatively few for polymers. Some of the efforts to incorporate materials with lower band gaps are discussed below.

4.4.2 Limited Photocurrent Generation

The excitons in most organic semiconductors (*singlet* excitons) have a lifetime of approximately 10^{-9} s (Greenham et al. 1997a). During that time, the exciton can diffuse over a length of 4 to 20 nm (Greenham et al. 1997a). If the exciton is not dissociated within that time, it will recombine and the energy will be lost. In addition, the exciton binding energies in these materials range from 0.1 to 1eV (Gregg and Hanna 2003; Pope and Swenberg 1999), which is too large to be overcome by the built-in electric field. Therefore, organic devices must have an interface between two disparate materials within one diffusion length of exciton generation in order to effectively separate the exciton into holes and electrons. In simple bilayer devices (see Figure 4.2), this presents a challenge because of the limited interfacial area between the electron-donating and -accepting layers. Many of the nanostructured devices have been designed to increase the interfacial surface area.

4.4.3 Poor Charge Transport

In most organic solar cells, the hole mobility in the polymer (e.g., polythiophene polymers $\cong 0.00001–0.1$ cm^2 V^{-1} s^{-1}) (Coakley et al. 2005; Mozer and

Sariciftci 2004; Mozer et al. 2005; Sirringhaus et al. 1998, 1999) is much lower than the electron mobility in the electron-accepting layer (e.g., TiO_2 ≅ 0.001–10 cm² V⁻¹ s⁻¹) (Hendry et al. 2004; Konenkamp 2000) so that the holes are often the bottleneck to fast transport. Most polymers are p-type hole-conducting materials, although some organic materials (e.g., perylene [Hoppe and Sariciftci 2004]) are n-type electron conductors. The mobility of the materials is dependent not only on the chemistry, but also on the nanomorphology (Choulis et al. 2003; Sirringhaus et al. 1999, 2000). This means that both the structure of the device and the preparation methods can influence the transport.

4.4.4 Additional Challenges

The inherent instability of organic materials, transitioning to large-scale manufacturing processes, and improving the theoretical understanding of device function and limits to performance are all areas that must be addressed in order for organic solar cells to penetrate the market.

4.5 DEVICE STRUCTURES

4.5.1 Dye-Sensitized Solar Cells

4.5.1.1 DSSC Structure

The device often associated with the photosynthetic process is the dye-sensitized solar cell (DSSC) first reported by O'Regan and Gratzel (1991). First introduced in 1991, these devices initially achieved relatively high power conversion efficiencies (η) of 7.1 to 7.9%, with open-circuit voltages (V_{oc}) of 0.65 to 0.7 V, short-circuit current densities (J_{sc}) of 1.1 to 1.3 mA/cm², and fill factors (*ff*) of 0.68 to 0.76. Due in part to the very promising first reports, the DSSC has been the subject of a great deal of research (~3,000 articles) and several commercial ventures (e.g., Konarka [2008], DyeSol [2008], and G24i [2008]) since that time.

Fabrication of DSSCs begins with a conducting glass substrate that acts as a transparent front electrode (Figure 4.6). The conducting layer is typically a coating of some form of tin oxide (SnO_2)—either indium tin oxide (ITO) or fluorine-doped tin oxide (FTO) (Qiao et al. 2006a). These

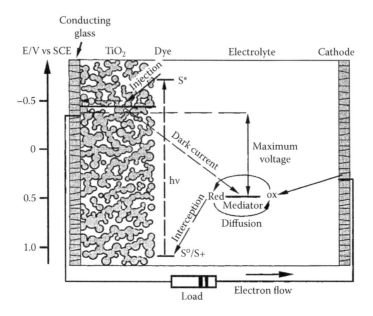

FIGURE 4.6
Structure and operation of a typical dye-sensitized solar cell. (With permission from Hagfeldt, A., and Gratzel, M., "Molecular Photovoltaics," *Acc. Chem. Res.* 33 [2000]: 269–277.)

transparent conducting oxides (TCOs) are used in photovoltaic devices made from a variety of materials.

The electron-accepting material in most DSSCs is titanium dioxide (titania or TiO_2). TiO_2 is nontoxic, low cost, and can be deposited simply by a variety of techniques. For DSSCs, a 5- to 20-μm thick (Gratzel 2003) nanocrystalline layer is formed by suspending TiO_2 nanoparticles with a diameter of 10 to 80 nm (Gratzel 2001) in a solvent such as water or acetic acid and spreading the viscous blend onto the substrate. The device is then annealed at 450 to 500°C to fuse the particles and form a porous nanocrystalline network of interconnected TiO_2 particles (see Figure 4.7). This interconnection allows electrical conduction to the electrode (Gratzel 2003).

As in photosynthesis, light absorption in DSSCs is performed by a monolayer of charge transfer dye that is deposited on the surface of the nanocrystalline TiO_2. The porosity of the TiO_2 is 50% and results in a surface area more than a thousand times greater than that of a flat surface of the same dimensions. The porosity enhances adsorption of the dye onto the surface, which in turn enhances light absorption. Upon absorption of a photon, the dye transfers an electron to the conduction band of the TiO_2,

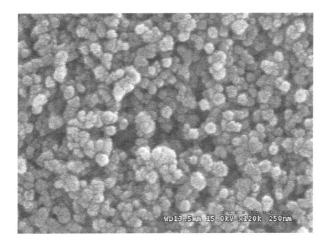

FIGURE 4.7
Scanning electron microscope image of a porous nanocrystalline network of interconnected TiO_2 particles.

but the dye itself is not responsible for charge transport (Li et al. 2006). Although many dyes have been investigated (including polypyridines, indolines, squaraines, perylenes, xanthenes, flavonoid anthocyanins, polyenes, coumarins, and polypyridyl derivatives [Burfeindt et al. 1996; Burke et al. 2007; Burrell et al. 2001; Campbell et al. 2004; Ehret et al. 2001; Hagfeldt and Gratzel 2000; Horiuchi et al. 2004; Howie et al. 2008; Mozer et al. 2006; Smestad 1998]), the most successful devices use dyes based on a ruthenium (Ru) complex (Figure 4.8).

An electrolyte solution restores the electrons to the dye through a redox reaction. The most common and successful electrolyte is an iodide/triiodide I^-/I_3^- solution (Saito et al. 2004). This electrolyte solution presents one of the greatest challenges to commercialization of DSSCs because the use of liquid electrolytes requires encapsulation to prevent evaporation or the reaction of water or oxygen with the electrolyte (Bai et al. 2008; Li et al. 2006). A back counterelectrode of gold or platinum on glass completes the circuit and closes the device.

4.5.1.2 DSSC Materials

The initial success of the first devices was due to the integration of the highly absorbing dye with the high surface area metal oxide. Since 1991, efforts have been made to improve the devices through the use of different materials.

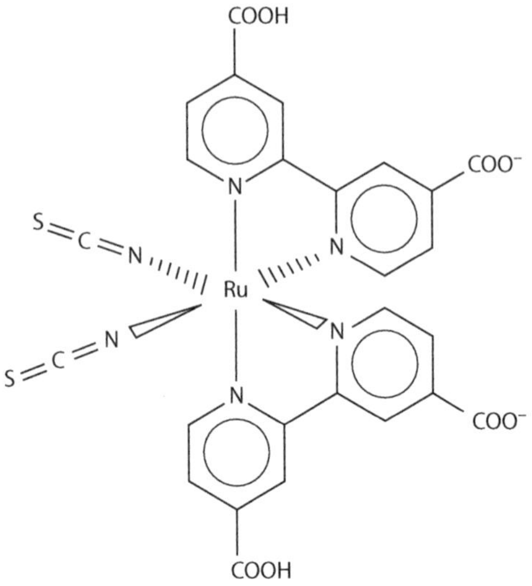

FIGURE 4.8
Chemical structure of the N3 ruthenium dye used in dye-sensitized solar cells. (With permission from Gratzel, M., "Dye-Sensitized Solar Cells," *J. Photochem. Photobiol. C* 4 [2003]: 145–53.)

Nanocrystalline TiO_2 has been the wide-band-gap semiconductor of choice since the fabrication of the first DSSC device. However, other metal oxides have been investigated. For example, zinc oxide (ZnO) (Kakiuchi et al. 2006; Keis et al. 2000; Quintana et al. 2007; Redmond et al. 1994; Rensmo et al. 1997; Tennakone et al. 1999) has a higher mobility and different morphologies can be synthesized more easily than for TiO_2 (Quintana et al. 2007). ZnO can be readily made into nanorods, wires, and tubes, which may lead to improved efficiencies. Nanorods and wires are discussed in Chapter 6. The best DSSC made to date with nanocrystalline ZnO had power conversion efficiency on the order of 4.1% (Kakiuchi et al. 2006)—significantly lower than the 10% that is readily obtained for TiO_2 devices. Other electron-accepting materials have been investigated, including Nb_2O_5 (Sayama et al. 1998) ($\eta \cong 2\%$) and tin oxide (SnO_2) (Ferrere et al. 1997) ($\eta \cong 1\%$), but none has approached the efficiency of TiO_2. Further improvement in device performance has been shown through the introduction of the concept of haze (Chiba et al. 2006), where larger (400 nm) particles were added to the TiO_2 electrodes to increase the ratio of diffuse transmittance to total optical transmittance.

Although many materials have been investigated for use as the dye in a DSSC, including blackberry juice (Cherepy et al. 1997; Gupta et al. 1999), and chlorophyll (Kay and Graetzel 1993; Kay et al. 1994), the best performance has been achieved using ruthenium complexes. Since 1993 (Nazeeruddin et al. 1993), the standard has been a dye referred to as N3 or *cis*-RuL$_2$-(NCS)$_2$, where L stands for 2,2'-bipyridyl-4,4'-dicarboxilic acid (Figure 4.8). The absorption onset of N3 occurs at approximately 800 nm, with a peak near 560 nm (Gratzel 2003), resulting in devices with a conversion efficiency of 10%. Only one variation of this dye, known as the black dye, or (tri(cyanato)-2,2'2''-terpyridyl-4,4'4''-tricarboxylate) Ru(II) (Nazeeruddin et al. 2001) has been found to provide better performance with an absorption onset at 920 nm and a peak at 630 nm (Figure 4.9). The highest confirmed efficiency for a DSSC is 11.1%, and this was achieved using the black dye (Chiba et al. 2006).

The iodide/triiodide I$^-$/I$_3^-$ electrolyte solution provides high efficiency, but as a liquid, it requires encapsulation and is not stable. A number of alternatives have been investigated to overcome these challenges (Li et al. 2006), with most based on the incorporation of a solid-state electrolyte to create a so-called solid-state dye-sensitized solar cell. The most promising solid-state hole conductors are generally one of two types. The first, p-type organic semiconductors such as copper iodide (CuI) (Meng et al. 2003;

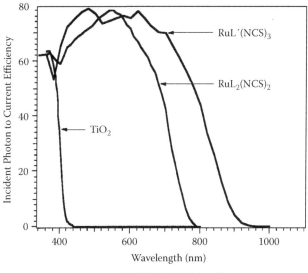

L = 4,4'-COOH-2,2'-bipyridine
L' = 4,4',4''-COOH-2,2':6',2''-terpyridine

FIGURE 4.9
IPCE curves for the N3 and "black" ruthenium dyes used in dye-sensitized solar cells. (With permission from Nazeeruddin, M. K., Pechy, P., Renouard, T., et al., "Engineering of Efficient Panchromatic Sensitizers for Nanocrystalline TiO$_2$-Based Solar Cells." *J. Am. Chem. Soc.* 123 [2001]: 1613–24.)

Tennakone et al. 1999), have achieved efficiencies of about 3.8% (Meng et al. 2003) but are still unstable. The second, polymer electrolytes such as 2,2′,7,7′-tetrakis-(n,N-di-p-methoxyphenyl-amine)-9,9′-spirobifluorene (or spiro-MeOTAD) (Johansson et al. 1997; Salbeck et al. 1997), have been introduced into DSSCs (Bach et al. 1998) and achieved efficiencies of greater than 4% (Schmidt-Mende et al. 2005).

4.5.2 Polymer Devices

4.5.2.1 Planar (Bilayer Heterojunction) Devices

As described above, the first modestly successful organic solar cells were fabricated in the 1980s by evaporating layers of two different organic semiconductors to form a heterojunction. In parallel with the work on DSSCs described above, investigation of the use of soluble semiconducting polymers in photovoltaic devices began in the early 1990s (Sariciftci et al. 1993). Since that time, a number of different bulk nanostructured materials have been incorporated into the devices as electron acceptors in planar or bilayer heterojunctions (Figure 4.10) in an effort to increase the efficiency.

One of the strongest electron acceptors is the buckminsterfullerene form of carbon, or C_{60}, which can accept up to six electrons per molecule (Ohsawa and Saji 1992). This material was first applied to photovoltaic devices by Sariciftci et al. (1993). Their device started with a glass substrate coated with indium tin oxide. ITO is a transparent conducting oxide that is popular in photovoltaic devices made from a variety of materials (Qiao et al. 2006a) and is commonly used in organic solar cells. Sariciftci et al. (1993) spin-coated a layer of poly[2-methoxy-5-(2′-ethyl-hexyloxy)-1,4-phenylene vinylene], or MEH-PPV, onto the ITO. The C_{60} was then

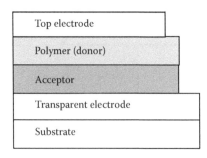

FIGURE 4.10

Schematic of a bilayer heterojunction organic solar cell incorporating bulk nanostructured materials as the electron-accepting layer.

evaporated on top, forming a bilayer p-n heterojunction. Unfortunately, these first C_{60} devices had low power conversion efficiencies (only 0.02%) under monochromatic illumination at 514.5 nm. These low efficiencies were likely due to narrow width of the interface between the polymer and the C_{60}, resulting in a limited interfacial area and low exciton separation.

In order to increase the interfacial area in C_{60}-based bilayer devices, Drees et al. (2002) used "thermally controlled interdiffusion." In this technique, 90-nm-thick films of MEH-PPV were deposited by spin coating. A 100 nm layer of C_{60} was sublimed on top (Figure 4.11a). The device was then placed on a hot plate at 250°C in an N_2 atmosphere for five minutes, causing the C_{60} and the MEH-PPV to interdiffuse (Figure 4.11b). Analysis of transmission electron microscopy (TEM) images and studies of the effect of polymer thickness on device performance (Drees et al. 2004) indicate that the C_{60} has diffused by tens of nanometers into the polymer. This interdiffusion greatly decreases the distance that the exciton must travel in

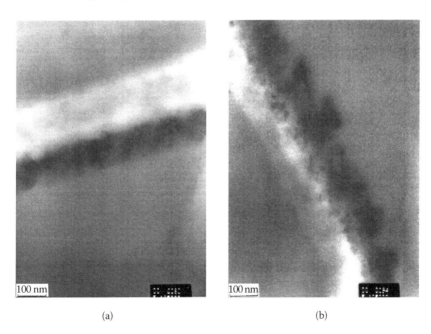

(a)   (b)

FIGURE 4.11

TEM images of an MEH-PPV/C_{60} bilayer heterojunction (a) before and (b) after heating. Heating causes interdiffusion of the C_{60} into the polymer to increase the interfacial area and facilitate charge separation and transport. (With permission from Drees, M., Davis, R. M., and Heflin, J. R., "Thickness Dependence, In Situ Measurements, and Morphology of Thermally Controlled Interdiffusion in Polymer-C_{60} Photovoltaic Devices, *Phys. Rev. B* 69 [2004]: 165320.)

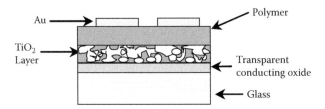

FIGURE 4.12
Schematic of a bilayer heterojunction solar cell utilizing porous nanocrystalline TiO_2 as the electron-accepting layer.

order to reach a material interface, and therefore increases the amount of exciton separation. The efficiency of these devices reached a maximum of 0.30% under monochromatic illumination at 470 nm (Drees et al. 2004).

The success of DSSCs led to the investigation of TiO_2 as the electron acceptor in polymer photovoltaic devices (Figure 4.12). In 1998, Savenije et al. (1998) deposited the TiO_2 onto an ITO-coated substrate using chemical vapor deposition (CVD). The polymer was again MEH-PPV, and they utilized a mercury (Hg) drop as the back contact. The power conversion efficiency was 0.15% under AM1.5 illumination. Shortly thereafter, other groups reported similar devices but with TiO_2 deposited by various means and using different materials for the back electrodes. In 1999, Arango et al. (1999) formed a nanocrystalline layer by spreading a viscous blend of TiO_2 nanoparticles in a water solution onto the substrate and then annealing it at 500°C to fuse the particles and form a nanocrystalline film (similar to Figure 4.7). Arango's device used calcium or aluminum back contacts. In 2001, Fan et al. (2001) formed a TiO_2 layer by using a sol-gel technique with titanium isopropoxide (TIP) as the titania precursor. Their device utilized a gold back contact and achieved a power conversion efficiency of 1.6% under monochromatic illumination at 500 nm.

Studies (Ravirajan et al. 2005) have shown that deep polymer/TiO_2 infiltration can be obtained by various approaches, such as *in situ* growth of TiO_2 from an organic precursor (van Hal et al. 2003) or spin coating of the polymer on ultrathin dip-coated TiO_2 films (Fan et al. 2001). It is generally agreed that the penetrating depth of polymer is dependent on the pore size. The pore size is in turn determined by the nanoparticle size. In addition, the infiltration depth has also been found to be a function of the polymer type. This infiltration greatly increases the interfacial area between the electron donor (polymer) and acceptor (TiO_2) and potentially leads to improved exciton separation.

TABLE 4.1

Photovoltaic Parameters for the Best Reported TiO_2 Devices with Different Polymers

Polymer	J_{SC} (mA cm^{-2})	V_{OC} (V)	FF	η (%)	Reference
P3UBT	0.45	0.67	0.29	0.10	Grant et al. (2002)
MEH-PPV	0.40	1.1	0.42	0.18	Breeze et al. (2001)
TPD(4M)-MEH-M3EH-PPV	0.96	0.86	0.50	0.41	Ravirajan et al. (2005)
P3HT	2.76	0.44	0.36	0.42	Kwong et al. (2004a)
PTEBS	0.17	1	0.8	0.17	Qiao et al. (2008)

A variety of different polymers have been used with TiO_2 and some of the most successful are listed in Table 4.1. In addition to polyphenylene vinylene, polymers such as MEH-PPV, and MEH-PPV with TPD groups (TPD is *N*,*N*'-diphenyl-*N*,*N*'-bis(3-methylphenyl)-(1,1'-biphenyl)-4,4'-diamine), the polymers utilized include several polythiophenes: poly(3-undecyl-2,2'-bithiophene) or P3UBT, poly(3-hexylthiophene) or P3HT, and sodium poly[2-(3-thienyl)-ethoxy-4-butylsulfonate or PTEBS.

PTEBS is especially interesting because it is water soluble. Using water as the solvent has numerous potential benefits. For example, acidic solutions of the PTEBS polymer develop a new absorption band in the near infrared (IR) (Figure 4.13) and films made from the self-acid form of the polymer show the same optical characteristics (Tran-Van et al. 2004). In addition to improved light harvesting from single-junction cells, this increased absorption also opens the possibility for building tandem-junction cells

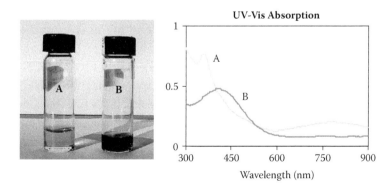

FIGURE 4.13

Absorption spectra of the PTEBS water-soluble polymer in (A) acidic and (B) basic solutions. Acidic solutions develop a new absorption band in the near-IR region. (See color insert following page 206.)

with layers made using both acidic and basic solutions in order to match a greater portion of the solar spectrum. Because water is part of the fabrication process, devices made from this polymer can show improved stability under atmospheric conditions. Obviously, water is environmentally friendly and nontoxic, which is not only compatible with the renewable aspects of solar energy, but also reduces the need for expensive waste containment and disposal. Furthermore, solvent evaporation rates have been shown to have a strong influence on film morphology and device performance (Kwong et al. 2004b, Strawhecker et al. 2001), and the evaporation of water can be carefully controlled using heat.

Other materials have been reported for nanostructured organic polymer bilayer heterojunctions. For example, zinc oxide (ZnO) nanorods can be readily grown using wet chemistry techniques and have yielded efficiencies ($\eta \cong 0.20\%$) (Peiro et al. 2006) comparable to those achieved with nanocrystalline TiO_2. Nanowires and nanorods are discussed more fully in Chapter 6.

4.5.2.2 Blended (Bulk Heterojunction) Devices

In order to overcome the limitations of bilayer devices and increase the interfacial surface area polymer solar cells, Halls et al. (1995) and Yu et al. (1995) introduced the bulk heterojunction in 1995. In a bulk (or dispersed) heterojunction, the electron acceptors are blended into the polymer to create a heterogeneous composite (Figure 4.14). By blending a p-type (electron donor) and an n-type (electron acceptor) material together in the solution and controlling the morphology of the devices, a high interfacial area throughout the bulk can be achieved resulting in enhanced exciton dissociation and charge transfer. The device fabricated by Halls et al. (1995) consisted of a blend of two different polymers (MEH-PPV and CN-PPV), while the device fabricated by Yu et al. (1995) was a blend of MEH-PPV and PCBM (a soluble form of C_{60}) and achieved an efficiency of 2.9%. Since that time, PCBM has become the first choice as an electron acceptor in bulk heterojunction devices, with the highest-efficiency single-junction cell having an independently confirmed efficiency of 4.8% (Schilinsky et al. 2006) under AM1.5 illumination at 100 mW/cm². These devices utilized P3HT.

As described above, in the electron acceptor material is distributed throughout the electron donor material bulk heterojunctions. In this way, excitons are always generated less than one diffusion length from an interface providing outstanding charge separation. However, the discontinuous

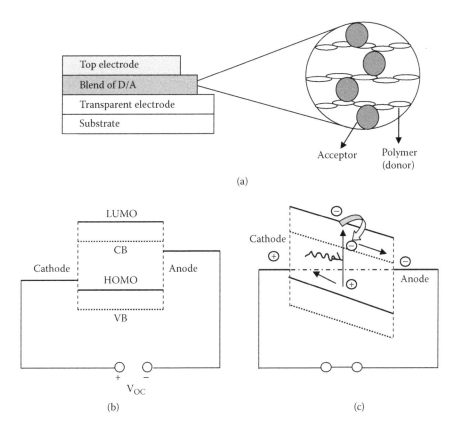

FIGURE 4.14

Schematic illustrations of the (a) structure as well as the band diagrams in (b) open-circuit mode and (c) short-circuit mode for bulk (dispersed) heterojunctions. In these devices, the nanostructured electron acceptors are blended into the polymer for enhanced charge separation.

nature of the electron acceptor material leads to a new problem. In order to move from the point of exciton separation to the electrode, an electron must pass through the polymer. This often results in recombination and limits the photocurrents.

To help overcome this challenge, Alivisatos et al. fabricated a cell using CdSe nanorods imbedded in the polymer (Figure 4.15) to achieve an overall efficiency of 1.7% in simulated AM1.5 light (Huynh et al. 2002). The longer the nanorods, the more electron transport was influenced by band conduction rather than by hopping, and the longer nanorods showed a much higher external quantum efficiency (EQE). Carroll et al. annealed a device containing PCBM so that the C_{60} formed single-crystal "nanowhiskers" (Reyes-Reyes et al. 2005b) oriented from the anode toward the cathode (Figure 4.16).

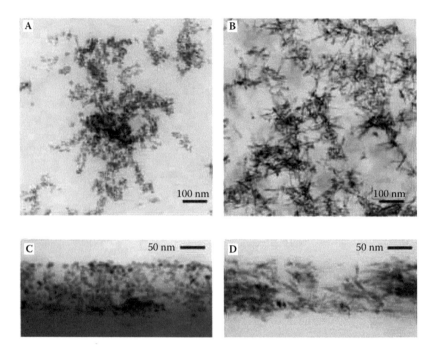

FIGURE 4.15

TEM images of thin-film CdSe nanocrystals in P3HT. (A, B) Top views of 7×7 nm nanocrystals and 7×60 nm nanocrystals, respectively. (C, D) The films in cross-section and the distribution of the nanocrystals in the film. The 7×60 nm nanocrystals are partially aligned perpendicular to the substrate plane. (With permission from Huynh, W. U., Dittmer, J. J., and Alivisatos, A. P., "Hybrid Nanorod-Polymer Solar Cells," *Science* 295 [2002]: 2425–27.)

These devices achieved efficiencies of approximately 5.2%, approximately 120% of the value before annealing. Integrating nanowires and nanorods into both bilayer and blended to polymer devices should allow for improved exciton separation while also increasing carrier mobility. Finding ways to manipulate these nanorod materials and build macroscale devices is the subject of extensive research and is discussed in Chapter 6.

In 2000, Forrest et al. introduced the double heterostructure organic solar cell (Peumans et al. 2000; Peumans and Forrest 2001) incorporating an exciton-blocking layer to prevent electron trapping and mitigate quenching effects caused by cathode deposition damage. These devices had an efficiency of 3.6%. Forrest et al. later extended this concept to build tandem cells made from a connected series of two stacked cells, each optimized to absorb in a specific region of the solar spectrum (Yakimov and Forrest 2002). This device had an efficiency of 5.7%.

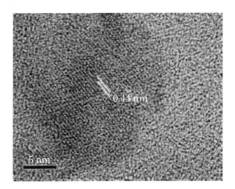

FIGURE 4.16
HRTEM image showing C_{60} "nanowhiskers" after annealing. (With permission from Reyes-Reyes, M., Kyungkon, K., Dewald, J., et al., "Meso-Structure Formation for Enhanced Organic Photovoltaic Cells," *Organic Lett.* 7 [2005b]: 5749–5752.)

In order to reduce the spectral mismatch leading to incomplete absorption of low-energy photons, the tandem-multijunction hybrid organic-inorganic solar cell has been studied (Chen et al. 2005; Dennler et al. 2006b; Gilot et al. 2007; Hadipour et al. 2007; Janssen et al. 2007; Kawano et al. 2006; J. Y. Kim et al. 2007; Peumans and Forrest 2001; Peumans et al. 2003; Uchida et al. 2004a; Xue et al. 2004; Yakimov and Forrest 2002). In a tandem cell consisting of multiple subcells in series, the open-circuit voltage (V_{oc}) is equal to the sum of the V_{oc}s of individual cells. Different subcells will absorb different regions in the solar spectra, enabling the absorption of a broader range than for a single-junction device. Typically, the wide-band-gap subcell will reside at the front side to harness high-energy photons, while low-band-gap cells are used on the back end to harvest low-energy photons. The interfacial middle electrode serves as both a protecting layer for bottom subcell and an electrical contact between the two subcells (via efficient electron-hole recombination). This middle electrode has been made by dc megnetron sputtering (e.g., ITO) (Kawano et al. 2006), thermal evaporation (e.g., Sm, Au, Ag, WO_3) (Gilot et al. 2007; Hadipour et al. 2007), dip coating (e.g., TiO_x) (J. Y. Kim et al. 2007), and spin coating (e.g., ZnO) (Gilot et al. 2007). The contact layer must be transparent so that the lower-energy photons can penetrate through, but does not need to have a low sheet resistance since it does not carry current in the lateral direction. Figure 4.17 shows an example of a tandem polymer solar cell, achieving a highest power conversion efficiency of about 6.5% (J. Y. Kim et al. 2007).

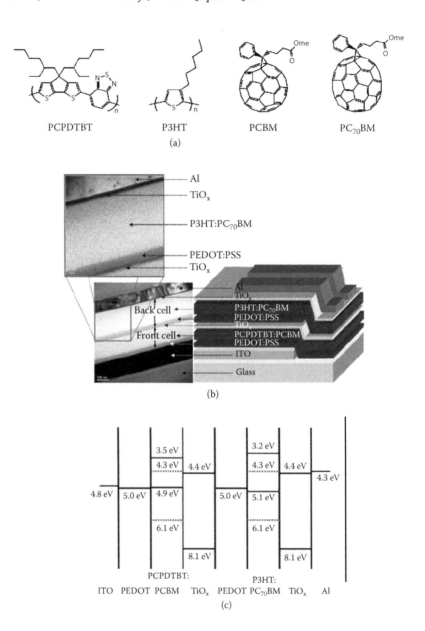

FIGURE 4.17

Tandem solar cell. (a) This device is based on two polymers with different absorption coefficients and two different C_{60} derivatives. (b) The device structure consists of two bulk heterojunction cells separated by a TiO_x layer. The front cell absorbs the high-energy photons. Low-energy photons pass through the device to be absorbed by the back cell. (c) A diagram showing the energy levels for each component material. (With permission from Kim, J. Y., Lee, K., Coates, N. E., et al., "Efficient Tandem Polymer Solar Cells Fabricated by All-Solution Processing," *Science* 317 [2007]: 222–25.)

4.6 NANOSCALE MATERIALS FOR POLYMER DEVICES

As described above, thanks to the development of nanotechnology and various nanoscale materials, polymer solar cells have evolved from a single-layer structure (homojunction) to a heterojunction geometry including bilayer and bulk heterojunctions.

4.6.1 Types of Nanomaterials

Electron acceptors have been fabricated from a number of materials, including TiO_2 (Arango et al. 2000; Breeze et al. 2001; Kwong et al. 2004a; Qiao et al. 2005a, 2006b), ZnO (Beek et al. 2004, 2005a,b), CdSe (Greenham et al. 1997a; Huynh et al. 1999), CdS (Greenham et al. 1996), carbon fullerenes and its derivatives (Aernouts et al. 2002; Brabec et al. 2002; Konkin et al. 2009; Krebs et al. 2009; Martens et al. 2003; Melzer et al. 2004; Mihailetchi et al. 2003; Munters et al. 2002; Nelson et al. 2004; Riedel et al. 2004; Sahin et al. 2009; Svensson et al. 2003; van Duren et al. 2002, 2004; Wienk et al. 2006a,b; Zhokhavets et al. 2003), and carbon nanotubes (Kymakis et al. 2003; Lee et al. 2001). Solvent-based conjugated polymers such as PPV [poly-(phenylene-vinylene)] (Piok et al. 2001) and its derivatives (e.g., MEH-PPV [Breeze et al. 2001], MDMO-PPV [Rispens et al. 2003], etc.) have been widely studied as photovoltaic materials in recent years (Mattoussi et al. 1998). Other groups have reported on results using the thiophene derivatives P3HT [poly(3-hexylthiophene)] (Kim et al. 2005a,b; Lee et al. 2001; Mozer and Sariciftci 2004; Zhokhavets et al. 2006a), P3OT [poly(3-octylthiophene)] (Gebeyehu et al. 2001; Kymakis and Amaratunga 2003; Landi et al. 2005; Qiao et al. 2006; Zafer et al. 2005; Zhokhavets et al. 2004), and PTEBS (Qiao et al. 2005a,b,c, 2006b).

4.6.1.1 Properties of Nanomaterials for Use in Photovoltaics

For heterojunction solar cell applications, nanomaterials need to have a higher electron affinity than the polymer. In other words, the conductance and valence bands of most semiconductor nanoparticle materials lie well below the related HOMO and LUMO of various polymers, making them energetically favorable for exciton dissociation and charge transfer at the interfaces. In addition, a high electron-accepting ability is also required for nanomaterials to be used in solar cells. For example, buckminsterfullerene

TABLE 4.2

The Conductance Bands, Valence Bands, and Band Gaps of Nanomaterials Commonly Used as Electron Acceptors

Nanoscale Materials	Conductance Band (eV)	Valence Band (eV)	Band Gap (eV)	Reference
C60	−3.83	−6.1	2.27	Sun and Sariciftci (2005)
PCBM-C61	−3.75	−6.1	2.35	Sun and Sariciftci (2005)
SWNT	−4.5			Kymakis et al. (2003), Kymakis and Amaratunga (2002)
BM-C60	−3.51	−6.1	2.59	Sun and Sariciftci (2005)
TiO_2	−4.2	−7.4	3.2	Breeze et al. (2001), Qiao et al. (2005a,b,c, 2006a,b)
CdS	−4.2	−6.45	2.25	Sun and Sariciftci (2005)
CdSe	−4.4	−6.1	1.7	Gratzel (2001), Greenham et al. (1997a, 1997b), Huynh et al. (1999, 2002)
CdTe	−4.12	−5.85	1.73	Gur et al. (2005)
SiC	−3.0	−6.0	3	Gratzel (2001)
SnO_2	−4.95	−8.75	3.8	Gratzel (2001)
WO_3	−4.6	−7.2	2.6	Gratzel (2001)
GaAs	−4.2	−5.6	1.4	Gratzel (2001)
GaP	−3.6	−5.85	2.25	Gratzel (2001)
ZnO	−4.4	−7.6	3.2	Beek et al. (2004)
Fe_2O_3	−4.85	−6.95	2.1	Gratzel (2001)

is capable of taking up to six electrons (Ohsawa and Saji 1992). Another property is high electron mobility, which works as a "speed limit" for electron transport.

4.6.1.2 Conductance and Valence Bands and Band Gaps for Typical Nanomaterials

The conductance bands, valence bands, and band gaps of nanomaterials commonly used as electron acceptors are shown in Table 4.2.

4.6.2 Conjugated Polymers

4.6.2.1 Commonly Used Polymers

Three classes of conjugated polymers, including poly(p-phenylenevinylene) (PPV) and its derivatives (CN-PPV [Yu and Heeger 1995], MEH-PPV [Breeze et al. 2001], MEH-CN-PPV [Becker et al. 1997], and MDMO-

PPV [Martens et al. 2003]), polythiophene (PT) and its derivatives (P3HT [Bettignies et al. 2005] and PTEBS [Qiao et al. 2005a,b,c, 2006a,b]), and the polyanilines (PAn [Tan et al. 2004]) have attracted attention and are being investigated widely for use in solar cells. The chemical structures of PPV and PT and their derivatives are shown in Figures 4.18 and 4.19, respectively. The chemical structure of PAn is shown in Figure 4.20.

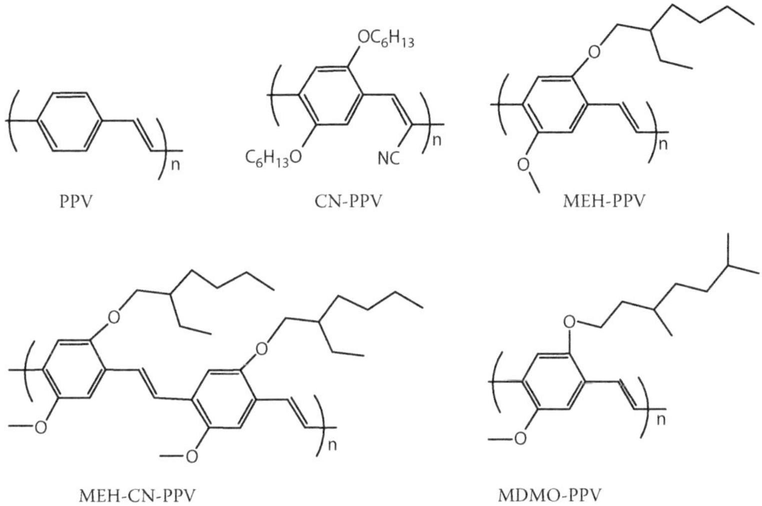

FIGURE 4.18
Chemical structure of poly(*p*-phenylenevinylene) (PPV) and its derivatives.

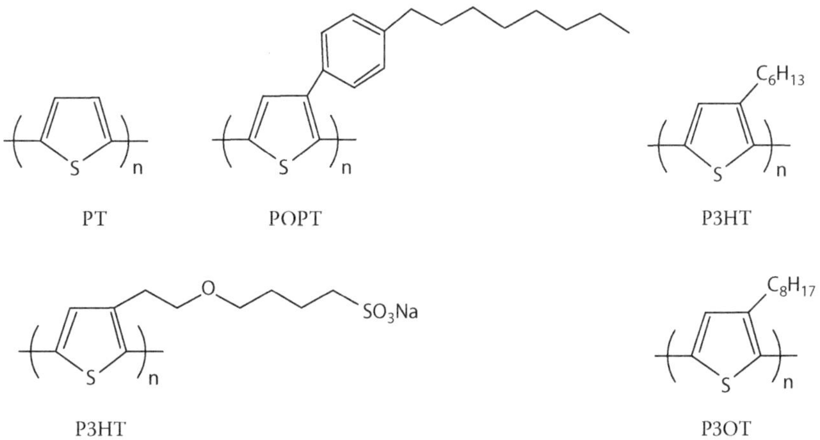

FIGURE 4.19
Chemical structure of polythiophene (PT) and its derivatives.

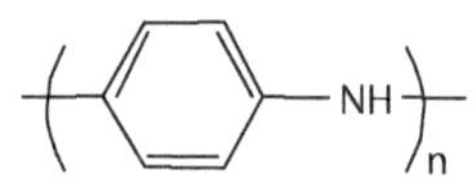

FIGURE 4.20
Chemical structure of the polyanilines (PAn).

Conjugated polymers are semiconducting due to their framework of alternating single and double carbon–carbon bonds. Single bonds are called σ-bonds, and double bonds include a σ-bond and a π-bond. The σ-bonds can be found in all conjugated polymers. However, the π-bonds are formed from the remaining out-of-plane p_z orbitals on the carbon atoms overlapping with neighboring p_z orbitals. The π-bonds are the source of the semiconducting properties of these polymers. First, the π-bonds are delocalized over the entire molecule; then, the overlap of p_z orbitals actually produces two orbitals, a bonding (π) orbital and an antibonding (π*) orbital. The lower-energy π-orbital serves as the highest occupied molecular orbitals (HOMOs), while the higher-energy π*-orbital forms the lowest unoccupied molecular orbitals (LUMOs). The difference in energy between the two levels produces the band gap that determines the optical properties of the material, such as photon absorption and emission.

Most conjugated polymers have a band gap between HOMO and LUMO in the range of 1.5 to 3 eV and a high absorption coefficient of ~10^5 cm^{-1}. Therefore, an incident visible-light photon has sufficient energy to excite an electron from the HOMO to LUMO of the conjugated polymers. This makes them well suited to absorb the visible light for photovoltaic devices. However, the optical absorption range is relatively narrow across the solar spectrum because most conjugated polymers only absorb light in the blue and green, and absorption in the red is difficult to accomplish. The incomplete light absorption in the solar spectrum limits the photocurrent generation.

4.6.2.2 Recently Developed Low-Band-Gap Polymers

As stated above, most conjugated polymers have a band gap larger than 1.9 eV, and accordingly only absorb light with wavelengths less than 650 nm. Recently, low-band-gap polymers (E_g < 1.8 eV) have been reported as an alternative for better light harvesting of the solar spectrum (Campos et al.

2005; Wang et al. 2004). Research has been done to synthesize low-band-gap polymers such as poly[5,7-bis-(3-octylthiophen-2-yl)thienopyrazine] (PB3OTP) (Sharma et al. 1995; Campos et al. 2005), poly-N-dodecyl-2,5,-bis(2′-thienyl)pyrrole,2,1,3-benzothiadiazole (PTPTB) (Brabec et al. 2002), side-chain-substituted poly(di-2-thienylthienopyrazine)s (PBEHTT and PTBEHT) (Wienk et al. 2006b), polymers based on alternating electron-donating 3,4,3′,4′-tetrakis[2-ethylhexyloxy]-2,2′-bithiophene and electron-deficient 2,1,3-benzothidiazole units along the chain (PBEHTB), and alternating polyfluorene copolymers with a green color (APFO-Green 1 or APFO Green 2) (Wang et al. 2004, 2005; Zhang et al. 2005). Some of these low-band-gap polymers have been reported to absorb from 300 to 850 nm (Wang et al. 2005).

4.6.2.3 HOMO/LUMO Levels and Band Gaps for Typical Polymers

The HOMO/LUMO levels and band gaps of various polymers, including PPV, PT, and recently developed low-band-gap polymers, are summarized in Table 4.3.

TABLE 4.3

HOMO/LUMO Levels and Band Gaps of Various Polymers

Conjugated Polymers	LUMO (eV)	HOMO (eV)	Band Gap (eV)	Reference
P3OT	−2.85	−5.25	2.4	Kymakis and Amaratunga (2002, 2003)
MEH-PPV	−3	−5.3	2.3	Breeze et al. (2001)
MDMO-PPV	−2.8	−5.0	2.2	Brabec et al. (2003)
PTEBS	−2.8	−5	2.2	Qiao et al. (2005a,b,c, 2006b)
P3HT	−3.2	−5.2	2.0	Kwong et al. (2004a)
PPE-PPV(DE21)	−3.6	−5.6	2.0	Sun and Sariciftci (2005)
APFO Green 2	−3.6	−5.6	2.0	Zhang et al. (2005)
PPE-PPV(DE69)	−3.56	−5.46	1.9	Sun and Sariciftci (2005)
PTBTB	−3.73	−5.5	1.77	Brabec et al. (2002), Dhanabalan et al. (2001b)
P3DDT	−3.55	−5.29	1.74	Sun and Sariciftci (2005)
PBEHTB	−3.6	−5.3	1.7	Wienk et al. (2006a)
PB3OTP	−2.75	−4.2	1.45	Campos et al. (2005)
PBEHTT	−3.6	−5.0	1.4	Wienk et al. (2006b)
APFO Green 1	−3.9	−5.14	1.24	Wang et al. (2004, 2005)
PTBEHT	−4.0	−5.2	1.2	Wienk et al. (2006b)

4.6.2.4 Properties of Polymers for Use in Photovoltaics

A singlet exciton is a bound electron-hole pair generated by photoexcitation. When illuminated under the light, electrons are pumped to the LUMO level by the absorbed photons, leaving the corresponding holes in the HOMO. However, electrostatic attraction keeps them together. The exciton binding energy of the conjugated polymers has been the subject of debate in the literature over the past decades. Reporters have proposed the values between a few k_bT (k_bT in the order of 10^{-5} eV) and 1 eV for the binding energy (Chandross et al. 1994; Lee et al. 1994; Marks et al. 1994). The common accepted value for the exciton binding energy of most conjugated polymers is about 0.3 to 0.4 eV (Barth and Bassler 1997; Marks et al. 1994). This means strong driving forces such as an electric field are needed to dissociate the photogenerated excitons into separate electrons and holes. These excitons need to be dissociated before the carriers can be transported and then collected at the electrodes. The diffusion range of singlet excitons of conjugated polymers is approximately 5 ~ 15 nm, and their radiative or nonradiative decays take place in the time of 100 ~ 1,000 ps (Greenham et al. 1997b).

Compared to those in the inorganic semiconductors, the charge mobilities are low (0.01 to 0.001 cm^2-V^{-1}-s^{-1}) (Zen et al. 2005) because charge transport takes place through hopping between the localized states in the polymer. In addition, the photocurrent is susceptible to temperature variation through hopping transport. This limits the useful thickness of the devices. Most conjugated polymers are vulnerable to degradation in the air with the presence of oxygen or moisture. For this reason, glove boxes are generally needed to make polymer solar cells (Dennler et al. 2006a).

The conjugated polymers described here are mainly p-type semiconductors in the sense that they can be partially oxidized to become p-doped (Wallace et al. 2000). A corresponding reduction or n-doping will destabilize the polymers (Brabec et al. 2003). Figure 4.21 shows an example for the oxidation

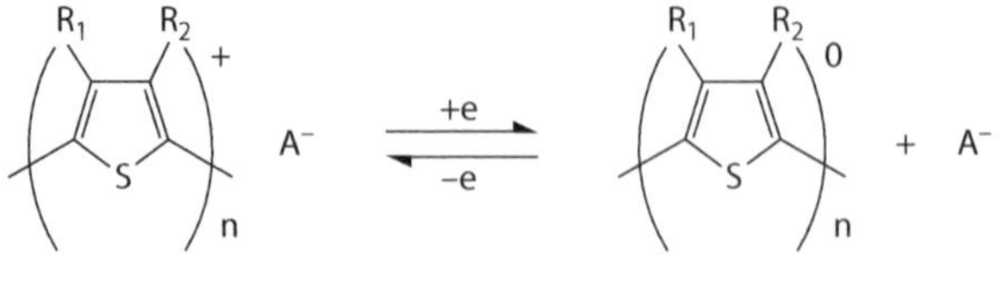

FIGURE 4.21
Oxidation and reduction of a polythiophene. P-type semiconductors can be partially oxidized to become p-doped. (With permission from Too, C. O., Wallace, G. G., Burrell, A. K., et al., "Photovoltaic Devices Based on Polythiophenes and Substituted Polythiophenes," *Synthetic Metals* 123 [2001]: 53–60.)

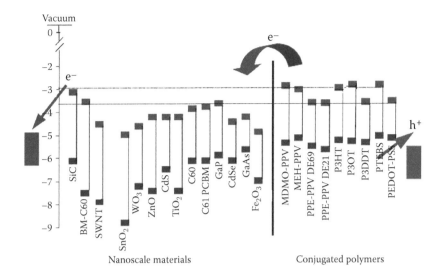

FIGURE 4.22

Matching of nanomaterials and polymer for use in solar cells based on energy band. The top squares represent the LUMO and conductance bands, and the bottom squares represent the HOMO and valence bands. The typical electrodes shown on the right are ITO and gold; those shown on the left are FTO and aluminum. (This figure is based on a concept presented in Gratzel, M., "Photoelectrochemical Cells," *Nature* 414 [2001]: 338–44.)

and reduction of a polythiophene (Too et al. 2001). N-type polymer semiconductors are also available (Huang et al. 1999; Janietz et al. 2001), however, less effort has been gone into synthesizing and characterizing them.

4.6.2.5 Energy Matching for Polymers and Nanomaterials

In heterojunction solar cells, the polymers work as p-type semiconductors and nanomaterials serve as n-type semiconductors. However, their energy structures must be energetically favorably matched to form an effective heterojunction. Figure 4.22 shows how a polymer and a nanomaterial match based on energy band structure, in which most of the common polymers and nanomaterials have been included.

4.7 NANOSCALE STRUCTURE

As discussed above, in order to achieve exciton separation, the exciton must be formed within one diffusion length of the donor-acceptor interface. Any farther and the exciton will recombine before reaching the interface. In an effort to overcome this challenge, other investigators have built devices with covalently attached supramolecules where the donor and acceptor molecules are in direct contact. Somewhat surprisingly, these dyads (Dhanabalan et al. 2001a; Eckert et al. 2000; Loi et al. 2003; Possamai et al. 2003) have yielded low efficiencies, implying that domain size can actually be too small and can result in enhanced recombination and poor-charge carrier transport. This indicates that an optimal domain size is needed in organic photovoltaics to balance exciton diffusion and charge transport.

The domain size depends on a number of factors. Most obvious is the relative concentrations of the donor and acceptor materials. For polymer/fullerene devices, the most efficient devices have high fullerene loading with weight ratios of 4:1 (fullerene:polymer) (Aernouts et al. 2002; Kroon et al. 2002; Munters et al. 2002; Shaheen et al. 2001). This seems to be necessary for electron transport.

Domain size also depends on the choice of solvent and solvent evaporation time (heating). Hoppe et al. (2004) conducted a comparison of toluene and chlorobenzene in MDMO-PPV:PCBM devices (MDMO-PPV is (poly[2-methoxy-53,7-dimethyloctyloxy)]-1,4-phenylenevinylene). The toluene resulted in grain sizes of 200 to 500 nm, while the chlorobenzene grain size was closer to 50 nm (Figure 4.23). The grain size when using toluene is too large to ensure exciton separation, and therefore the

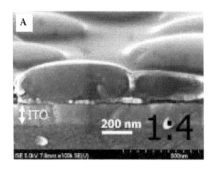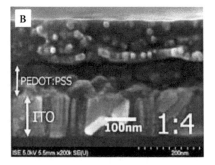

FIGURE 4.23
Effect of solvent on domain size. Domain size in MDMO-PPV:PCBM devices for (A) toluene and (B) chlorobenzene solvents. Toluene results in grain sizes of 200 to 500 nm, while the chlorobenzene grain size is closer to 50 nm (Figure 4.23). The grain size when using toluene is too large to ensure exciton separation, and therefore the devices made using chlorobenzene showed a higher efficiency. (With permission from Hoppe, H., Niggemann, M., Winder, C., et al., "Nanoscale Morphology of Conjugated Polymer/Fullerene-Based Bulk-Heterojunction Solar Cells," *Adv. Funct. Mater.* 14 [2004]: 1005–11.)

devices made using chlorobenzene showed a higher efficiency. Other studies (Kwong et al. 2004a) indicate that solvents with lower vapor pressure result in better mixing of the polymer and the electron acceptor material, and thus improved efficiency. Correspondingly, solvent evaporation times can impact the morphology and efficiency of the devices (Halls et al. 2000; Strawhecker et al. 2001).

In Section 5.2.2, the use of annealing to form C_{60} nanowhiskers (Reyes-Reyes et al. 2005b) was discussed. Other studies (Camaioni et al. 2002; Dittmer et al. 2000; Hoppe and Sariftci 2004; Padinger et al. 2003; Schmidt-Mende et al. 2001) have shown that annealing can change the nanostructure of the device and can either enhance or degrade the efficiency of the device, depending on how long the annealing takes place and how much change there is in the size of the domains.

4.8 SUMMARY

The materials and devices of organic polymer solar cells and dye-sensitized solar cells have been discussed. The nanoscale donor and acceptor materials are very important in achieving highly efficient organic polymer solar cells and dye-sensitized solar cells. In the former device, conjugated polymers with a high absorption coefficient and high electron affinity are typically used as donors, including PPV, PT, and PAn and their derivatives. In the latter, Ru-based complexes and other all-organic dyes, including polypyridines, indolines, squaraines, perylenes, xanthenes, flavonoid anthocyanins, polyenes, coumarins, and polypyridyl derivatives, are currently used. Acceptor materials, including metal oxide (TiO_2, ZnO, etc., in the form of particle, rod, or wire), fullerene, and carbon nanotubes, are being studied. The electrical and optical properties of these materials affect directly the photon absorption, exciton generation, exciton dissociation, charge transfer, and charge transport.

4.9 FUTURE PERSPECTIVE

Since visible light only accounts for 50% of the solar energy, near-infrared light needs to be harvested for significant improvement of device efficiency. Low-band-gap polymers and dyes and near-infrared quantum dots are strongly needed in the future. Also, efficient multiexciton generation using quantum dots, quantum rods, and quantum wires is currently gaining extensive attention and is expected to play a role in increasing the PV device efficiency. Another approach to increasing device efficiency is to incorporate films with a meso-macroporous structure (Ito et al. 2000). In addition, in order to increase use of excitons generated in the donor materials, new materials with longer exciton diffusion lengths are also worthy for future research. Most of these future approaches are discussed in other chapters of this book (e.g., nanowires, etc.).

REFERENCES

Aernouts, T., Geens, W., Poortmans, J., et al. 2002. Extraction of bulk and contact components of the series resistance in organic bulk donor-acceptor-heterojunctions. *Thin Solid Films* 403–4:297–301.

Arango, A. C., Carter, S. A., and Brock, P. J. 1999. Charge transfer in photovoltaics consisting of interpenetrating networks of conjugated polymer and TiO2 nanoparticles. *Appl. Phys. Lett.* 74:1698–700.

Arango, A. C., Johnson, L. R., Bliznyuk, V. N., et al. 2000. Efficient titanium oxide/conjugated polymer photovoltaics for solar energy conversion. *Adv. Mater.* 12:1689.

Bach, U., Lupo, D., Comte, P., et al. 1998. Solid-state dye-sensitized mesoporous TiO2 solar cells with high photon-to-electron conversion efficiencies. *Nature* 395:583–85.

Bai, Y., Cao, Y., Zhang, J., et al. 2008. High-performance dye-sensitized solar cells based on solvent-free electrolytes produced from eutectic melts. *Nature Mater.*, Vol. 7, n. 8, pp. 626–630.

Barth, S., and Bassler, H. 1997. Intrinsic photoconduction in PPV-type conjugated polymers. *Phys. Rev. Lett.* 79:4445–48.

Becker, H., Lux, A., Holmes, A. B., et al. 1997. PL and EL quenching due to thin metal films in conjugated polymers and polymer LEDs. *Synthetic Metals* 85:1289–90.

Beek, W. J. E., Wienk, M. M., and Janssen, R. A. J. 2004. Efficient hybrid solar cells from zinc oxide nanoparticles and a conjugated polymer. *Adv. Mater.* 16:1009–13.

Beek, W. J. E., Slooff, L. H., Wienk, M. M., et al. 2005a. Hybrid ZnO:polymer bulk heterojunction solar cells from a ZnO precursor. In *Organic Photovoltaics VI. Proceedings of SPIE—The International Society for Optical Engineering*, San Diego, CA, Vol. 5938, 1–11.

Beek, W. J. E., Wienk, M. M., and Janssen, R. A. J. 2005b. Hybrid bulk heterojunction solar cells: Blends of ZnO semiconducting nanoparticles and conjugated polymers. In *Organic Photovoltaics VI. Proceedings of SPIE—The International Society for Optical Engineering*, San Diego, CA, Vol. 5938, 1–9.

Bettignies, R. d., Leroy, J., Firon, M., et al. 2005. Study of P3HT:PCBM bulk heterojunction solar cells: Influence of components ratio and of the nature of electrodes on performances and lifetime. In *Organic Photovoltaics VI. Proceedings of SPIE—The International Society for Optical Engineering*, San Diego, CA, Vol. 5938, 59380C1–14.

Brabec, C. J., Dyakonov, V., Parisi, J., et al. 2003. *Organic photovoltaics: Concepts and realization.* Ed. R. Hull, R. M. J. Osgood, and J. Parisi. Springer Series in Materials Science, Vol. 60. New York: Springer-Verlag.

Brabec, C. J., Winder, C., Sariciftci, N. S., et al. 2002. A low-bandgap semiconducting polymer for photovoltaic devices and infrared emitting diodes. *Adv. Funct. Mater.* 12:709–12.

Breeze, A. J., Schlesinger, Z., and Carter, S. A. 2001. Charge transport in TiO2/MEH-PPV polymer photovoltaics. *Phys. Rev. B* 64:1–9.

Burfeindt, B., Hannappel, T., Storck, W., et al. 1996. Measurement of temperature-independent femtosecond interfacial electron transfer from an anchored molecular electron donor to a semiconductor as acceptor. *J. Phys. Chem.* 100:16463–65.

Burke, A., Schmidt-Mende, L., Ito, S., et al. 2007. A novel blue dye for near-IR dye-sensitised solar cell applications. *Chem. Commun.* 2007:234–36.

Burrell, A. K., Officer, D. L., Plieger, P. G., et al. 2001. Synthetic routes to multiporphyrin arrays. *Chem. Rev.* 101:2751–96.

Camaioni, N., Ridolfi, G., Casalbore-Miceli, G., et al. 2002. The effect of a mild thermal treatment on the performance of poly(3-allkylthiophene)/fullerene solar cells. *Adv. Mater.* 14:1735.

Campbell, W. M., Burrell, A. K., Officer, D. L., et al. 2004. Porphyrins as light harvesters in the dye-sensitised TiO2 solar cell. *Coordination Chem. Rev.* 248:1363–79.

Campos, L. M., Tontcheva, A., Gunes, S., et al. 2005. Extended photocurrent spectrum of a low band gap polymer in a bulk heterojunction solar cell. *Chem. Mater.* 17:4031–33.

Chamberlain, G. A. 1983. Organic solar cells: A review. *Solar Cells* 8:47–83.

Chandross, M., Mazumdar, S., Jeglinski, S., et al. 1994. Excitons in poly(para-phenylenevinylene). *Phys. Rev. B* 50:14702–5.

Chen, C.-W., Lu, Y.-J., Wu, C.-C., et al. 2005. Effective connecting architecture for tandem organic light-emitting devices. *Appl. Phys. Lett.* 87:241121.

Cherepy, N. J., Smestad, G. P., Gratzel, M., et al. 1997. Ultrafast electron injection: Implications for a photoelectrochemical cell utilizing an anthocyanin dye-sensitized TiO2 nanocrystalline electrode. *J. Phys. Chem. B* 101:9342–51.

Chiba, Y., Islam, A., Watanabe, Y., et al. 2006. Dye-sensitized solar cells with conversion efficiency of 11.1%. *Jpn. J. Appl. Phys. Part 2 Lett.* 45:638–40.

Choulis, S. A., Nelson, J., Kim, Y., et al., 2003. Investigation of transport properties in polymer/fullerene blends using time-of-flight photocurrent measurements. *Appl. Phys. Lett.* 83:3812–14.

Coakley, K. M., Srinivasan, B. S., Ziebarth, J. M., et al. 2005. Enhanced hole mobility in regioregular polythiophene infiltrated in straight nanopores. *Adv. Funct. Mater.* 15:1927–32.

Dennler, G., Lungenschmied, C., Neugebauer, H., et al. 2006a. A new encapsulation solution for flexible organic solar cells. *Thin Solid Films* 511:349–53.

Dennler, G., Prall, H.-J., Koeppe, R., et al. 2006b. Enhanced spectral coverage in tandem organic solar cells. *Appl. Phys. Lett.* 89:073502.

Dhanabalan, A., Knol, J., Hummelen, J. C., et al. 2001a. Design and synthesis of new processible donor-acceptor dyad and triads. *Synthetic Metals* 119:519–22.

Dhanabalan, A., van Duren, J. K. J., van Hal, P. A., et al. 2001b. Synthesis and characterization of a low bandgap conjugated polymer for bulk heterojunction photovoltaic cells. *Adv. Funct. Mater.* 11:255–62.

Dittmer, J. J., Lazzaroni, R., Leclere, P., et al. 2000. Crystal network formation in organic solar cells. *Solar Energy Mater. Solar Cells* 61:53–61.

Drees, M., Davis, R. M., and Heflin, J. R. 2004. Thickness dependence, in situ measurements, and morphology of thermally controlled interdiffusion in polymer-C_{60} photovoltaic devices. *Phys. Rev. B* 69:165320.

Drees, M., Premaratne, K., Graupner, W., et al. 2002. Creation of a gradient polymer-fullerene interface in photovoltaic devices by thermally controlled interdiffusion. *Appl. Phys. Lett.* 81:4607–9.

DyeSol. 2008. www.dyesol.com.

Eckert, J.-F., Nicoud, J.-F., Nierengarten, J.-F., et al. 2000. Fullerene-oligophenylenevinylene hybrids: Synthesis, electronic properties, and incorporation in photovoltaic devices. *J. Am. Chem. Soc.* 122:7467–79.

Ehret, A., Stuhl, L., and Spitler, M. T. 2001. Spectral sensitization of TiO2 nanocrystalline electrodes with aggregated cyanine dyes. *J. Phys. Chem.* 105:9960–65.

Fan, Q., McQuillin, B., Bradley, D. D. C., et al. 2001. A solid state solar cell using sol-gel processed material and a polymer. *Chem. Phys. Lett.* 347:325–30.

Fang, P. H., Hirata, M., and Hirata, M. 1970. *Investigation of organic semiconductor for photovoltaic application,* NTIS, Springfield, VA.

Ferrere, S., Zaban, A., and Gregg, B. A. 1997. Dye sensitization of nanocrystalline tin oxide by perylene derivatives. *J. Phys. Chem. B* 101:4490–93.

G24i. 2008. www.g24i.com.

Gebeyehu, D., Brabec, C. J., Sariciftci, N. S., et al. 2001. Hybrid solar cells based on dye-sensitized nanoporous TiO2 electrodes and conjugated polymers as hole transport materials. *Synthetic Metals* 125:279–87.

Ghosh, A. K., and Feng, T. 1973. Rectification, space-charge-limited current, photovoltaic and photoconductive properties of Al/tetracene/Au sandwich cell. *J. Appl. Phys.* 44:2781–88.

Gilot, J., Wienk, M. M., and Janssen, R. A. J. 2007. Double and triple junction polymer solar cells processed from solution. *Appl. Phys. Lett.* 90:143512.

Grant, C. D., Schwartzberg, A. M., Smestad, G. P., et al. 2002. Characterization of nanocrystalline and thin film TiO2 solar cells with poly(3-undecyl-2,2′-bithiophene) as a sensitizer and hold conductor. *J. Electroanal. Chem.* 522:40–48.

Gratzel, M. 2001. Photoelectrochemical cells. *Nature* 414:338–44.

Gratzel, M. 2003. Dye-sensitized solar cells. *J. Photochem. Photobiol. C* 4:145–53.

Greenham, N. C., Peng, X., and Alivisatos, A. P. 1996. Charge separation and transport in conjugated-polymer/semiconductor-nanocrystal composites studied by photoluminescence quenching and photoconductivity. *Phys. Rev. B* 54:17628–37.

Greenham, N. C., Peng, X., and Alivisatos, A. P. 1997a. A CdSe nanocrystal/MEH-PPV polymer composite photovoltaic. In *AIP Conference Proceedings* N. 404, 295–302. Future Generation Photovoltaic Technologies. First NREL Conference, March 24–26. Denver, CO.

Greenham, N. C., Peng, X., and Alivisatos, A. P. 1997b. Charge separation and transport in conjugated polymer/cadmium selenide nanocrystal composites studied by photoluminescence quenching and photoconductivity. *Synthetic Metals* 84:545–46.

Gregg, B. A. 2005. The photoconversion mechanism of excitonic solar cells. *MRS Bull.* 30:20–22.

Gregg, B. A., and Hanna, M. C. 2003. Comparing organic to inorganic photovoltaic cells: Theory, experiment, and simulation. *J. Appl. Phys.* 93:3605–14.

Gupta, T. K., Cirignano, L. J., Shah, K. S., et al. 1999. Characterization of screen-printed dye-sensitized nanocrystalline TiO2 solar cells. In *Solar Optical Materials XVI. Proceedings of the SPIE—The International Society for Optical Engineering,* Solar Optical Materials XVI, July 22. Denver, CO. Vol. 3789, 149–57.

Gur, I., Fromer, N. A., Geier, M. L., et al., 2005. Air-stable all-inorganic nanocrystal solar cells processed from solution. *Science* 310:462–65.

Hadipour, A., Boer, B. d., and Blom, P. W. M. 2007. Solution-processed organic tandem solar cells with embedded optical spacers. *J. Appl. Phys.* 102:074506.

Hagfeldt, A., and Gratzel, M. 2000. Molecular photovoltaics. *Acc. Chem. Res.* 33:269–77.

Halls, J. J. M., Arias, A. C., MacKenzie, J. D., et al. 2000. Photodiodes based on polyfluorene composites: Influence of morphology. *Adv. Mater.* 12:498–502.

Halls, J. J. M., Walsh, C. A., Greenham, N. C., et al. 1995. Efficient photodiodes from interpenetrating polymer networks. *Nature* 376:498–500.

Harima, Y., Yamashita, K., and Suzuki, H. 1984. Spectral sensitization in an organic p-n junction photovoltaic cell. *Appl. Phys. Lett.* 45:1144–45.

Hendry, E., Wang, F., Shan, J., et al. 2004. Electron transport in TiO2 probed by THz time-domain spectroscopy. *Phys. Rev. B* 69:81101.

Hoppe, H., Niggemann, M., Winder, C., et al. 2004a. Nanoscale morphology of conjugated polymer/fullerene-based bulk-heterojunction solar cells. *Adv. Funct. Mater.* 14:1005–11.

Hoppe, H., and Sariciftci, N. S. 2004b. Organic solar cells: An overview. *J. Mater. Res.* 19:1924–45.

Horiuchi, T., Miura, H., Sumioka, K., et al. 2004. High efficiency of dye-sensitized solar cells based on metal-free indoline dyes. *J. Am. Chem. Soc.* 126:12218–19.

Howie, W. H., Claeyssens, F., Miura, H., et al. 2008. Characterization of solid-state dye-sensitized solar cells utilizing high absorption coefficient metal-free organic dyes. *J. Am. Chem. Soc.* 130:1367–75.

Huang, W., Meng, H., Yu, W. L., et al. 1999. A novel series of p-n diblock light-emitting copolymers based on oligothiophenes and 1,4-bis(oxadiazolyl)-2,5-dialkyloxybenzene. *Macromolecules* 32:118–26.

Huynh, W. U., Dittmer, J. J., and Alivisatos, A. P. 2002. Hybrid nanorod-polymer solar cells. *Science* 295:2425–27.

Huynh, W. U., Peng, X., and Alivisatos, A. P. 1999. CdSe nanocrystal rods/poly(3-hexylthiophene) composite photovoltaic devices. *Adv. Mater.* 11:923–27.

Ito, S., Ishikawa, K., Wen, C. J., et al. 2000. Dye-sensitized photocells with meso-macroporous TiO2 film electrodes. *Bull. Chem. Soc. Jpn.* 73:2609–14.

Janietz, S., Anlauf, S., and Wedel, A. 2001. New n-type rigid rod full aromatic poly(l,3,4-oxadiazole)s and their application in organic devices. *Synthetic Metals* 122:11–14.

Janssen, A. G. F., Riedl, T., Hamwi, S., et al. 2007. Highly efficient organic tandem solar cells using an improved connecting architecture. *Appl. Phys. Lett.* 91:073519.

Johansson, N., dos Santos, D. A., Guo, S., et al. 1997. Electronic structure and optical properties of electroluminescent spiro-type molecules. *J. Chem. Phys.* 107:2542–49.

Kakiuchi, K., Hosono, E., and Fujihara, S. 2006. Enhanced photoelectrochemical performance of ZnO electrodes sensitized with N-719. *J. Photochem. Photobiol. A* 179:81–86.

Kawano, K., Ito, N., Nishimori, T., et al. 2006. Open circuit voltage of stacked bulk heterojunction organic solar cells. *Appl. Phys. Lett.* 88:073514.

Kay, A., and Graetzel, M. 1993. Artificial photosynthesis. 1. Photosensitization of titania solar cells with chlorophyll derivatives and related natural porphyrins. *J. Phys. Chem.* 97:6272–77.

Kay, A., Humphry-Baker, R., and Graetzel, M. 1994. Artificial photosynthesis. 2. Investigations on the mechanism of photosensitization of nanocrystalline TiO2 solar cells by chlorophyll derivatives. *J. Phys. Chem.* 98:952–59.

Keis, K., Lindgren, J., Lindquist, S.-E., et al. 2000. Studies of the adsorption process of Ru complexes in nanoporous ZnO electrodes. *Langmuir* 16:4688–94.

Kim, J. Y., Lee, K., Coates, N. E., et al. 2007a. Efficient tandem polymer solar cells fabricated by all-solution processing. *Science* 317:222–25.

Kim, K., Liu, J., Namboothiry, M. A. G., et al. 2007b. Roles of donor and acceptor nanodomains in 6% efficient thermally annealed polymer photovoltaics. *Appl. Phys. Lett.* 90:163511.

Kim, Y., Choulis, S. A., Nelson, J., et al. 2005a. Composition and annealing effects in polythiophene/fullerene solar cells. *J. Mater. Sci.* 40:1371–76.

Kim, Y., Cook, S., Choulis, S. A., et al. 2005b. Effect of electron-transport polymer addition to polymer/fullerene blend solar cells. *Synthetic Metals* 152:105–8.

Klimov, V. 2003. Discovery of pheophytin function in the photosynthetic energy conversion as the primary electron acceptor of photosystem II. *Photosynthesis Res.* 76:247–53.

Konarka. 2008. www.konarka.com.

Konenkamp, R. 2000. Carrier transport in nanoporous TiO2 films. *Phys. Rev. B* 61:11057–64.

Konkin, A. L., Sensfuss, S., Roth, H.-K., et al. 2009. LESR study on PPV-PPE/PCBM composites for organic photovoltaics. *Synthetic Metals*, Vol. 148, 199–204.

Krebs, F. C., Carle, J. E., Cruys-Bagger, N., et al. 2009. Lifetimes of organic photovoltaics: Photochemistry, atmosphere effects and barrier layers in ITO-MEHPPV:PCBM-aluminum devices. *Solar Energy Mater. Solar Cells*, Vol. 86, N. 4, 499–516.

Kroon, J. M., Wienk, M. M., Verhees, W. J. H., et al. 2002. Accurate efficiency determination and stability studies of conjugated polymer/fullerene solar cells. *Thin Solid Films* 403–4:223–28.

Kwong, C. Y., Choy, W. C. H., Djurisic, A. B., et al. 2004a. Poly(3-hexylthiophene): TiO2 nanocomposites for solar cell applications. *Nanotechnology* 15:1156–61.

Kwong, C. Y., Djurisic, A. B., Chui, P. C., et al. 2004b. Influence of solvent on film morphology and device performance of poly(3-hexylthiophene):TiO2 nanocomposite solar cells. *Chem. Phys. Lett.* 384:372–75.

Kymakis, E., Alexandrou, I., and Amaratunga, G. A. J. 2003. High open-circuit voltage photovoltaic devices from carbon-nanotube-polymer composites. *J. Appl. Phys.* 93:1764–68.

Kymakis, E., and Amaratunga, G. A. J. 2002. Single-wall carbon nanotube/conjugated polymer photovoltaic devices. *Appl. Phys. Lett.* 80:112–14.

Kymakis, E., and Amaratunga, G. A. J. 2003. Photovoltaic cells based on dye-sensitisation of single-wall carbon nanotubes in a polymer matrix. *Solar Energy Mater. Solar Cells* 80:465–72.

Landi, B. J., Raffaelle, R. P., Castro, S. L., et al. 2005. Single-wall carbon nanotube-polymer solar cells. *Progress Photovoltaics Res. Appl.* 13:165–72.

Lee, C. H., Yu, G., Moses, D., et al. 1994. Picosecond transient photoconductivity in poly(P-phenylenevinylene). *Phys. Rev. B* 49:2396–407.

Lee, S. B., Katayama, T., Kajii, H., et al. 2001. Electrical and optical properties of conducting polymer-C60-carbon nanotube system. *Synthetic Metals* 121:1591–92.

Li, B., Wang, L., Kang, B., et al. 2006. Review of recent progress in solid-state dye-sensitized solar cells. *Solar Energy Mater. Solar Cells* 90:549–73.

Loi, M. A., Denk, P., Hoppe, H., et al. 2003. Long-lived photoinduced charge separation for solar cell applications in phthalocyanine-fulleropyrrolidine dyad thin films. *J. Mater. Chem.* 13:700–4.

Lyons, L. E., and Newman, O. M. G. 1971. Photovoltages in tetracene films. *Austr. J. Chem.* 24:13–23.

Marks, R. N., Halls, J. J. M., Bradley, D. D. C., et al. 1994. The photovoltaic response in poly(p-phenylene vinylene) thin-film devices. *J. Phys. Condensed Matter* 6:1379–94.

Martens, T., D'Haen, J., Munters, T., et al. 2003. Disclosure of the nanostructure of MDMO-PPV:PCBM bulk hetero-junction organic solar cells by a combination of SPM and TEM. *Synthetic Metals* 138:243–47.

Mattoussi, H., Radzilowski, L. H., Dabbousi, B. O., et al. 1998. Electroluminescence from heterostructures of poly(phenylene vinylene) and inorganic CdSe nanocrystals. *J. Appl. Phys.* 83:7965–74.

Melzer, C., Koop, E. J., Mihailetchi, V. D., et al. 2004. Hole transport in poly(phenylene vinylene)/methanofullerene bulk-heterojunction solar cells. *Adv. Funct. Mater.* 14:865–70.

Meng, Q.-B., Takahashi, K., Zhang, X.-T., et al. 2003. Fabrication of an efficient solid-state dye-sensitized solar cell. *Langmuir* 19:3572–74.

Mihailetchi, V. D., van Duren, J. K. J., Blom, P. W. M., et al. 2003. Electron transport in a methanofullerene. *Adv. Funct. Mater.* 13:43–46.

Mozer, A. J., and Sariciftci, N. S. 2004. Negative electric field dependence of charge carrier drift mobility in conjugated, semiconducting polymers. *Chem. Phys. Lett.* 389:438–42.

Mozer, A. J., Sariciftci, N. S., Pivrikas, A., et al. 2005. Charge carrier mobility in regioregular poly(3-hexylthiophene) probed by transient conductivity techniques: A comparative study. *Phys. Rev. B* 71:35214.

Mozer, A. J., Wada, Y., Jiang, K. J., et al. 2006. Efficient dye-sensitized solar cells based on a 2-thiophen-2-yl-vinyl-conjugated ruthenium photosensitizer and a conjugated polymer hole conductor. *Appl. Phys. Lett.* 89:043509.

Mukherjee, T. K. 1970. Photoconductive and photovoltaic effects in dibenzothiophene and its molecular complexes. *J. Phys. Chem.* 74:3006–14.

Munters, T., Martens, T., Goris, L., et al. 2002. A comparison between state-of-the-art 'gilch' and 'sulphinyl' synthesised MDMO-PPV/PCBM bulk hetero-junction solar cells. *Thin Solid Films* 403–4:247–51.

Nazeeruddin, M. K., Kay, A., Rodicio, I., et al. 1993. Conversion of light to electricity by cis-X2bis(2,2′-bipyridyl-4,4′-dicarboxylate)ruthenium(II) charge-transfer sensitizers (X = Cl-, Br-, I-, CN-, and SCN-) on nanocrystalline titanium dioxide electrodes. *J. Am. Chem. Soc.* 115:6382–90.

Nazeeruddin, M. K., Pechy, P., Renouard, T., et al. 2001. Engineering of efficient panchromatic sensitizers for nanocrystalline TiO_2-based solar cells. *J. Am. Chem. Soc.* 123:1613–24.

Nelson, J., Choulis, S. A., and Durrant, J. R., 2004. Charge recombination in polymer/fullerene photovoltaic devices. *Thin Solid Films* 451–52:508–14.

Ohsawa, Y., and Saji, T. 1992. Electrochemical detection of C60(6-) at low-temperature. *J. Chem. Soc. Chem. Commun.* Issue 10, 781–82.

O'Regan, B., and Gratzel, M. 1991. A low-cost, high-efficiency solar-cell based on dye-sensitized colloidal TiO2 films. *Nature* 353:737–40.

Padinger, F., Rittberger, R. S., and Sariciftci, N. S. 2003. Effects of postproduction treatment on plastic solar cells. *Adv. Funct. Mater.* 13:85–88.

Peiro, A. M., Ravirajan, P., Govender, K., et al. 2006. Hybrid polymer/metal oxide solar cells based on ZnO columnar structures. *J. Mater. Chem.* 16:2088–96.

Peumans, P., Bulovic, V., and Forrest, S. R. 2000. Efficient photon harvesting at high optical intensities in ultrathin organic double-heterostructure photovoltaic diodes. *Appl. Phys. Lett.* 76:2650–52.

Peumans, P., and Forrest, S. R. 2001. Very-high-efficiency double-heterostructure copper phthalocyanine/C[sub 60] photovoltaic cells. *Appl. Phys. Lett.* 79:126–28.

Peumans, P., Uchida, S., and Forrest, S. R. 2003. Efficient bulk heterojunction photovoltaic cells using small-molecular-weight organic thin films. *Nature* 425:158–62.

Piok, T., Schroeder, R., Brands, C., et al. 2001. Photocarrier generation quantum yield for ionically self-assembled monolayers. *Synthetic Metals* 121:1589–90.

Pope, M., and Swenberg, C. E. 1999. *Electronic processes in organic crystals and polymers*. New York: Oxford University Press.

Possamai, A., Camaioni, N., Ridolfi, G., et al. 2003. A fullerene-azothiophene dyad for photovoltaics. *Synthetic Metals* 139:585–88.

Qiao, Q., Beck, J., and McLeskey, J. T. 2005a. Photovoltaic devices from self-doped polymers. In *Organic Photovoltaics VI*, Vol. 5938, 59380E-1. San Diego, CA.

Qiao, Q., and McLeskey, J. T. 2005b. Water-soluble polythiophene/nanocrystalline TiO_2 solar cells. *Appl. Phys. Lett.* 86:153501.

Qiao, Q., Su, L. Y., Beck, J., et al. 2005c. Characteristics of water-soluble polythiophene: TiO2 composite and its application in photovoltaics. *J. Appl. Phys.* 98:094906.

Qiao, Q., Beck, J., Lumpkin, R., et al. 2006a. A comparison of fluorine tin oxide and indium tin oxide as the transparent electrode for P3OT/TiO2 solar cells. *Solar Energy Mater. Solar Cells* 90:1034–40.

Qiao, Q., Kerr, W. C., Beck, J., et al., 2006b. Optimization of photovoltaic devices from layered PTEBS and nanocrystalline TiO_2. *ECS Trans.* 1:1–6.

Qiao, Q., Xie, Y., and McLeskey, J. J. T. 2008. Organic/inorganic polymer solar cells using a buffer layer from all-water-solution processing. *J. Phys. Chem. C* 112:9912–16.

Quintana, M., Edvinsson, T., Hagfeldt, A., et al. 2007. Comparison of dye-sensitized ZnO and TiO2 solar cells: Studies of charge transport and carrier lifetime. *J. Phys. Chem. C* 111:1035–41.

Ravirajan, P., Haque, S. A., Durrant, J. R., et al. 2005. The effect of polymer optoelectronic properties on the performance of multilayer hybrid polymer/TiO2 solar cells. *Adv. Funct. Mater.* 15:609–18.

Redmond, G., Fitzmaurice, D., and Graetzel, M. 1994. Visible light sensitization by cis-bis(thiocyanato)bis(2,2′-bipyridyl-4,4′-dicarboxylato)ruthenium(II) of a transparent nanocrystalline ZnO film prepared by sol-gel techniques. *Chem. Mater.* 6:686–91.

Rensmo, H., Keis, K., Lindstrom, H., et al. 1997. High light-to-energy conversion efficiencies for solar cells based on nanostructured ZnO electrodes. *J. Phys. Chem. B* 101:2598–601.

Reucroft, P. J., Kronick, P. L., and Hillman, E. E. 1968. Research directed toward the study of materials for organic photovoltaic cells. In *Research directed toward the study of materials for organic photovoltaic cells*, NTIS, Springfield, MA. 69.

Reucroft, P. J., Kronick, P. L., and Hillman, E. E. 1969. Photovoltaic effects tetracene crystals. *Mol. Crystals Liquid Crystals* 6:247–54.

Reyes-Reyes, M., Kim, K., and Carroll, D. L. 2005a. High-efficiency photovoltaic devices based on annealed poly(3-hexylthiophene) and 1-(3-methoxycarbonyl)-propyl-1-phenyl-(6,6)C-61 blends. *Appl. Phys. Lett.* 87, 83506-1.

Reyes-Reyes, M., Kyungkon, K., Dewald, J., et al. 2005b. Meso-structure formation for enhanced organic photovoltaic cells. *Organic Lett.* 7:5749–5752.

Riedel, I., Parisi, J., Dyakonov, V., et al. 2004. Effect of temperature and illumination on the electrical characteristics of polymer-fullerene bulk-heterojunction solar cells. *Adv. Funct. Mater.* 14:38–44.

Rispens, M. T., Meetsma, A., Rittberger, R., et al. 2003. Influence of the solvent on the crystal structure of PCBM and the efficiency of MDMO-PPV:PCBM 'plastic' solar cells. *Chem. Commun.* Issue 17, 2116–18.

Sahin, Y., Alem, S., de Bettignies, R., et al. 2005. Development of air stable polymer solar cells using an inverted gold on top anode structure. *Thin Solid Films.* Vol. 476, N. 2, 340–343.

Saito, Y., Kubo, W., Kitamura, T., et al. 2004. I-/I-3(-) redox reaction behavior on poly(3,4-ethylenedioxythiophene) counter electrode in dye-sensitized solar cells. *J. Photochem. Photobiol. A* 164:153–57.

Salbeck, J., Yu, N., Bauer, J., et al. 1997. Low molecular organic glasses for blue electroluminescence. In *International Conference on Electroluminescence of Molecular Materials and Related Phenomena,* Fukuoka, Japan, May 21–24, 209–215.

Sariciftci, N. S., Braun, D., Zhang, C., et al. 1993. Semiconducting polymer-buckminsterfullerene heterojunctions: Diodes, photodiodes, and photovoltaic cells. *Appl. Phys. Lett.* 62:585–87.

Savenije, T. J., Warman, J. M., and Goossens, A. 1998. Visible light sensitisation of titanium dioxide using a phenylene vinylene polymer. *Chem. Phys. Lett.* 287:148–53.

Sayama, K., Sugihara, H., and Arakawa, H. 1998. Photoelectrochemical properties of a porous Nb2O5 electrode sensitized by a ruthenium dye. *Chem. Mater.* 10:3825–32.

Schilinsky, P., Waldauf, C., and Brabec, C. J. 2006. Performance analysis of printed bulk heterojunction solar cells. *Adv. Funct. Mater.* 16:1669–72.

Schmidt-Mende, L., Bach, U., Humphry-Baker, R., et al. 2005. Organic dye for highly efficient solid-state dye-sensitized solar cells. *Adv. Mater.* 17:813–15.

Schmidt-Mende, L., Fechtenkotter, A., Mullen, K., et al. 2001. Self-organized discotic liquid crystals for high-efficiency organic photovoltaics. *Science* 293:1119–22.

Shaheen, S. E., Brabec, C. J., Sariciftci, N. S., et al. 2001. 2.5% efficient organic plastic solar cells. *Appl. Phys. Lett.* 78:841–43.

Shaheen, S. E., Ginley, D. S., and Jabbour, G. E. 2005. Organic-based photovoltaics: Toward low-cost power generation. *MRS Bull.* 30:10–19.

Sharma, G. D., Sangodkar, S. G., and Roy, M. S. 1995. Study on electrical and photoelectrical behaviour of undoped and doped furazano[3,4-b]piperazine (FP) thin-film devices. *Synthetic Metals* 75:201–7.

Sirringhaus, H., Brown, P.J., Friend, R. H., et al. 1999. Two-dimensional charge transport in self-organized, high-mobility conjugated polymers. *Nature* 401:685–88.

Sirringhaus, H., Tessler, N., and Friend, R. H. 1998. Integrated optoelectronic devices based on conjugated polymers. *Science* 280:1741–44.

Sirringhaus, H., Wilson, R. J., Friend, R. H., et al. 2000. Mobility enhancement in conjugated polymer field-effect transistors through chain alignment in a liquid-crystalline phase. *Appl. Phys. Lett.* 77:406–8.

Smestad, G. 1998. *Nanocrystalline solar cell kit: Recreating photosynthesis.* Ed. A. Huseth and K. Shanks. Madison, WI: Institute for Chemical Education.

Strawhecker, K. E., Kumar, S. K., Douglas, J. F., et al. 2001. The critical role of solvent evaporation on the roughness of spin-cast polymer films. *Macromolecules* 34:4669–72.

Sun, S.-S., and Sariciftci, N. S. 2005. Organic photovoltaics: Mechanisms, materials, and devices. Boca Raton, FL: Taylor and Francis Group.

Svensson, M., Zhang, F. L., Veenstra, S. C., et al. 2003. High-performance polymer solar cells of an alternating polyfluorene copolymer and a fullerene derivative. *Adv. Mater.* 15:988.

Tan, S. X., Zhai, J., Xue, B. F., et al. 2004. Property influence of polyanilines on photovoltaic behaviors of dye-sensitized solar cells. *Langmuir* 20:2934–37.

Tang, C. W. 1986. Two-layer organic photovoltaic cell. *Appl. Phys. Lett.* 48:183–85.

Tennakone, K., Perera, V. P. S., Kottegoda, I. R. M., et al. 1999. Dye-sensitized solid state photovoltaic cell based on composite zinc oxide/tin (IV) oxide films. *J. Phys. D* 32:374–79.

Too, C. O., Wallace, G. G., Burrell, A. K., et al. 2001. Photovoltaic devices based on polythiophenes and substituted polythiophenes. *Synthetic Metals* 123:53–60.

Tran-Van, F., Carrier, M., and Chevrot, C. 2004. Sulfonated polythiophene and poly (3,4-ethylenedioxythiophene) derivatives with cations exchange properties. *Synthetic Metals* 142:251–58.

Uchida, S., Xue, J., Rand, B. P., et al. 2004. Organic small molecule solar cells with a homogeneously mixed copper phthalocyanine: C_{60} active layer. *Appl. Phys. Lett.* 84:4218–20.

van Duren, J. K. J., Loos, J., Morrissey, F., et al. 2002. In-situ compositional and structural analysis of plastic solar cells. *Adv. Funct. Mater.* 12:665–69.

van Duren, J. K. J., Yang, X. N., Loos, J., et al. 2004. Relating the morphology of poly(p-phenylene vinylene)/methanofullerene blends to solar-cell performance. *Adv. Funct. Mater.* 14:425–34.

van Hal, P. A., Wienk, M. M., Kroon, J. M., et al. 2003. Photoinduced electron transfer and photovoltaic response of a MDMO-PPV: TiO2 bulk-heterojunction. *Adv. Mater.* 15:118.

Wallace, G. G., Dastoor, P. C., Officer, D. L., et al. 2000. Conjugated polymers: New materials for photovoltaics. *Chem. Innovation* 30:14.

Wang, X. J., Perzon, E., Delgado, J. L., et al. 2004. Infrared photocurrent spectral response from plastic solar cell with low-band-gap polyfluorene and fullerene derivative. *Appl. Phys. Lett.* 85:5081–83.

Wang, X. J., Perzon, E., Oswald, F., et al. 2005. Enhanced photocurrent spectral response in low-bandgap polyfluorene and C-70-derivative-based solar cells. *Adv. Funct. Mater.* 15:1665–70.

Wienk, M. M., Struijk, M. P., and Janssen, R. A. J. 2006a. Low band gap polymer bulk heterojunction solar cells. *Chem. Phys. Lett.* 422:488–91.

Wienk, M. M., Turbiez, M. G. R., Struijk, M. P., et al. 2006b. Low-band gap poly(di-2-thienylthienopyrazine): Fullerene solar cells. *Appl. Phys. Lett.* Issue 15, 153511. 88.

Xue, J., Uchida, S., Rand, B. P., et al. 2004a. Asymmetric tandem organic photovoltaic cells with hybrid planar-mixed molecular heterojunctions. *Appl. Phys. Lett.* 85:5757–59.

Xue, J., Uchida, S., Rand, B. P., et al. 2004b. 4.2% efficient organic photovoltaic cells with low series resistances. *Appl. Phys. Lett.* 84:3013–15.

Yakimov, A., and Forrest, S. R. 2002. High photovoltage multiple-heterojunction organic solar cells incorporating interfacial metallic nanoclusters. *Appl. Phys. Lett.* 80:1667–69.

Yu, G., Gao, J., Hummelen, J. C., et al. 1995. Polymer photovoltaic cells—Enhanced efficiencies via a network of internal donor-acceptor heterojunctions. *Science* 270:1789–91.

Yu, G., and Heeger, A. J. 1995. Charge separation and photovoltaic conversion in polymer composites with internal donor-acceptor heterojunctions. *J. Appl. Phys.* 78:4510–15.

Zafer, C., Karapire, C., Sariciftci, N. S., et al. 2005. Characterization of N, N'-bis-2-(1-hydoxy-4-methylpentyl)-3, 4, 9, 10-perylene bis (dicarboximide) sensitized nanocrystalline TiO2 solar cells with polythiophene hole conductors. *Solar Energy Mater. Solar Cells* 88:11–21.

Zen, A., Saphiannikova, M., Neher, D., et al. 2005. Comparative study of the field-effect mobility of a copolymer and a binary blend based on poly(3-alkylthiophene)s. *Chem. Mater.* 17:781–86.

Zhang, F. L., Perzon, E., Wang, X. J., et al. 2005. Polymer solar cells based on a low-bandgap fluorene copolymer and a fullerene derivative with photocurrent extended to 850 nm. *Adv. Funct. Mater.* 15:745–50.

Zhokhavets, U., Erb, T., Hoppe, H., et al. 2006. Effect of annealing of poly(3-hexylthiophene)/fullerene bulk heterojunction composites on structural and optical properties. *Thin Solid Films* 496:679–82.

Zhokhavets, U., Gobsch, G., Hoppe, H., et al. 2004. A systematic study of the anisotropic optical properties of thin poly(3-octylthiophene)-films in dependence on growth parameters. *Thin Solid Films* 451–52:69–73.

Zhokhavets, U., Goldhahn, R., Gobsch, G., et al. 2003. Anisotropic optical properties of conjugated polymer and polymer/fullerene films. *Thin Solid Films* 444:215–20.

5

Recent Progress in Quantum Well Solar Cells

Keith W. J. Barnham, I. M. Ballard, B. C. Browne,
D. B. Bushnell, J. P. Connolly, N. J. Ekins-Daukes,
M. Fuhrer, R. Ginige, G. Hill, A. Ioannides,
D. C. Johnson, M. C. Lynch, M. Mazzer,
J. S. Roberts, C. Rohr, and T. N. D. Tibbits

5.1 INTRODUCTION

This chapter reviews recent work by the Quantum Photovoltaic Group (QPV) at Imperial College London and their collaborators in the application of quantum wells (QWs) to photovoltaics (PV). The group has been working on the application of such nanostructures to PV for nearly two decades, but this chapter will focus on work since the last review.[1] The QPV group work on the quantum dot concentrator (QDC) is described elsewhere in this book (Chapter 9).

We will discuss the potential advantages of QW cells over lattice-matched (LM) and metamorphic (MM) multijunction solar cells in high-optical-concentration (CPV) systems. A wide variety of such high-concentration systems are currently being developed in response to concerns about global warming and stimulated by feed-in tariffs in a growing number of countries.[2] The concentrators are designed to bring down the cost of the III-V multijunction technology, which has already been deployed in space, to levels below flat-panel PV.

We will introduce the quantum well cell and discuss recent progress in exploiting the radiative domination of recombination at concentrator current levels by photon recycling, the performance of the QWSC as a concentrator cell, and new deep-well structures of particular application to multijunction solar cells. A new spin-out company from Imperial College London, QuantaSol Ltd., is close to commercializing the tandem QWSC.

5.2 COMPARISON OF STRAIN-BALANCED QUANTUM WELL CELLS WITH BULK CELLS

The highest-efficiency single-junction solar cells under both 1-sun conditions and concentration are GaAs cells. Both single-junction records have held for two decades.[3] The GaAs band gap (1.42 eV) is, however, rather high for terrestrial applications, as optimal efficiency in an AM1.5 direct spectrum requires a band gap around 1.1 eV at both 1 sun and high concentration.[4] Since those records were established, the main effort to raise efficiency has gone into developing tandem- and triple-junction cells. As will be discussed in Section 5.5, for terrestrial concentrator applications multijunction would also benefit from lowering the band gap of the conventional GaAs cell. The problem in doing so, which III-V cell designers have had to face for at least two decades, is that, while there are *higher-band-gap* alloys such as GaInP and AlGaAs, that can be grown lattice matched to GaAs, there are no LM binary or ternary III-V compounds with a *lower band gap* than GaAs. To increase the GaInP/GaAs/Ge triple-junction cell efficiency, considerable effort is going into studying the quaternary GaInNAs that can be grown lattice matched to GaAs.[5] However, it demonstrates poor minority carrier lifetimes, resulting in insufficient current to avoid limiting the multijunction performance. The approach favored by most triple-junction manufacturers has therefore been to grow on a relaxed InGaAs "virtual" substrate, which increases the lattice constant and lowers the band gap of both the middle and top junction cells. As will be discussed below, this metamorphic approach achieves lower band gaps, but at the expense of a number of residual dislocations, however effective the buffering of the virtual substrate.[6]

The strain-balanced quantum well solar cell (SB-QWSC), which is illustrated schematically in Figure 5.1a, is an alternative approach to the problem of lowering the GaAs band gap.[7] The low-band-gap, higher-lattice-constant (a_2) alloy $In_xGa_{1-x}As$ wells (with In composition $x \sim 0.1$ to 0.25) are compressively strained. The higher-band-gap, lower-lattice-constant (a_1) alloy $GaAs_{1-y}P_y$ barriers (with P composition $y \sim 0.1$) are in tensile strain.

We find that the *zero-stress* condition gives the best material quality[8] and enables at least sixty-five wells to be grown without relaxation.[9]

One feature of the SB-QWSC that is of particular significance for concentrator cells is that the dark current of the GaAs/GaAsP/InGaAs SB-QWSC has ideality factor $n = 1$ at current levels above around 200 suns concentration.[1] We believe this is a result of the complete absence of dislocations in zero-stress material. It is in contrast to the situation with the best InGaAs metamorphic cells, which have $n \sim 1.2$ at concentrator current levels.[10]

There are two contributions to the $n = 1$ dark current. The first is *radiative recombination*, which can be described in terms of the generalized Planck model.[11] The highest probability for radiative recombination occurs between the electron in the lowest energy state in the conduction band well and the hole at the top of the valence band well; i.e., the recombination results in a photon corresponding to energy E_a in Figure 5.1b. The bulk p- and n-regions of the cell, which are formed from higher-band-gap material such as GaAs, are transparent to such photons. Hence, if the SB-QWSC structure is grown on some type of mirror, such as a distributed Bragg reflector (DBR), the phenomenon of photon recycling can be observed,[12] as discussed further in Section 5.4.

The higher-band-gap p- and n-regions are important in another respect, to be discussed in the next section. There is a second contribution to the $n = 1$ dark current, in addition to the radiative recombination from the quantum wells, namely, the *ideal Shockley injection current.* The magnitude of this contribution is determined by the minority carrier recombination in the GaAs p- and n-regions (Figure 5.1b), and hence is suppressed relative to the radiative recombination across the lower energy gap, E_a. A back mirror in combination with the total internal reflection at the top surface of the cell can, in principle, trap a major fraction of the radiative recombination from the wells in the cell apart from the loss through the top surface escape cone. The light trapping cannot be complete if incident sunlight is to enter the cell and detailed balance requires some photon loss.[12] However, a significant fraction of the radiative loss can be trapped by a back mirror and, if reabsorbed in the wells, contribute to the output current and enhance efficiency.[13] The first practical demonstrations of photon recycling are discussed in Section 5.4.

The SB-QWSC approach also enables one to achieve absorption edges and effective band gaps that can only be achieved by conventional bulk cells in metamorphic systems, which of necessity involve some residual dislocations. The further the lattice constant of the relaxed bulk system is from the

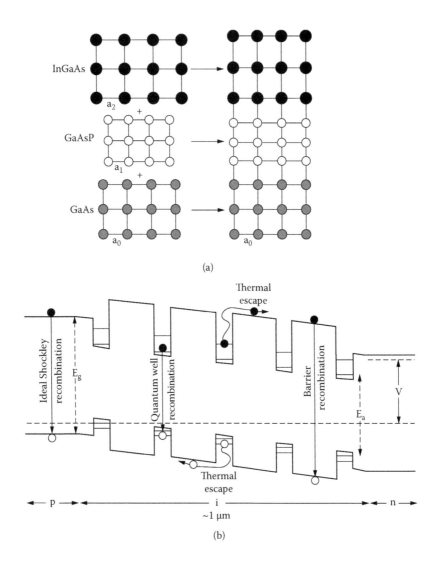

(a)

(b)

FIGURE 5.1

(a) Schematic of SB-QWSC with compressively strained $In_xGa_{1-x}As$ wells and tensile-strained $GaAs_{1-y}P_y$ barriers. (b) Energy band edge diagram of p-i-n SB-QWSC. Note that the band gap of the GaAs in the n- and p-regions is significantly higher than the absorption threshold E_a in the QWs. Hence they are transparent to radiative recombination from the bottom of the wells, and recombination due to ideal Shockley injection is suppressed compared to radiative recombination.

lattice constant of the substrate, the more the residual dislocations. Hence, the comparison of a dislocation-free strain-balanced multiquantum well (SB-MQW) with a relaxed bulk alloy with the same band edge will be dominated by the extra recombination centers in the MM case.

We have recently answered one important question about QW solar cells: whether an MQW system is superior in its photovoltaic properties to bulk material with the same thickness and band edge, if the material quality of both is similar. As part of a study of thermophotovoltaic cells[14] we studied the performance of MQW cells with lattice-matched QWs wells and barriers formed from the quaternary InGaAsP lattice matched to InP substrates. We compared these MQW cells with similar double-heterostructure (DH) cells with the same emitter and base, but with an i-region composed entirely of a lattice-matched InGaAsP quaternary with the same absorption edge as the QWs and the thickness equal to the *total thickness* of the QWs and the barrier. Both the MQW and DH cells were lattice matched and of similar material quality. It is important that the thickness of the i-region was the same in both cases, namely, 1 μm of bulk InGaAsP in the DH case compared to 1 μm of transparent barrier plus absorbing wells in the MQW case.

Despite having considerably less low-band-gap material in the i-region than the DH cell, under narrow band illumination in the QW region the short-circuit current (I_{sc}) is higher in the QW cell than in the DH cell, as shown in Figure 5.2, resulting from enhanced QE at the absorption threshold.[14]

The dark current of the MQW is lower than for the DH structure with the same band edge.[14] Furthermore, the open-circuit voltage (V_{oc}) is higher for the MQW even when the short-circuit current is corrected for the extra absorption, as shown in Figure 5.2. We conclude that in addition to the other advantages discussed in this chapter, an MQW has better current, voltage, and efficiency properties than the same thickness of bulk absorber with the same band edge and similar material quality. In this test both cells were operating at relatively low bias where nonradiative recombination dominates.

One more important advantage of the QW cell is the ability to tune the absorption threshold by changing the width and depth of the quantum well without introducing dislocations. This is a particularly useful property for the optimization of monolithic, multijunction cells given the series current constraint, as will be discussed in Section 5.5.

A number of research groups are investigating the replacement of quantum wells in a p-i-n cell by epitaxially grown quantum dots (for example,

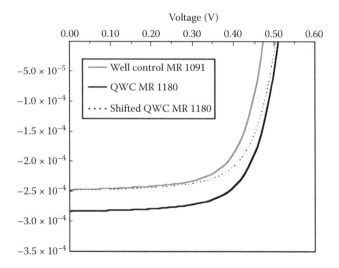

FIGURE 5.2

Measured current and voltage characteristics of a GaInAsP quaternary MQW cell (MR1180) compared to a quaternary DH cell (MR1091) with the same absorption threshold and i-region width under 1,500 nm narrow band illumination. Also shown is the QW cell MR1180 light IV shifted down to the same short-circuit current as MR1091 (dotted line). (Reprinted with permission from Rohr, C. et al. American Institute of Physics. *J. Appl. Phys.* 2006. 100:14510.)

Luque et al.[15] and Hubbard et al.[16]). We have investigated this option. Given the practically achievable areal density of QDs and their lower absorption coefficient (compared to QWs), it is very difficult to incorporate enough dot layers in a p-i-n structure to increase the photocurrent sufficiently to overcome the voltage loss. The latter is likely to be higher in a QD cell than a QW cell due to unavoidable nonradiative recombination arising from relaxation resulting from residual, unbalanced strain in the QDs.

QD material of sufficient quality may be achievable by stress balancing, but the effective absorption coefficient appears to be at least an order of magnitude too small. A feature of both strain-balanced and strained epitaxially grown QDs is the presence of a wetting layer, which is, in fact, a *quantum well* of around 1.8 nm width. Quantum efficiency data with epitaxial dots in p-i-n structures[15,16] show features that correspond to absorption at the level to be expected in such shallow wells at an energy just below that of the bulk regions. The longer wavelength signal from the QDs is a further order of magnitude down. The short-circuit current enhancement observed in Hubbard et al.[16] is primarily due to the QW wetting layer rather than the QDs.

5.3 RADIATIVE RECOMBINATION DOMINANCE AT CONCENTRATOR CURRENT LEVELS

We have developed a detailed simulation model SOL at Imperial for both the spectral response and the dark current behavior of LM, strained, and strain-balanced MQW quantum wells.[1,13,17] Figures 5.3a and b show typical examples of fits to the internal quantum efficiency (IQE) and dark current of a fifty-shallow-well SB-QWSC. The cell is a p-i-n diode with an i-region containing fifty QWs that are 7 nm wide of compressively strained $In_xGa_{1-x}As$ with $x \sim 0.1$ inserted into tensile-strained $GaAs_{1-y}P_y$ barriers with $y \sim 0.1$.

The IQE in the neutral regions of the cell is fitted in the conventional way in terms of minority carrier diffusion lengths in p- and n-regions.[17] The energy levels in the quantum wells are calculated on a first-principles quantum mechanical approach, and either the absorption level is adjusted to fit the data at the first continuum in the well or allowance is made for reabsorption following reflection at the back substrate, as described in Section 5.5.

The shape and height of the dark current in Figure 5.3b in the low-bias ideality $n \sim 2$ region is well reproduced by a model in terms of one parameter, an effective lifetime defined as follows. SOL solves for the variation in the carrier distributions $n(x)$ and $p(x)$ with position x throughout the i-region using the QW density of states based on the energy levels in the QE calculation, assuming the depletion approximation holds.[17] This approach gives similar results to a self-consistent calculation up to the voltages at which the $n = 1$ contribution dominates. From carrier densities a recombination rate is determined assuming the Shockley–Hall–Read (SHR) approach.[17] This requires the nonradiative lifetimes of the carriers. The evidence suggests we can equate the electron and hole lifetimes, so we are left with two parameters that depend on material quality, the carrier lifetimes in the barrier and well, τ_B and τ_W, respectively. As bulk material of comparable band gap to the QW and barrier does not exist, we assume that, for low P and In fractions, the well and barrier quality are sufficiently similar that lifetimes are equal. This leaves one effective lifetime for SHR recombination, $\tau = \tau_W = \tau_B$. We find that reasonable fits can be obtained to the shape and height of the SRH dark current over a wide range of quantum well samples with the single parameter τ.

As discussed in Section 5.2, there are two distinct contributions to the $n = 1$ current. First, the *ideal Shockley injection current* assumes no

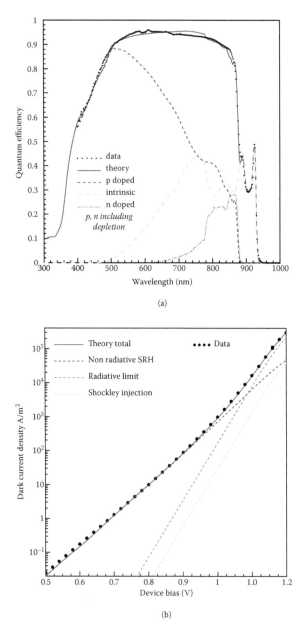

FIGURE 5.3

(a) Measured internal quantum efficiency (IQE) of a fifty-shallow-well SB-QWSC and the fit made with program SOL. The separate contributions of p-, intrinsic, and n-regions are also shown. (b) Measured dark current for the cell in (a) showing the single-parameter fit to the nonradiative SRH contribution in the $n \sim 2$ region and the predictions for the two contributions to the $n = 1$ region based on IQE fits in the bulk and well region of (a).

recombination in the depletion region but does assume the radiative and nonradiative recombination of injected minority carriers with majority carriers in the field-free regions. This contribution depends in a standard way on the minority carrier diffusion lengths, doping levels, and the surface recombination in the neutral regions. We can estimate this current from the minority carrier parameters obtained when fitting the spectral response in the neutral regions, which are the p- and n-regions in Figure 5.3a.

The second contribution to the $n = 1$ current results from the recombination of carriers in the QWs and barriers in the depletion region. Like the ideal Shockley injection current, this is expected to have both radiative and nonradiative contributions. However, we assume that the nonradiative contribution in the i-region is described by the SHR $n \sim 2$ model discussed above. The *radiative recombination* contribution to the QW recombination can be estimated by a detailed balance argument. This relates the photons absorbed to the photons radiated, as discussed in Nelson et al.[11] and references cited therein. The radiated spectrum as a function of photon energy, E, in an electrostatic field F, $L(E,F)dE$, is determined by the *generalized Planck equation*:

$$L(E,F)dE = \frac{2\pi n^2 L_W}{h^3 c^2} \frac{\alpha(E,F)E^2}{e^{(E-\Delta E_F)/k_B T} - 1} dE \qquad (5.1)$$

We integrate this spectrum over the energy and the cell geometry as described in Nelson et al.[11] to give a total radiative current. This will depend on the *quasi-Fermi level separation* ΔE_F and the absorption coefficient $\alpha(E,F)$ as a function of energy and field. In general, we assume that $\Delta E_F = eV$ where V is the diode bias. However, in Section 5.5 we discuss evidence that in deep-well samples, $\Delta E_F < eV$.

The absorption coefficient $\alpha(E,F)$ is calculated from first principles as discussed above when fitting the spectral response of the QWs, assuming unity quantum efficiency for escape from the wells.

It is important to note that the important parameters for both $n = 1$ contributions, the ideal Shockley injection (minority carrier diffusion lengths in p- and n-regions), and the QW radiative current levels (absorption coefficient $\alpha(E,F)$) are therefore determined by the spectral response fits in the bulk and QW regions, respectively. Hence, the fit in Figure 5.3b depends on only one parameter, and yet reproduces the data over six decades of dark current. This suggests we have a good understanding of the factors influencing the dark current. We have compared a wide range of well numbers and well depths, and as long as we can fit the internal quantum efficiency, we have found good agreement with the measured dark current, apart from the deep-well $n = 1$ contribution discussed in Section 5.5.[18] These quantitative results confirm the observation discussed in Section 5.2, that radiative recombination dominates over the ideal Shockley injection.

5.4 DARK CURRENT REDUCTION AND V_{OC} ENHANCEMENT BY PHOTON RECYCLING

As a result of the dislocation-free material, the high-band-gap bulk regions of the cell, and the application of SOL to a wide range of samples as discussed above, we know that radiative recombination dominates in SB-QWSCs at high current levels. This makes possible the exploitation of *radiative recycling,* which we first observed as a reduction of the dark current and confirmed by a model for the electroluminescence.[13,19]

We studied three SB-QWSC designs that varied in the number and depth of the QWs, each grown on both a distributed Bragg reflector (DBR) and a GaAs substrate. To ensure that the SB-QWSCs were similar, a pre-grown DBR wafer and an n-type GaAs substrate were placed down- and upstream, respectively, in the metal organic vapor phase epitaxy (MOVPE) reactor and both overgrown in the same gas flow. The QWs consisted of three depths of $In_xGa_{1-x}As$ wells, with $x = 0.11$, 0.17 and $x = 0.22$. The widths of the wells and barriers ($GaAs_{1-y}P_y$, where $y = 0.1$) were adjusted to satisfy the zero-stress condition.[8]

The DBRs consisted of 20.5 periods of alternating layers of material with low ($Al_{0.8}Ga_{0.2}As$) and high ($Al_{0.13}Ga_{0.87}As$) refractive indices. To reduce series resistance, intermediate layers of $Al_{0.5}Ga_{0.5}As$ were grown between each DBR layer. Layer thicknesses were designed to double the pass of normal incidence light and incident sunlight within ~20° of normal. This resulted in a significant enhancement of the IQE in the QW, as can be seen in Figure 5.4 for two-shallow fifty-well devices. The DBR also acts as a good reflector for radiative recombination at angles greater than 60°, as can be seen in Figure 5.5.

The dark currents of ~15 fully metallized mesa structures were measured using the four-point method and pulsed current sourcing to reduce heating. All measurements were taken at 25°C using a Peltier cooled stage.

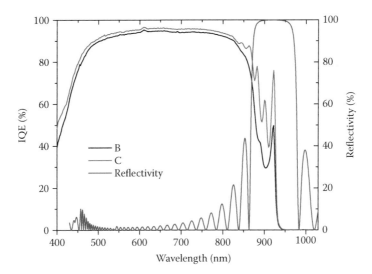

FIGURE 5.4

Internal quantum efficiency (IQE) of a fifty-shallow-well SB-QWSC overgrown on either a GaAs substrate (device B) or a distributed Bragg reflector (DBR) on a GaAs wafer (device C). Both wafers were loaded side by side in the MOVPE reactor. The modeled reflectivity of the DBR is also shown.

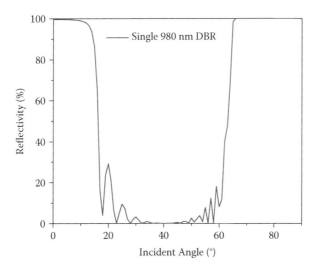

FIGURE 5.5

Calculated angular reflectivity for the DBR with peak reflectivity at a wavelength of 980 nm. Reflectivity is high for near-normal incidence sunlight as well as at angles greater than 60°.

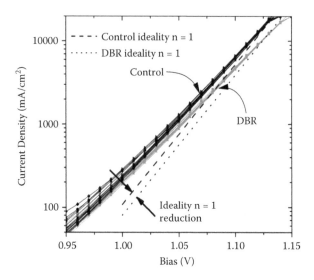

FIGURE 5.6

Measured dark currents at high bias of ~15 fully metalized devices from fifty-shallow-well SB-QWSCs grown on a DBR or control GaAs substrate. The predictions are shown of the electroluminescence (EL) model. (Reprinted with permission from Johnson, D. C. et al. 2007. American Institute of Physics. *Appl. Phys. Lett.* 90:213505.)

As shown for the shallow-well samples, for example, in Figure 5.6, the dark current is significantly lower in all the DBR cells at high bias.

The measurements were fit with a two-exponential model:

$$J = J_{01} \cdot e^{q(V-J\cdot R_s)/(n_1 \cdot k \cdot T)} + J_{02} \cdot e^{q(V-J\cdot R_s)/(n_2 \cdot k \cdot T)} \tag{5.2}$$

where J_{01} and J_{02} are reverse saturation currents, n_1 and n_2 are ideality factors, R_S is the series resistance, k is Boltzmann's constant, and T is the temperature. The ideality factor n_1 is fixed at 1. The first term in Equation 5.1 dominates at high bias and corresponds to QW radiative recombination, which, as discussed in Section 5.3, dominates over the ideal Shockley injection.

Fits to the data were made by varying J_{01}, J_{02}, n_2, and R_S. The average J_{01} values obtained from ~15 devices for each design are listed in Table 5.1. The ideality $n = 1$ reverse saturation current is reduced by between 18 and 33%, depending on design.

To confirm that this reduction in J_{01} is a result of photon recycling, a model has been developed to provide for the shape and the integral of the electroluminescence (EL) emitted as a function of high forward bias in the device. The net photon flux exiting the surface of the device is described

TABLE 5.1

Measured Reverse Saturation Currents and Predictions from EL Model

			EL Model	Measured	
Design No.	In Fraction X	Number of Wells	J_{01} 10^{-21}A	J_{02} 10^{-13}A	J_{01} 10^{-21}A
1 control	0.11	50	1.7	3.2 ± 0.4	1.5 ± 0.2
1 DBR	0.11	50	1.2	2.9 ± 0.7	1.0 ± 0.2
2 control	0.17	30	6.8	4.5 ± 0.5	8.9 ± 0.3
2 DBR	0.17	30	5.1	5.1 ± 0.2	6.7 ± 0.1
3 control	0.22	30	1.8	18	34 ± 1
3 DBR	0.22	30	1.3	13	28 ± 2

Source: Reprinted with permission from Johnson, D. C. et al. 2007. American Institute of Physics. *Appl. Phys. Lett.* 90:213505.

using the generalized Planck equation (Equation 5.1). Integrating the photon flux density over energy, one obtains the radiative dark current in absolute units for a given bias or quasi-Fermi-level separation, assuming radiative recombination dominates. This method requires calculation of device absorptivity, which is determined by a complex refractive index calculated from the measured internal quantum efficiency. A matrix method is used to calculate the absorption of a stack of thin films, including antireflection coatings, the window layer, active layer, and DBR, taking into account multiple passes of photons and interference effects in the structure. Further details are given in Johnson et al.[19]

EL spectra were measured by driving 1-mm-sized concentrator devices into forward bias while maintaining the device at 25°C with the Peltier stage. Emission spectra are measured using a computer-controlled monochromator and charge-coupled device. Figure 5.7 compares the measured and modeled EL spectra of a SB-QWSC grown on a DBR (design 1 in Table 5.1). In both spectra one can see Fabry-Perot cavity oscillations, and the EL model fits the shape of these experimental measurements quite well.

The EL calculation is in absolute units. Hence, by integrating the predicted EL spectrum over energy, we can calculate the expected radiative contribution to the dark current and compare with the measured dark currents. Examples of the predictions of this EL model for the $n = 1$ radiative contributions to the dark current are shown for the control and DBR devices in Figure 5.6.

The reduction of the dark current is evident in both the measurements and the calculated values from our model. Table 5.1 compares the reverse

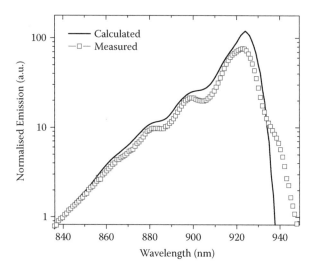

FIGURE 5.7

Measured electroluminescence (EL) spectrum for the fifty-shallow-well device grown on a DBR showing evidence for Fabry–Perot cavity oscillations. The full line is the EL model discussed in the text and in Johnson et al.[19] (Reprinted with permission from Johnson, D. C. et al. 2007. American Institute of Physics. *Appl. Phys. Lett.* 90:213505.)

saturation currents predicted from the EL model with the average values fitted to dark current measurements for all three device designs. The model quite accurately describes the ideality $n = 1$ component of the measured dark currents in relative terms. Further details can be found Johnson et al.[19] and references therein.

In order to demonstrate that the reduction in dark current due to photon recycling leads to efficiency enhancement, 1-mm concentrator cells were processed from the shallow fifty-well devices and extensively studied under concentrated light. Measurements were made using a shuttered system, with the devices mounted on a Peltier element temperature-controlled copper stage and illuminated through a high-speed Uniblitz shutter by a Xe light source.

The procedure used is to measure the external quantum efficiency (EQE) of the cell and convolute the EQE with the known standard spectrum (e.g., AM1.5D low AOD) to find the short-circuit current density (J_{sc}) that the cell should have in this spectrum. The cell position is adjusted until the J_{sc} achieves this value illuminating within the designated area, and the light IV measured as described above. The cell is then moved closer to the source, the position reoptimized, the light IV repeated, and the concentration defined as the ratio of the new J_{sc} to the standard J_{sc}.

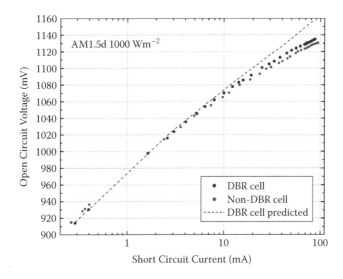

FIGURE 5.8

Open-circuit voltage as a function of short-circuit current, which is assumed to be proportional to the concentration of the light, for fifty-shallow-well concentrator cells with and without a DBR. (From T. N. D. Tibbits, I. Ballard, K. W. J. Barnham, D. C. Johnson, M. Mazzer, J. S. Roberts, R. Airey, and N. Foan, in *Proceedings of the 4th World Conference on Photovoltaic Energy Conversion (WCPEC4)*, Hawaii, 2006, IEEE. p. 861.)

A similar pair of shallow-well SB-QWSC concentrator devices was measured with uniform illumination over the mesa at the Fraunhofer ISE standards laboratory, producing performance parameters very similar to those measured with the procedure described above when allowance is made for the different areas illuminated in the two systems.

Pairs of fifty-shallow-well wafers grown as above side by side on GaAs substrates and DBRs were processed as 1-mm concentrator cells and measured under this system.[20] Figure 5.8 shows how the V_{OC}s of both a DBR SB-QWSC and a non-DBR SB-QWSC vary with solar concentration (assumed to be proportional to short-circuit current), and clearly shows an enhancement of the V_{OC} in the DBR device as the concentration is increased corresponding to the fall in dark current. We believe this is the first time in which the V_{OC} of an illuminated solar cell has been shown to increase as a result of direct enhancement of photon recycling by adding a DBR.[20,21]

With the increase in J_{SC} and V_{OC} due to the DBR, we achieved an efficiency of 27.0% (AM1.5 low-AOD spectrum at 328 suns).[20,21] Subsequently this has been increased significantly above 27% at 500 suns[22] and is now close to the world single-junction record in AM1.5D of (27.8 ± 1.0)% at 216 suns.[3]

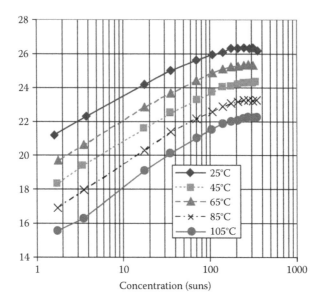

FIGURE 5.9
Measured AM1.5D efficiency of a fifty-shallow-well concentrator cell as a function of concentration (assumed proportional to short-circuit current) at different device temperatures measured on the system described in the text and in Ballard et al.[23]

The efficiency of a similar fifty-shallow-well DBR device has been measured as a function of concentration at different temperatures using the apparatus described above as further discussed in Ballard et al.[23] The results are shown in Figure 5.9. A reduction in efficiency with increasing temperature is expected because of the shrinkage of the band gap. However, it can be seen that the reduction is smaller at higher concentrations. The behavior is also expected, as the dark current ideality factor reduces from $n \sim 2$ at low current levels to $n = 1$ at high concentrator current levels.[2]

We can be more quantitative about the temperature dependence at high current levels, as we have shown in Section 5.4 that SOL can predict the absolute dark current levels of the two $n = 1$ components. The level of the *radiative recombination* contribution will be determined from detailed-balance by the level of absorption in the well and the *ideal Shockley injection* recombination by the minority carrier diffusion lengths in the bulk p- and n-regions.

In Figure 5.10 we show the measurements of the reverse saturation current density of the fifty-shallow-well DBR cell in Figure 5.9 compared with the absolute predictions of SOL for the radiative recombination and ideal Shockley injection contributions determined from the IQE fits. It can be

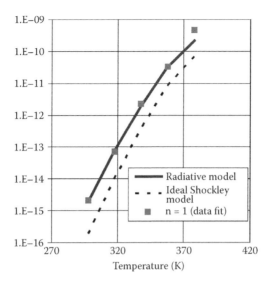

FIGURE 5.10

Measured reverse saturation current density of the ideality $n = 1$ dark current from a fit to the dark current with Equation 5.2 of the shallow fifty-well DBR cell in Figure 5.9. The measurements are shown as a function of temperature compared with the absolute prediction of SOL for the radiative recombination and ideal Shockley contributions.

seen that the radiative recombination contribution reproduces the measured data in level and temperature dependence. The ideal Shockley injection contribution is significantly lower due the higher band gap in the bulk regions of the cell. Were a higher proportion of the EL to be recycled, as discussed in Section 5.3, then the saturation current density would be lower and the efficiency higher.

5.5 DEEP-WELL SB-QWSC AND TANDEM QUANTUM WELL CELLS

The current world record PV cells are triple-junction cells grown by Spectrolab[3,5] consisting of two variants of the GaInP/GaAs tandem on active Germanium (Ge) substrates. The first has a high-band-gap GaInP top cell grown lattice matched (LM) to the GaAs bottom cell. In the second approach, the metamorphic (MM) cell, a relaxed InGaAs buffer layer (often called a *virtual substrate*), is grown on the Ge substrate to increase the lattice constant so that InGaAs with lower band gap than GaAs can

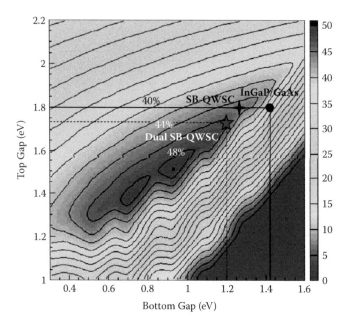

FIGURE 5.11

Ideal efficiency contour plot for a tandem cell under 500× concentration in AM1.5D low AOD. Dark lines show the band gaps of GaInP and GaAs. The red cross shows the band gap of a deep-well SB-QWSC and the blue star a tandem cell with this lower cell and a QW top cell. (From B. Browne, A. Ioannides, J. Connolly, K. Barnham, J. Roberts, R. Airey, G. Hill, G. Smekens, and J. Van Begin. 2008. *Proceedings of the 33rd IEEE Photovoltaic Specialists Conference,* IEEE, San Diego, CA.) (See color insert following page 206.)

be grown. However good the buffer layer, there will always be significantly more dislocations in this cell than with a good GaAs substrate, and as noted in Section 5.2, the lowest ideality factor of the Spectrolab InGaAs cell is $n = 1.2$.[10]

Electrical connection between the subcells is provided by two tunnel junctions. As the cells are in series the same current passes through all three cells and is limited by the current generated in the poorer cell. In the LM case this turns out to be the GaAs middle cell in most spectra, again because of the relatively high GaAs band gap. In both the LM and MM cases the Ge cell overproduces current in many spectra of interest. Therefore, optimizing the Ge-based triple-junction cell simplifies to considering the optimum top and middle cell band gaps. The parameter space can be represented as contours on a two-dimensional plot of the two band gaps for a given spectrum. Figure 5.11 shows the tandem plot taken from Browne et al.,[24] which is similar to the triple-junction plot in King et al.[6]

The strong correlation between the band gaps implied by the contours reflects the series current constraint and the need to ensure the same current in the top and middle cells.

It is clear from Figure 5.11 that reducing the bottom cell band gap of the GaInP/GaAs tandem as indicated by the SB-QWSC point has the potential to significantly increase the tandem- or triple-junction efficiency. Furthermore, an MQW top cell (dual SB-QWSC) would make it possible to *independently optimize both band gaps* without introducing dislocations and potentially to get close to the summit of the contours. In contrast, growth on virtual substrates as in MM cells results in the band gap reductions of the top and bottom cells being correlated. The MM cells can be optimized along *diagonal lines* for the two options of ordered and disordered top cells,[5] but only at the expense of adding residual dislocations.

The ability of a double QWSC, whether as a tandem or on an active Ge substrate, to independently optimize the top and middle cell absorption edges, as, for example, in Figure 5.11, is a very significant advantage for the QW approach. Note also that in different spectra, for different temperatures of operation and for nonideal transmission coefficients in a high-concentration system, the maximum point of the contour plot will be different from that in Figure 5.11. The QW approach gives one the ability to optimize the energy harvest over a year for spectral and temperature variation, and concentrator transmission, simply by changing the number or depth of the quantum wells. This advantage will be worth at least 1 to 2% absolute in terms of average electrical energy harvested over one year. Both the Imperial QPV group and QuantaSol have the aim to demonstrate this advantage as their highest priority.

The first stage of the tandem optimization, namely, the ability to produce a bottom cell of good quality with an absorption edge longer than 1 μm or below 1.2 eV, with good material quality, has been achieved. Some interesting physics has emerged in these deep strain-balanced wells. A study along the lines of the analysis of the contributions to the $n = 1$ dark current discussed in Section 5.3 has been conducted as a function of well depth. Figure 5.12 shows the experimental J_{01} values determined by fits with Equation 5.2 to the dark currents measured in fully metalized diodes, for different well depths in samples with between thirty and fifty SB wells. The well depth is given by the position of the first exciton peak.

It can be seen from Figure 5.12 that the ideal Shockley contribution to the $n = 1$ dark current is indeed independent of well depth as expected, since

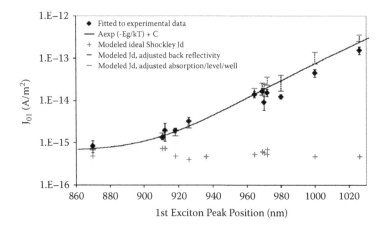

FIGURE 5.12

Experimental values of J_{01} plotted as a function of band edge given by the first exciton position compared to modeled predictions described in the text. Note the modeling predictions for the ideal Shockley component of J_{01} for each device.

the contribution is determined by the GaAs band gap in all cases. On the basis of the detailed balance condition described by Equation 5.1, SOL predicts two values for the J_{01} of the $n = 1$ contribution, depending on whether, as discussed in Section 5.3, one raises the absorption parameter in the well to fit the measured quantum efficiency at the first continuum or fixes the absorption at the theoretical value and fits a back-reflectivity. In the latter case the photon recycling (allowed for in this calculation) will make a more significant contribution, thus reducing the predicted $n = 1$ level.

This ambiguity of interpretation is indicated by the error bars in Figure 5.12. However, whichever limit one takes, the experimental values of J_{01} obtained from the fits to the dark currents are below the prediction for all the deep-well data points above 970 nm. We interpret this reduction as being further evidence in support of the observation made in low-well-number devices, that radiative recombination is reduced in QW solar cells.[11,25] We parameterize this radiative recombination suppression in terms of a reduction in the quasi-Fermi-level separation ΔE_F from the expected value of eV, where V is the bias. The quasi-Fermi-level reductions that explain the lower J_{01} values in Figure 5.12 are presented in Figure 5.13.

A further interesting effect has been observed in SB-QWSCs of medium well depth and low well number. A change in shape of the electroluminescence from the QW as a function of voltage has been observed at

(a)

(b)

FIGURE 1.9
Images of ground-mounted PV systems in (a) Halbergmoss, Germany (Courtesy of O. Stern and O. Mayer) and (b) Aichi Airport, Japan (From Ichiro Araki et al. 2009. "Bifacial PV system in Aichi Airport-site demonstration research plant for new energy power generation," *Solar Energy Materials & Solar Cells*. Elsevier. 911–916.)

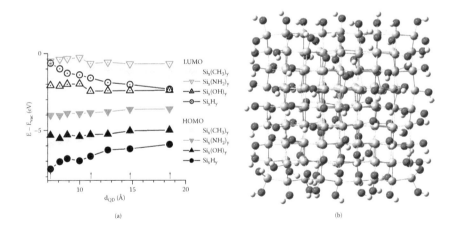

(a)

(b)

FIGURE 3.18
(a) HOMO and LUMO levels of Si core approximants completely terminated with functional groups. Computations were carried out with the B3LYP hybride DF and an all-electron Gaussian 6-31G(d) molecular orbital basis set. (From König, D., Rudd, J., Green, M. A., and Conibeer, G., "Role of the Interface for the Electronic Structure of Si Quantum Dots," *Phys. Rev. B* 78 [2008a]: 035339 1–9, and König, D., Rudd, J., Green, M. A., and Conibeer, G., "Critical Discussion of Computation Accuracy," EPAPS article to [57], 2008b, ftp://ftp.aip.org/epaps/phys_rev_b/E-PRBMDO-78-101827/EPAPS_Rev3. pdf) OH-, NH$_2$-, and CH$_3$-groups emulate SiO$_2$, Si$_3$N$_4$, and SiC as a dielectric matrix. (b) Example of optimized Si$_{165}$(OH)$_{100}$ approximant, corresponding to a Si NC diameter of 18.5 Å. O, Si, and H atoms are shown in dark grey, grey, and light grey, respectively.

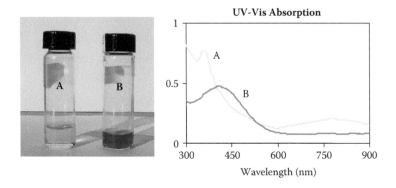

FIGURE 4.13

Absorption spectra of the PTEBS water-soluble polymer in (A) acidic and (B) basic solutions. Acidic solutions develop a new absorption band in the near-IR region.

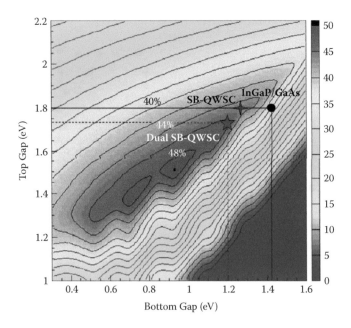

FIGURE 5.11

Ideal efficiency contour plot for a tandem cell under 500× concentration in AM1.5D low AOD. Dark lines show the band gaps of GaInP and GaAs. The red cross shows the band gap of a deep-well SB-QWSC and the blue star a tandem cell with this lower cell and a QW top cell. (From B. Browne, A. Ioannides, J. Connolly, K. Barnham, J. Roberts, R. Airey, G. Hill, G. Smekens, and J. Van Begin. 2008. *Proceedings of the 33rd IEEE Photovoltaic Specialists Conference,* San Diego, CA.)

FIGURE 6.13
Picture of large area Si NW film grown by CVD (VLS mechanism) on a stainless steel foil.

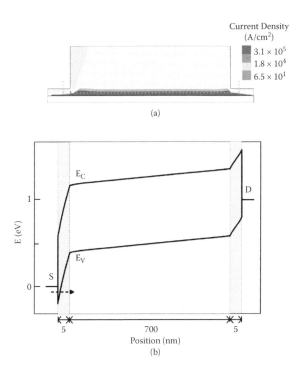

FIGURE 7.7
(a) Cross-sectional contour plot showing the current density in a back-gated Ge FET with Ni contacts, and with l_{diff} = 30 nm, V_{GS} = −4 V, and V_{DS} = −1 V. The conduction in the Ge channel takes place primarily near the bottom oxide interface, where the holes accumulate. (b) Conduction (E_C) and valence (E_V) energy band profile along the dashed path shown in the upper panel. Note that the applied V_{DS} = −1 V bias falls primarily across the metal-semiconductor barrier at the source contact, which in turn renders the device current highly dependent on the tunneling through the Schottky barrier at the source contact, and relatively insensitive to the carrier mobility in the channel. (Reproduced from Liu, E. S., Jain, N., Varahramyan, K. M., et al., "Role of Metal-Semiconductor Contact in Nanowire Field-Effect Transistors," to appear in *IEEE Transactions on Nanotechnology*, 2010.)

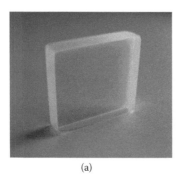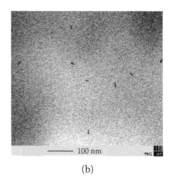

(a) (b)

FIGURE 9.5

(a) Photograph of a homogeneous LSC containing nanorods under UV illumination. (b) TEM image of CdSe/CdS nanorods in a polymer nanocomposite LSC.

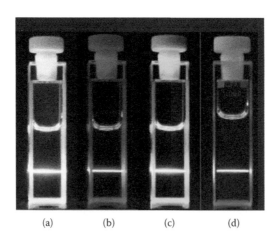

(a)   (b)   (c)   (d)

FIGURE 10.14

Photographs of the upconversion luminescence in 1 wt% colloidal solutions of nanocrystals in dimethylsulfoxide excited at 10,270 cm^{-1} (invisible) with a laser power density of 5.9 kW/cm^2. (a) Total upconversion luminescence of the NaYF4:20% Yb^{3+}, 2% Er^{3+} sample. (b, c) The same luminescence through green and red color filters, respectively. (d) Total upconversion luminescence of the NaYF4: 20% Yb^{3+}, 2% Tm^{3+} sample. (Courtesy of Prof. M. Haase.)

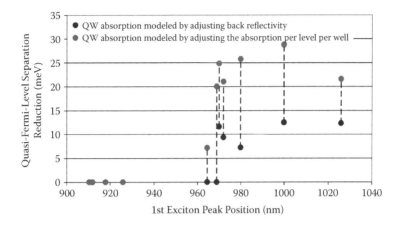

FIGURE 5.13

The quasi-Fermi-level reduction that would be required to explain the radiative dark current reduction in Figure 5.12, plotted as a function of the lowest energy exciton peak.

biases above about 1 V. Examples for one and ten medium-depth wells (In ~ 0.17, exciton ~ 970 nm) are shown in Figure 5.14. There is also some evidence for exciton suppression at high bias in fifty-well samples.[26] As shown in Figure 5.14, the bulk GaAs EL signal around 870 nm does not change shape, but there is evidence that the EL signal at the exciton peak falls and the peak widens with increasing voltage as bias increases above around 1 V bias.

As in the depth dependence of J_{01} in Figure 5.12, this effect corresponds to a suppression of radiative recombination at high bias. We are studying two possible explanations for this effect: first, that there is a temperature difference between the carriers and the lattice, and second, that this is some form of screening effect at high carrier concentrations in the well.[26] The first of these explanations would have the advantage of providing a thermodynamic force between well and barrier.[27] This could explain the observed nonzero quasi-Fermi-level separation suppression that has been identified at lower biases and in the deep-well samples in Figure 5.13.

5.6 CONCLUSIONS

QWSCs have some interesting novel properties that we have only recently uncovered:

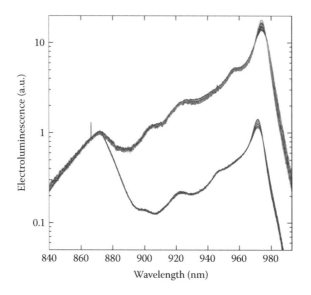

FIGURE 5.14

Electroluminescence from one and ten medium-well-depth SB-QWSCs. The single-well device (bottom) was biased between 1.0 and 1.1 V, and the ten-well device (top) was biased between 1.016 and 1.066 V. (From Führer, M. F. et al. 2008. *Proceedings of the 33rd IEEE Photovoltaic Specialists Conference*, San Diego, CA.)

- The radiative dominance of the ideality $n = 1$ dark current at high concentrator currents can now be understood in terms of the excellent material quality (better than MM) and ideal Shockley injection current into high-band-gap neutral regions.
- Radiative recombination means one can recycle waste photons adding approximately a few percent absolute to efficiencies.
- The 50+ deep wells ideal for tandem- and triple-junction applications appear to display a similar radiative recombination suppression to that observed in the electroluminescence of low-well-number devices. The change in shape of the electroluminescence signal at the first exciton may be a related phenomenon.
- A double QW tandem- or triple-junction cell would make it possible to tailor a cell for spectral and temperature conditions and for concentrator transmission variations, so as to optimize the electrical energy harvesting over a year.
- QDs have problems of significantly lower absorption and residual dislocations as replacement for QWs. However, colloidal QDs have a significant potential in luminescent concentrators (Chapter 9), which are complementary to the high-concentrations system currently

being developed, for which the SB-QWSC in tandem- and multi-junction configurations is particularly suited.

REFERENCES

1. K. W. J. Barnham, I. Ballard, A. Bessière, A. J. Chatten, J. P. Connolly, N. J. Ekins-Daukes, D. C. Johnson, M. C. Lynch, M. Mazzer, T. N. D. Tibbits, G. Hill, J. S. Roberts, and M. A. Malik. 2006. Quantum well solar cells and quantum dot concentrators. In *Nanostructured materials for solar energy conversion*, ed. T. Soga, chap. 16, 517. Amsterdam: Elsevier.
2. K. W. J. Barnham, M. Mazzer, and B. Clive. 2006. *Nature Mater.* 5:161.
3. M. A. Green, K. Emery, Y. Hisikawa, and W. Warta. 2007. *Prog. Photovolt. Res. Appl.* 15:425.
4. J. S. Ward, M. W. Wanlass, K. A. Emery, and T. J. Coutts. 1993. In *Proceedings of the 23rd IEEE Photovoltaic Specialists Conference,* Louisville, p. 650.
5. D. J. Friedman, J. F. Geisz, S. R. Kurtz, and J. M. Olson. 1998. In *2nd World Conference and Exhibition on Photovoltaic Energy Conversion,* Vienna, p. 3.
6. R. R. King, D. C. Law, K. M. Edmondson, C. M. Fetzer, G. S. Kinsey, H. Yoon, R. A. Sherif, and N. H. Karam. 2007. *Appl. Phys. Lett.* 90:183516.
7. N. J. Ekins-Daukes, K. W. J. Barnham, J. G. Connolly, J. S. Roberts, J. C. Clark, G. Hill, and M. Mazzer. 1999. *Appl. Phys. Lett.* 75:4195.
8. N. J. Ekins-Daukes, K. Kawaguchi, and J. Zhang. 2002. *Crystal Growth Design* 2:287.
9. M. C. Lynch, I. M. Ballard, D. B. Bushnell, J. P. Connolly, D. C. Johnson, T. N. D. Tibbits, K. W. J. Barnham, N. J. Ekins-Daukes, J. S. Roberts, G. Hill, R. Airey, and M. Mazzer. 2005. *J. Mater. Sci.* 40:1445.
10. R. R. King, D. C. Law, K. M. Edmondson, C. M. Fetzer, R. A. Sherif, G. S. Kinsey, D. D. Krut, H. L. Cotal, and N. H. Karam. 2006. In *Proceedings of the 4th World Conference on PV Energy Conversion*, Hawaii, p. 760.
11. J. Nelson, J. Barnes, N. Ekins-Daukes, B. Kluftinger, E. Tsui, K. Barnham, C. T. Foxon, T. Cheng, and J. S. Roberts. 1997. *J. Appl. Phys.* 82:6240.
12. A. Marti, J. L. Balenzategui, and R. F. Reyna. 1997. *J. Appl. Phys.* 82:4067.
13. J. P. Connolly, D. C. Johnson, I. M. Ballard, K. W. J. Barnham, M. Mazzer, T. N. D. Tibbits, J. S. Roberts, G. Hill, and C. Calder. 2007. In *Proceedings of the International Conference on Solar Concentrators for the Generation of Electricity or Hydrogen (ICSC-4)*, Escorial, Spain, p. 21.
14. C. Rohr, P. Abbott, I. Ballard, J. P. Connolly, K. W. J. Barnham, M. Mazzer, C. Button, G. Hill, J. S. Roberts, G. Clarke, and R. Ginige. 2006. *J. Appl. Phys.* 100:114510.
15. A. Luque, A. Marti, N. Lopez, E. Antolin, E. Canovas, C. Stanley, C. Farmer, and P. Diaz. 2006. *J. Appl. Phys.* 99:094503.
16. S. M. Hubbard, C. D. Cress, C. G. Bailey, R. P. Raffaelle, S. G. Bailey, and D. M. Wilt. 2007. *Appl. Phys. Lett.* 92:123512.
17. J. P. Connolly, I. M. Ballard, K. W. J. Barnham, D. B. Bushnell, T. N. D. Tibbits, and J. S. Roberts. 2004. In *Proceedings of the 19th European Photovoltaic Solar Energy Conference,* Paris, p. 355.

18. B. Browne, J. Connolly, I. Ballard, K. Barnham, D. Bushnell, M. Mazzer, M. Lynch, R. Ginige, G. Hill, C. Calder, and J. Roberts. 2007. In *Proceedings of the 22nd European Photovoltaic Solar Energy Conference*, Milan, p. 782.
19. D. C. Johnson, I. M. Ballard, K. W. J. Barnham, J. P. Connolly, M. Mazzer, A. Bessière, C. Calder, G. Hill, and J. S. Roberts. 2007. *Appl. Phys. Lett.* 90:213505.
20. D. C. Johnson, I. M. Ballard, K. W. J. Barnham, M. Mazzer, T. N. D. Tibbits, J. S. Roberts, G. Hill, and C. Calder. 2006. In *Proceedings of the 4th World Conference on Solar Energy Conversion (WCPEC4)*, Hawaii, p. 26.
21. T. N. D. Tibbits, I. Ballard, K. W. J. Barnham, D. C. Johnson, M. Mazzer, J. S. Roberts, R. Airey, and N. Foan. 2006. In *Proceedings of the 4th World Conference on Photovoltaic Energy Conversion (WCPEC4)*, Hawaii, p. 861.
22. J. G. J. Adams, K. W. J. Barnham, J. P. Connolly, G. Hill, J. S. Roberts, T. N. D. Tibbits, M. Geen, and M. Pate. 2008. In *Proceedings of the 33rd IEEE Photovoltaic Specialists Conference*, San Diego, paper 21.
23. I. M. Ballard, A. Ioannides, K. W. J. Barnham, J. Connolly, D. C. Johnson, M. Mazzer, and R. Ginige. 2007. In *Proceedings of the 22nd European Photovoltaic Solar Energy Conference*, Milan, p. 761.
24. B. Browne, A. Ioannides, J. Connolly, K. Barnham, J. Roberts, R. Airey, G. Hill, G. Smekens, and J. Van Begin. 2008. In *Proceedings of the 33rd IEEE Photovoltaic Specialists Conference*, San Diego, paper 144.
25. A. Bessière, J. P. Connolly, K. W. J. Barnham, I. M. Ballard, D. C. Johnson, M. Mazzer, G. Hill, and J. S. Roberts. 2005. In *Proceedings of the 31st IEEE Photovoltaic Specialists Conference*, Orlando, FL, p. 699.
26. M. F. Führer, J. P. Connolly, M. Mazzer, I. M. Ballard, D. C. Johnson, K. W. J. Barnham, A. Bessière, J. S. Roberts, R. Airey, C. Calder, G. Hill, T. N. D. Tibbits, M. Pate, and M. Geen. 2008. In *Proceedings of the 33rd IEEE Photovoltaic Specialists Conference*, San Diego, paper 253.
27. M. Mazzer, K. W. J. Barnham, N. J. Ekins-Daukes, D. B. Bushnell, J. C. Connolly, G. Torsello, D. Diso, S. Tundo, M. Lomascolo, and A. Licciulli. 2003. In *Proceedings of the 3rd World Conference on Photovoltaic Energy Conversion*, Osaka, Japan, p. 2661.

6

Nanowire- and Nanotube-Based Solar Cells

Loucas Tsakalakos

6.1 INTRODUCTION

Wire-like nano- and microstructures are increasingly of great interest for photovoltaics. This geometry provides several distinct advantages for PV that may in the future lead to low-cost, high-efficiency modules. For example, it has been shown that nanowire/tube arrays possess excellent antireflective and light-trapping/absorption properties compared to planar thin-film or bulk materials (Tsakalakos et al. 2007a; Xi et al. 2007). Nanowires/tubes also provide a direct path for transport of minority and majority charge carriers, hence providing the potential to minimize losses associated with surface/interface recombination (Haraguchi et al. 1992; Chung et al. 2000). Doping of nanowires can be controlled during their synthesis (Cui et al. 2000), and both homojunctions and heterojunctions have been shown (Haraguchi et al. 1992; Peng et al. 2004; Dick et al. 2007; Paladugu et al. 2008). Regarding nanowire synthesis, it is indeed possible to grow nanowire arrays over large areas using well-established, scalable processes such as chemical vapor deposition (including plasma enhanced and metal-organic based), wet etching, and solution-based chemical synthesis. The above features make (quasi-) one-dimensional nanostructures such as nanorods, wires, and tubes particularly attractive for photovoltaics. The elongated geometry also allows the nanorod/tube structures to be dispersed in an interpenetrating network such that they may be used as transparent conducting layers. These applications of nanorods (NRs), nanowires (NWs), and nanotubes (NTs) will be explored in more detail below.

6.2 BACKGROUND

6.2.1 Nanowire and Nanotube Synthesis

One-dimensional nanostructures have been synthesized in numerous compositions, including metals (Valizedeh et al. 2006; Navas et al. 2008), elemental semiconductors (Cui et al. 2000; Heath and LeGoues 1993; Morales and Lieber 1998), compound semiconductors (Persson et al. 2004), ceramics such as oxides (Park and Yong 2004; Fang et al. 2004), carbides (Tsakalakos et al. 2005, 2007b), nitrides (Duan and Lieber 2000; Zhao et al. 2005), and polymers (Kemp et al. 2007). Due to this tremendous wealth of compositional space, a wide variety of applications have been envisioned in the fields of electronics, optoelectronics, photonics, chemistry, mechanics, and biotechnology. Furthermore, multiple techniques have been employed in the synthesis of one-dimensional nanostructures, including wet (Peng et al. 2004) and dry etching (Melosh et al. 2003; Chang et al. 2007); vapor (Kong et al. 1998; Huang et al. 2001b), liquid (Greene et al. 2003), or critical point phase growth (Hanrath and Korgel 2003); electrochemical (Toimil Morales et al. 2001b); and by lithographic means (Z. Li et al. 2005). Each method provides advantages and disadvantages with regard to control of the nanostructure geometrical features (mean size and distribution), orientation control with respect to the substrate, stoichiometry, ability to form homo- or heterojunctions, and ability to deposit on large area substrates.

Perhaps the most widely used synthetic method for 1D nanostructures is chemical vapor deposition (CVD). The origins of this growth process can be traced to the work of Wagner and Ellis (1964) at Bells Labs in the early 1960s, who first showed that elongated crystals of silicon with lateral dimensions ranging from submicron to centimeter dimensions could be grown using the vapor-liquid-solid (VLS) mechanism of single-crystal growth. In this prototypical reaction, Au is used as a catalyst for the decomposition of silane or dichlorosilane precursor gas (often provided with a simultaneous hydrogen, helium, or argon flow). Upon decomposition of the Si precursor as Si, atoms dissolve in the liquid Au catalyst, forming a binary liquid phase that is expected to be present above the eutectic point of the Au-Si equilibrium phase diagram (Figure 6.1). Upon further addition of Si into the liquid alloy particle supersaturation occurs, at which point a Si single-crystal nucleates either homogeneously (if unsupported) or heterogeneously from the surface of a substrate supporting the particle

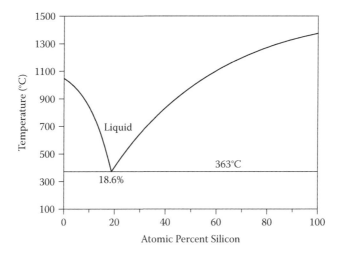

FIGURE 6.1
The Au-Si eutectic alloy phase diagram. (With permission from Au-Si Phase Diagram. ASM Alloy Phase Diagram Center. http://www.asminternational.org/ASMEnterprise/APD)

(Figure 6.2). The lateral size of the nucleated crystal is proportional to that of the original metal particle (Cui et al. 2001). Growth of the crystal continues by further supply of precursor gas to the particle and ends when the supply is stopped or when the catalyst particle falls off.

The crystal growth rate, *V*, for the VLS mechanism is impacted by several parameters, such as temperature (*T*), precursor flow rate, and diameter (*d*). This latter parameter can be understood by the Gibbs–Duheim effect and is qualitatively derived as (Y. Wu et al. 2002)

$$\sqrt[n]{V} = \frac{\Delta\mu_o}{kT}\sqrt[n]{b} - \frac{4\Omega\alpha_{VS}}{kT}\sqrt[n]{b}\frac{1}{d}$$

in which $\Delta\mu_o$ is the effective difference between the chemical potentials of growth in the nutrient phase and the nanowire at a plane interface. It has been shown that the growth rate decreases as the diameter nanowire is reduced. Recent work has also shown that the surface energies of the nanowires are critical in determining their surface morphology (Ross et al. 2005), and diffusion of atoms on the surface and in between nanowires also plays a critical role in the ensemble growth of nanowire arrays (Hannon et al. 2006). The minimal diameter that can be grown by VLS has been addressed by thermodynamic arguments and shown that there is no inherent minimum diameter for nanowire growth by the VLS

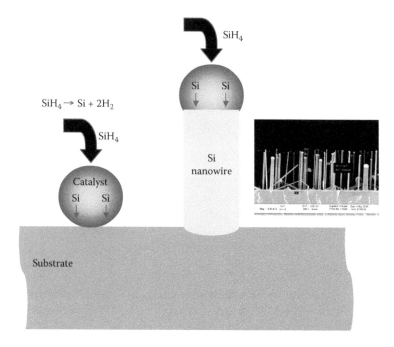

FIGURE 6.2

Schematic of prototypical nanowire growth (silicon) on a substrate and a corresponding scanning electron microscope (SEM) image of epitaxially grown Si NWs by VLS on a <111> Si substrate.

mechanism, yet kinetics may ultimately limit the growth of very fine wires (Tan et al. 2003). Silicon nanowires as small as ~3 nm have been demonstrated experimentally by VLS growth (Wu et al. 2004). Achieving such small diameters requires careful control of the catalyst particle, yet also of the relative longitudinal and lateral growth rates. It is generally known that higher temperatures lead to an enhanced sidewall or lateral deposition rate that can lead to nanowires with a tapered morphology (Wang et al. 2006), though other mechanisms, such as catalyst etching, can also contribute to tapering of nanowires (Bae et al. 2008).

A critical question for application of VLS-grown nanowires and tubes to photovoltaic applications is whether the metal catalyst is incorporated into the nanowire crystal. It is well known that even small levels of certain metal impurities can lead to deep-level trap states in a semiconductor crystal that reduces the minority lifetime (Sze 1969). Based on equilibrium arguments, i.e., consideration of the segregation coefficients, it has been postulated that one can expect a residual concentration of Au in Si nanowires of the order of 10^{15} cm^{-3} (Kayes et al. 2006), even though the equilibrium phase diagram suggests complete immiscibility of solid Au and Si at room temperature (Massalski 1990). Such concentrations are enough to significantly reduce minority carrier lifetimes in the bulk (Bemski 1958; Schmid and Reiner 1982). Experimental data measuring such concentrations are quite limited owing to the difficulty in measuring such low concentrations in small-volume nanostructures. Some groups have reported the presence of catalyst metal in semiconducting nanowires by secondary ion mass spectrometry (SIMS) or laser-enhanced atom probe (LEAP) methods (Putnam et al. 2008; Allen et al. 2008), whereas others have not measured the presence of catalyst metal particles in nanowires but have measured other impurities, such as carbon that is postulated to be related to dopant precursor gases (e.g., from trimethylboron) (Prosa et al. 2008). Since the growth rates for VLS are significantly higher than the thin-film growth rates, often by two to three orders of magnitude, and growth is often system dependent, it is difficult at present to arrive at a universal conclusion to this issue.

There are several related growth mechanisms that rely on a catalyst-mediated nanoparticle seed. The vapor-solid-solid (VSS) mechanism has been shown to occur in the synthesis of Ti-catalyzed Si nanowires (Kamins et al. 2001), as well as certain III-V nanowire systems (Persson et al. 2004). In this case, the nanoparticle remains solid (as dictated by the phase diagram) and diffusion to the growth front is dominated by surface diffusion on the nanoparticle. Another mechanism that has been postulated for certain nanowire material-synthetic method combinations is the oxide-mediated growth mechanism (Zhang et al. 2003).

Carbon nanotubes (CNTs) are also typically grown by a catalytic CVD process. Catalysts used for CNT growth include Fe, Ni, and Co. However, rather than leading to a well-defined crystal structure as in the case of inorganic nanowires, the nanotube chirality cannot be precisely controlled. Indeed, the exact mechanism of CNT growth by catalytic CVD is not known, though it has been addressed theoretically (Reich et al. 2005; Bolton et al. 2006) and experimentally via *in situ* electron microscopy measurements (Hofmann et al. 2007). Preferential control of semiconducting CNTs has been shown by modifying the precursor energy with plasmas (Li et al. 2004).

Plasma-enhanced chemical vapor deposition (PECVD) is a related CVD method that has been used to synthesize nanowires and nanotubes. The nanostructure morphologies obtained by PECVD are generally different from those obtained by thermal CVD since the temperature of growth is often lower and the presence of the plasma can influence the growth morphology. Silicon nanowires synthesized by PECVD are generally less aligned and have a wider diameter distribution (Hofmann et al. 2003). However, it is possible to grow Si NWs using catalysts such as indium that lead to shallow, impurity states, since the plasma is able to minimize oxidation of the catalyst by creating atomic hydrogen (Sunkara et al. 2001; Iacopi et al. 2007). CNTs grown by PECVD tend to be better aligned owing to the presence of the electric field, yet they tend to form bundled fiber morphologies (Chhowalla et al. 2001; Griffiths et al. 2007).

Metal-organic chemical vapor deposition is another method widely used to grow III-V thin films and nanowires. Both catalyst-mediated (Persson 2004) and catalyst-free (Motohisa et al. 2004) growth of III-V nanowires has been demonstrated. This method is particularly suited for growth of nanowires containing heterostructures in either the longitudinal direction (Paladugu et al. 2008) or radial direction (Mohan et al. 2006). Nanowires are particularly useful for heterostructure growth since buildup of strain energy associated with lattice mismatch and subsequent thermal expansion mismatch is better relieved in nanowires than in thin films that are by definition biaxially constrained; i.e., the critical thickness for introduction of strain energy-relieving defects such as dislocations is essentially increased (Alizadeh et al. 2004; Ertekin et al. 2005), hence allowing for the growth of nearly defect-free (nonequilibrium planar and linear defects) crystals (Tambe et al. 2008). It is for this reason that is also possible to grow III-V nanowires on substrates that are not possible in thin-film form, e.g., GaAs on Si (Bao et al. 2008). Jagadish and coworkers have shown that the density of defects in III-V nanowires is reduced to nearly zero by increasing the precursor flow rate (and hence NW growth rate), which is opposite of what is encountered in conventional bulk and thin-film epitaxial crystal growth (Joyce et al. 2009).

Pulsed laser ablation (PLA), a technique that is well known for the deposition of thin-film materials (Chrisey and Hubler 2003; Ashfold et al. 2004), was among the early methods used to grow one-dimensional nanostructures. PLA is particularly noted as a means of growing multicomponent materials such as perovskite oxide ceramics, since the laser enables rapid ablation of all constituents in the target material and subsequent deposition on the growth substrate (Singh and Kumar et al. 1998). Smalley and coworkers used PLA to grow single-wall carbon nanotubes (SWNTs) and multiwall carbon nanotubes (MWNTs) that led to pioneering and Nobel Prize–winning work in understanding the fundamental properties of CNTs (Guo et al. 1995). Subsequent *in situ* spectroscopy analysis of CNT synthesis by PLA showed that indeed the CNTs form by nucleation of metal catalyst in the gas phase followed by subsequent nucleation and growth of CNTs that are then transported to colder regions of the reactor and deposited in a random fashion on a suitable substrate or on the reactor walls (Puretzky et al. 2000). Lieber and coworkers subsequently applied PLA to the synthesis of semiconducting nanowire materials (Morales and Lieber 1998). As in the case of CNTs, metal catalyst powder is mixed with a precursor powder to form a compact pellet that is then used as the PLA source. CNTs and NWs nucleate in the gas phase. It is apparent that PLA is not well suited, therefore, for controlled growth of nanostructures on substrates, though combining PLA with CVD does allow for controlled synthesis of multilayer nanowires such as Si-SiGe heterostructures, as shown in Figure 6.3 (Y. Wu et al. 2002).

The concept of catalytic growth has also been applied to the use of critical point synthesis of nanowires. In this method, precursor materials (e.g., organosilane or organogermane) that also contain catalyst metal are placed in a reactor that is heated under high pressure to drive the reaction products to the critical point in the phase diagram (Holmes et al. 2000). This allows for fabrication of a high volume of nanowires in a random morphology (Figure 6.4), though it is difficult to create one-dimensional nanostructures on substrates using this approach. III-V nanowires have also been obtained using this so-called supercritical fluid-liquid-solid (SFLS) process (Schricker et al. 2006).

Wet etching has recently been applied to the fabrication of nanowire materials. One such method is metal-assisted galvanic etching of silicon (Peng et al. 2004), leading to large-scale arrays of vertically aligned silicon NWs (Figure 6.5). The mechanism for formation of the Si nanowires has been definitively explained as such (Benoit et al. 2008): (1) a silicon wafer or thin film is placed in a bath containing HF and $AgNO_3$; (2) Ag nanocrystals precipitate onto the Si surface as nanoparticles; (3) the nanocrystals locally catalyze oxidation of the Si surface, the oxide of which is subsequently etched by the HF; (4) Ag dendrites continue to deposit in the nanopores created by the local catalysis; and (5) a simple Ag etch removes metal from the nanopores, leaving an array of nanowires. There is a clear crystallographic anisotropy to the etching process, with <100> Si leading to vertically aligned nanowires and etching of <111> Si yielding to slanted nanowire arrays (Figure 6.6). This is also manifested in the etching of polycrystalline Si films (Figure 6.7). It

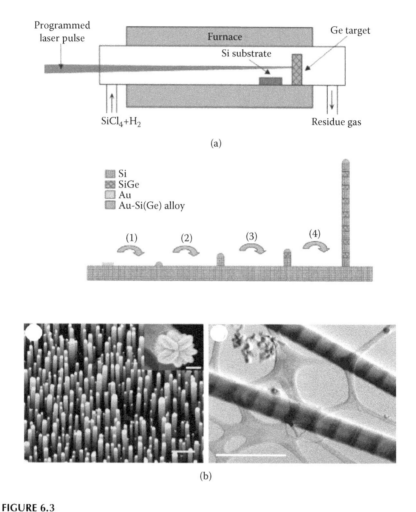

FIGURE 6.3

(a) Schematic of a combined CVD and PLA system for Si-SiGe heterostructure nanowire synthesis and (b) representative SEM and TEM images of such nanostructures. (With permission from Y. Wu et al. [2002]. Copyright of the American Chemical Society.)

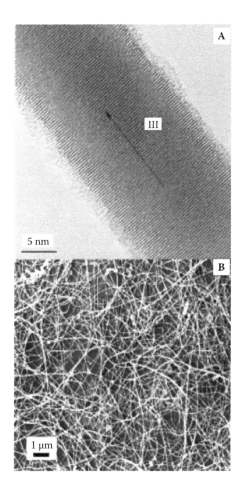

FIGURE 6.4

TEM and SEM images of GaAs nanowires synthesized by supercritical fluid solid synthesis showing random deposition of nanowires. (With permission from Schricker, A. D., Davidson, F. M., Wiacek, R. J., and Korgel, B. A., "Space Charge Limited Currents and Trap Concentrations in GaAs Nanowires," *Nanotechnology* 17 [2006]: 2681–88. Copyright of the American Institute of Physics.)

3 µm

1 µm

FIGURE 6.5
Si NWs fabricated by metal-assisted wet etching on <100> Si substrates.

should be noted that to date this method has been demonstrated only for silicon.

Electrochemically assisted wet etching is another method of forming nanowire/pillar arrays on conducting materials. This is particularly useful for microcrystalline silicon thin films, for which the reflectance may be reduced using such an approach (Rappich et al. 2000).

Electrochemical deposition (ECD) is another widely used nanowire fabrication technique (Whitney et al. 1993; Yin et al. 2001). This is almost exclusively coupled with a nanoscale dielectric template into which nanowires are electrochemically deposited. Typical templates include track-etched polycarbonate (Schonenberger et al. 1997) or anodized aluminum oxide (AAO) (Masuda and Fukuda et al. 1995; Routkevitch et al. 1996) as well as nanoscale templates formed by block copolymer nanolithography (Thurn-Albrecht et al. 2000). ECD has been applied to metallic systems, including Au (Liu et al. 2006) and Ni (Whitney et al. 1993), and magnetic intermetallic Co-Pt or Fe-Pt nanowires (Huang et al. 2002).

FIGURE 6.6
Si NWs fabricated by metal-assisted wet etching on <111> Si substrates showing titled orientation of the nanostructures.

Semiconducting systems such as CdS (Xu et al. 2000) and CdTe (CdTe NWs) have been deposited in nanotemplates. Nanotemplates can be fabricated on substrates (Rabin et al. 2003), or may be free standing (Whitney et al. 1993). Regardless, of particular importance is the electrical contact to the underlying metal layer such that uniform deposition in the nanopores can occur. In general, nanowires grown by ECD contain a polycrystalline microstructure, though in specific materials systems it is possible to post-anneal the nanowires and form single-crystalline (or primarily single-crystalline) nanowires (Toimil Morales et al. 2001).

Solution-based chemical synthesis without the use of electrochemistry has also been used to fabricate small aspect ratio nanorods (defined here as having an aspect ratio of approximately less than 10) as well as nanowires and nanotubes. II-VI semiconductors such as CdSe and CdTe are synthesized by surfactant-mediated growth that promotes growth along a specific crystallographic axis, while nanorod sidewall crystal faces preferentially adsorb the surfactant molecules (Peng et al. 2000). These nanorods nucleate in solution,

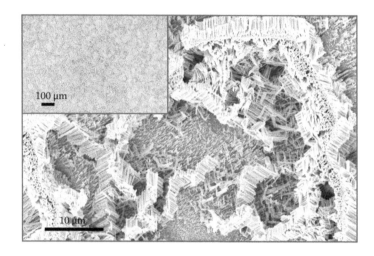

FIGURE 6.7
Si NWs fabricated by metal-assisted wet etching from a polycrystalline Si film on alumina showing multiple orientations of the nanostructures due to the various crystal planes available to the HF:AgNO$_3$ etch solution.

leading to a colloidal suspension of nanorods. Specific materials systems, e.g., ZnO, can, however, be seeded on a template layer to yield well-aligned nanowire arrays (Greene et al. 2003). Perovskite oxide nanowires/rods such as barium and strontium titanate have also been synthesized by solution phase decomposition methods (Urban et al. 2002). It must be noted, however, that elemental semiconductors such as Si and Ge have not been grown using a solution-grown (nonsupercritical) or ECD technique, since aqueous precursors are not possible (due to oxidation) and would thus require exotic organic solvents amenable to high-temperature synthesis.

Dry etching methods are also available for fabrication of nanowires. Vertically aligned arrays of nanowires can be fabricated by application of nanolithography to form a suitable etch mask material (e.g., a metal such as Ni) followed by reactive ion etching (Lewis et al. 1998). Similarly, nanosphere lithography (Figure 6.8) can be used to form nanoscale metal patterns on a substrate by self-assembly of nano/microspheres on the surface followed by evaporation of the metal through the interstices of the nominally ordered nanosphere array (Hulteen and Van Duyne 1995; Cheung et al. 2006). After removal of the nano/microspheres by wet processing, reactive ion etching is used to create the nanowire/pillar arrays. In-plane nanowires can be formed by applying electron beam lithography (Vieu et al. 2000), standard photolithography (Tong et al. 2009), or nanoimprint lithography (Melosh et al. 2003; Jung et al. 2004) followed by dry etching (Figure 6.9).

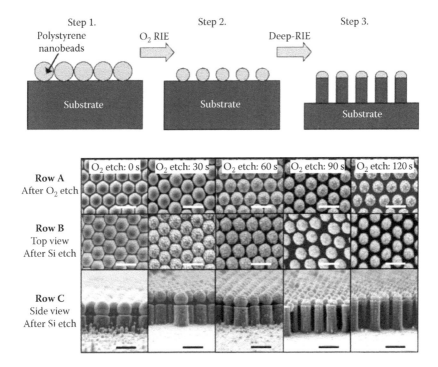

FIGURE 6.8
Use of nanosphere lithography to form etch masks for reactive ion etching of Si nanopillars. (With permission from Cheung, C. L., Nikoli, R. J., Reinhardt, C. E., and Wang, T. F., "Fabrication of Nanopillars by Nanosphere Lithography," *Nanotechnology* 17 [2006]: 1339–43. Copyright of the American Institute of Physics.)

Physical vapor deposition (PVD) methods such as sputtering or thermal/electron beam evaporation can be modified to produce aligned to slanted nanorod/wire/pillar arrays. This is achieved by the so-called glancing angle deposition (GLAD) method, in which the vapor source is applied at a high angle coupled with rotation of the substrate (which is kept at a relatively low temperature) such that a self-shadowing of the atom flux occurs. In this manner one can create relatively well-defined nanopillars (typically polycrystalline) or cork-screw-type structures that are either vertically aligned or at an angle to the substrate (Zhao et al. 2003). Carbon nanotubes were also first demonstrated using PVD, in particular arc melting and related methods (Iijima and Ichihashi 1993).

Finally, molecular beam epitaxy (MBE) and related processes (e.g., chemical beam epitaxy [CBE]) have also been used to form nanowire arrays of Si (Schubert et al. 2004), GaAs, and other III-V materials (Z. H. Wu et al. 2002; Jensen et al. 2004; Bertnessa et al. 2006; Debnath et al. 2007). MBE

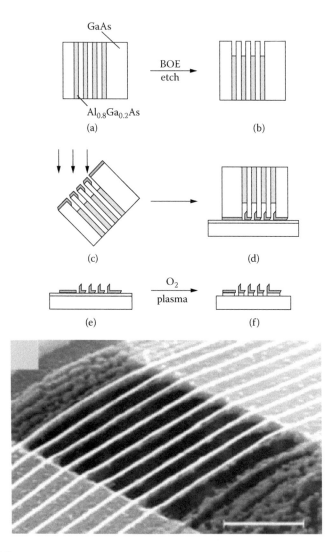

FIGURE 6.9

Schematic of the SNAP process showing (a) growth of multi-quantum wells on a GaAs substrate, (b) selective etching of the Al0.8Ga0.2As at the edge of the wafer, (c) oblique angle deposition of a metal film on the etched edges, (d) and (e) transfer of the metal film to the device substrate of interest containing a thin film of the nanowire materials to be formed, (f) formation of the nanowire features in the thin film by dry etching. A representative in-plane nanowire array formed by SNAP is also shown. (With permission from Melosh, N. A., Boukai, A., Diana, F., et al., "Ultrahigh-Density Nanowire Lattices and Circuits," *Science* 300 [2003]: 112–15. Copyright of American Association for the Advancement of Science [AAAS].)

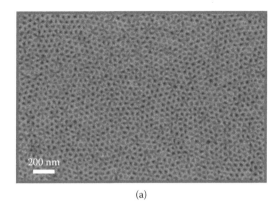

(a)

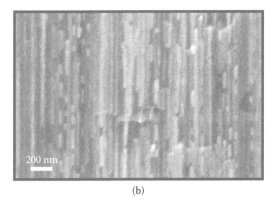

(b)

FIGURE 6.10
(a) Plan view and (b) cross-sectional view of ~18 nm diameter InSb NWs fabrication by electrodeposition in an anodized aluminum oxide nanotemplate.

has been extensively applied to the growth of epitaxial thin films by creating an atomic flux in a Knudsen cell, often with multiple cells to form multicomponent stoichiometries, which allows for precise control of the growth rate down to the submonolayer level (Panish and Temkin 1989). While MBE is not particularly suited to mass production, it allows for the fabrication of novel crystal structures and heterostructures. By use of the catalytic growth processes described above, one can form nanowires in a manner similar to that for CVD or metal-organic chemical vapor deposition (MOCVD).

6.2.2 Nanowire and Nanotube Applications

One-dimensional nanostructures are currently being explored for a vast array of potential applications beyond photovoltaics. In the field of electronics, NWs and NTs are being considered as replacements for conventional transistors owing to the fact that their mobility and other scaled performance parameters, such as transconductance and subthreshold slope, perform on par or, in many cases, better than those of Complementary Metal Oxide Semiconductor (CMOS) fabricated devices (Marchi et al. 2006). A key technical challenge with this concept is related to the integration of such nanostructures over large area wafers with precise control of position, alignment, and ultimately electronic properties. In the area of vacuum microelectronics NWs and NTs are being explored as field emission devices in logic (Kang et al. 2006) or display applications (Teo et al. 2004) since their high aspect geometry is well suited to efficient extraction of electrons from exposed tips by the Fowler-Nordheim field emission mechanism (Bonard et al. 1998; Frederick et al. 1999). The superconducting properties of nanowires have also been explored (Michotte et al. 2003).

Optoelectronic applications (beyond PV) are another topic of intense research for NTs and NWs. Nanowires have been demonstrated as lasers and light-emitting diodes (LEDs) with potential for enhanced efficiency and light output. Lasers have been fashioned by optical pumping to yield UV emission from GaN or ZnO NWs (Huang et al. 2001b), as well as by electrically excited lasing from CdS nanowire devices in the optical regime (Duan et al. 2003). LEDs have been demonstrated in nanowires (Huang et al. 2005), and electroluminescence in the IR from CNT devices has been observed (Misewich et al. 2003).

In the area of energy technologies, nanowires and tubes are being explored for batteries and thermoelectrics. Silicon nanowires have been shown to produce a higher-energy density in Li-ion batteries while increasing the electrode lifetime since the structural changes that occur by Li intercalation into the Si lattice are better tolerated in nanowires than in bulk electrodes (Chan et al. 2008). High discharge current densities were also shown for composite gold-cobalt oxide nanowire electrodes fabricated by a virus-templated method (Nam et al. 2006). Nanowires of Bi_2Te_3 (Sander et al. 2002), Si/SiGe (D. Li et al. 2005), and other compositions have also been explored as thermoelectrics, as it has been postulated that thermal conductivity (k) can be reduced in quantum confined 1D nanowires by sidewall scattering of phonons that also allows effective electronic charge transport for a high electrical conductivity (σ). This is of central importance to the problem of thermoelectrics since the relevant figure of merit is the ZT value that is proportional to the ratio of σ/k. To date nanowires have not been effectively utilized to produce a full thermoelectric device, though promising results toward such a device have been reported (Wang et al. 2005).

6.3 STATE OF THE ART

6.3.1 Nanowire Solar Cells

The device physics of nanowire photovoltaic devices employing p-n junctions has been analyzed in detail by Kayes et al. (2005) for the case of a single nanowire with top illumination; i.e., no light trapping or other sub-wavelength optical effect was assumed. This author has also analyzed the expected efficiency for silicon nanowire devices and compared two fundamental geometries accessible in a nanowire geometry: conformal p-n junctions and longitudinal p-n junction devices (Tsakalakos et al. 2008). A minority carrier lifetime of 1 μs, a Shockley–Read–Hall trap density of 10^{18} cm^{-3}, and a surface recombination velocity of 10^5 cm/s were assumed in the calculations, which were performed in a standard device physics package. Lengths ranging from 10^0 to 10^3 microns were explored, as were radii ranging from 10^{-1} to 10^1 microns. Longitudinal nanowire devices showed a maximum efficiency of ~13% for the design space explored, whereas for conformal nanowire devices the maximum efficiency was 21.5%. The lower efficiency for the longitudinal devices is related to a higher probability for surface recombination. Therefore, it can be concluded that in general, a conformal device configuration is preferable for nanowire devices, and by assuming realistic recombination parameters and defect densities within the nanowires, a more realistic expected efficiency is in the range of 15 to 18% for silicon nanowire solar cells.

These results agree with the broad conclusions of the study by Kayes et al. (2005), who discuss further qualitative insights into the device operation. It can be expected that the J_{sc} is independent of nanowire diameter as long as the diameter is less than minority carrier diffusion long, after which J_{sc} decreases. J_{sc} increases with nanowire length, reaching a constant value once the length is greater than the optical thickness of the nanowire material. V_{oc} has an opposite trend in that it decreases with nanowire/rod length yet increases with nanowire diameter. It is also very sensitive to the density of mid-gap traps (Shockley–Read–Hall type) in the depletion region, degrading rapidly above a specific material-dependent trap concentration.

Key design parameters are that the radius of the nanowire be approximately equal to the minority carrier diffusion length for optimal performance, the doping should be relatively high on both sides of the junction to minimize the depletion width, and the trap density within the depletion region should be as low as possible. These general design rules suggest that the semiconductor recombination parameters are known in nanowires; only recently have measurements on the minority carrier diffusion length and lifetime been performed on specific materials. These data have shown that in large-diameter nanowires the diffusion length is relatively long (Kelzenberg et al. 2008), i.e., on the order of a few microns, yet in smaller-diameter wires (80 to 200 nm) the diffusion length is short due to surface recombination (Allen et al. 2008). This suggests that relatively large-diameter wires are required, yet this leads to a trade-off between optimal lateral charge transport properties and achieving good antireflective properties in nanowire arrays, another key feature of nanowire solar cells (see below). Finally, the length of the nanowire should approximately be equal to the optical thickness of the materials in question, though it is noted this is the case for a single nanowire device and not the case of a nanowire array where strong light trapping occurs (Tsakalakos 2007a). These design rules are also of general relevance to organic-based devices that utilize nanowires, with the benefit being the same, namely, that it is possible with nanowire devices to effectively decouple light absorption from charge transport by using a conformal geometry that allows for lateral extraction of charge. A review of the major nanowire- and nanotube-based solar cell designs is provided below.

6.3.1.1 Hybrid Nanowire/Rod Devices

Among the first solar cells to be demonstrated with a 1D nanostructure were hybrid organic-inorganic devices by Alivisatos and coworkers (Huynh et al. 2002). CdSe nanocrystals were embedded in a regioregular P3HT organic semiconductor matrix and showed a maximum power conversion efficiency of 1.7% (Figure 6.11). It was shown that the external quantum efficiency of the devices increased dramatically as the aspect ratio was increased from 1 to 8.5, and the authors argued this was related to enhanced charge transport through the random nanorod network. Of particular significance is that these devices have the potential to offer flexible photovoltaic modules.

Integration of organic matrices with aligned nanowire arrays is also a strategy of active research. Si nanowire arrays have been integrated with

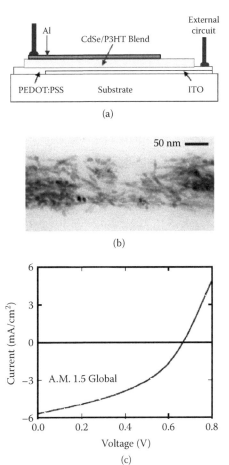

FIGURE 6.11
Hybrid organic-inorganic nanorod solar cells demonstrated by Alivisatos and coworkers showing (a) the structure of these devices, (b) the typical microstructure of such layers as viewed in a TEM, and (c) a typical I-V characteristic under AM1.5 illumination with an efficiency of ~1.7%. (With permission from Huynh, W. U., Dittmer, J. J., and Alivisatos, A. P., "Hybrid Nanorod-Polymer Solar Cells," *Science* 295 [2002]: 2425–27. Copyright of the American Association for the Advancement of Science [AAAS].)

organics such as P3HT. Planar devices were fabricated as a first step in demonstrating such a device concept, with efficiency of 1.6%, V_{oc} of 704 mV, and J_{sc} of 4.22 mA/cm^2 (Alet et al. 2006). Si nanowire-based devices were fabricated by Goncher and Solanki (2008), showing a power conversion efficiency of less than 0.1%. Hybrid PV devices based on InP nanowires and P3HT have also been demonstrated by Novotny et al. (2008), who showed that the forward bias current is increased by six to seven orders of magnitude in the nanowire devices, and found a relatively low ideality factor of 1.31, with the photovoltaic effect yielding a fill factor of 44%. The InP nanowires were grown directly on indium tin oxide (ITO).

6.3.1.2 Inorganic NWs

All inorganic nanowire devices have also been the subject of growing research interest in recent years. The impetus in using all-inorganic structure is the potential concern regarding the long-term stability of organic semiconductors, an area that is currently being studied in detail. Alivisatos and coworkers demonstrate an all-inorganic nanorod solar cell with an efficiency of ~2.9%. CdTe nanorods were spin-cast onto ITO/glass substrates to yield a dense layer of nanorods, followed by spin casting of CdSe nanorods and evaporation of a thin Al top contact. The charge-separating heterojunction is formed at the interface between the two nanorod layers, and it was shown that the device performance improved by sintering of the bilayer nanocrystal films. These solar cells were also found to be stable after exposure to ambient atmosphere and light for up to thirteen thousand hours.

Large area silicon nanowire-based solar cells grown by chemical vapor deposition were demonstrated by Tsakalakos et al. (2007a) on stainless steel substrates. There are several advantages in using Si NWs for photovoltaics. The most important feature of such devices is the fact that one can create conformal p-n junctions on the nanowires in order to allow for efficient lateral extraction of charge carriers, as discussed above. Furthermore, it has been shown that NW arrays possess excellent antireflective and light-trapping optical properties (Tsakalakos et al. 2007b), hence allowing for enhanced absorption of incoming solar photons in a broadband fashion. This is clearly evident in Figure 6.12, in which the effective optical absorption is increased across the full spectrum compared to a planar thin film. This can be explained as being due to subwavelength scattering, followed by strong light trapping within the nanowire ensemble. In general, aligned nanowire arrays lead to a higher absorption than randomly oriented

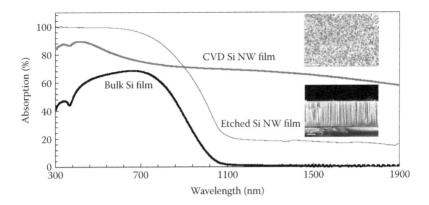

FIGURE 6.12
Optical absorption spectra of a solid Si film, a wet-etched Si NW film of same thickness, and a CVD-grown (randomly oriented) Si NW film on glass showing enhanced absorption of the Si NW films above and below the Si bandgap (~1,100 nm).

nanowire films since the latter tend to scatter/diffract/refract a higher portion of the light out of the top of the structure, which is evident in a higher total reflection (Tsakalakos et al. 2007b; Street et al. 2008; Muskens et al. 2009). Silicon nanowires are also of interest for PV since Si is the second most abundant material in the earth's crust (Skinner 1979), making it an appealing choice for low-cost, environmentally friendly photovoltaic modules. Si NWs can be grown over large areas using CVD, hence making them strong candidates for low-cost PV devices.

The growth of Si NWs on alternative, low-cost substrates is particularly appealing for terrestrial PV applications. Goncher et al. (2006) demonstrated Si NW growth on transparent indium tin oxide/glass substrates. Growth on metal substrates, which also enables flexible modules, is particularly challenging, as multiple materials constraints are placed on the grower. This includes (1) formation of nanoscale catalyst islands, (2) assurance that the nanocatalyst will not react with underlying substrate material, and (3) limiting reactions between the nanowire material and the substrate. It is also desirable to form an ohmic contact between the nanowires and the metal substrate. It was found that attempts to grow Si NWs on stainless steel substrates did not produce nanowires due to the formation of iron silicide. Silicon oxide was found to be an effective diffusion barrier but does not allow electrical back contact for the nanowire, though after an extensive materials evaluation Ta_2N thin films were found to act as an effective diffusion barrier that enabled the growth of high-density Si NW arrays (Figure 6.13).

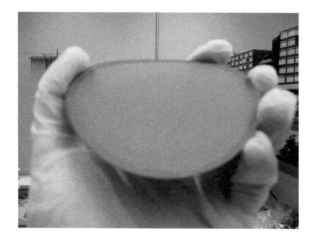

FIGURE 6.13

Picture of large area Si NW film grown by CVD (VLS mechanism) on a stainless steel foil. (See color insert following page 206.)

Using this basic materials technology, it is possible to fabricate large area solar cells in the following manner (details can be found in Tsakalakos 2007a):

1. p-Type single-crystalline silicon (c-Si) NWs are grown by CVD on large area substrates (stainless steel or Si).
2. The nanowires are partially oxidized at >700°C.
3. The top portion of the nanowire is etched in buffered HF after application and etch back of a photoresist layer in order to leave a dielectric film at the base of the nanowire to minimize shunts.
4. A conformal n-type hydrogenated amorphous Si (a-S:H) thin film is deposited on the nanowire array in order to form a p-n junction.
5. The NW array is coated with a thin, conformal layer of ITO to electrically tie together the nanowires on the top side of the device.
6. Metal fingers are shadow evaporated on the top of the nanowire array.

A schematic of the device structure is shown in Figure 6.14, along with the most promising device data. A power conversion efficiency (under an AM1.5-simulated light spectrum) on the order of 0.1% was obtained with a promising J_{sc} of 2 mA/cm^2 and a broad external quantum efficiency curve. A particular concern was whether the presence of Ta at the back contacts led to incorporation of Ta in the NWs or a reaction layer that could limit the device performance. Detailed transmission electron microscopy (TEM)

(a)

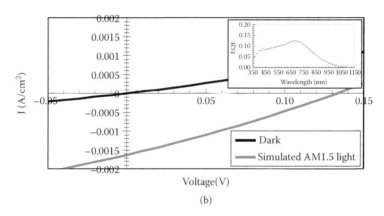

(b)

FIGURE 6.14

(a) Cross-sectional SEM and inset of CVD-grown amorphous-crystalline Si NW solar cell. (b) Representative J-V data with an inset of the device EQE.

analysis did not yield any evidence of such a reaction layer (Figure 6.15), and it is believed that at the process temperatures employed, Ta diffusion into the NWs is minimal.

Such hybrid a-Si/c-Si nanowire devices were also fabricated on bulk Si wafers into which large area nanowire arrays were fashioned using metal-assisted galvanic wet etching (Tsakalakos et al. 2008). The best-performing devices obtained power conversion efficiencies of ~1.3%, whereas control samples had an efficiency of ~13%, and based on external quantum efficiency data it was concluded that front surface recombination limited device performance. The nanowire array density is very high for wet-etched nanowires, and as a result it was shown by microscopy studies

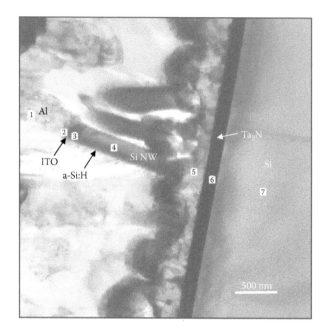

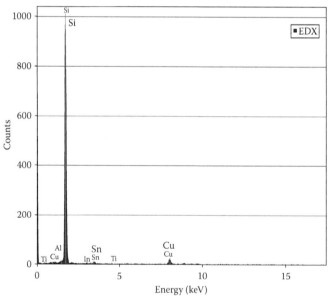

FIGURE 6.15

Cross-sectional TEM image of a CVD-grown amorphous-crystalline Si NW solar cell on a Ta_2N-coated Si substrate, and energy-dispersive spectroscopy (EDS) data from location 5 (base of NW) just above the Ta_2N film showing no evidence of a reaction layer or Ta contamination.

(Figure 6.16) that the a-Si:H film did not coat the full length of the nano-wires in the array. Due to the harsh chemical treatment of the nanowire surface, this lack of suitable a-Si passivation led to a significant residual density of nonradiatively recombining surface states that degraded device performance. This is the case even though for both the CVD grown and wet-etched devices the optical reflectance was reduced compared to control samples by one to two orders of magnitude (Figure 6.17).

Silicon nanowire solar cells have also been demonstrated by several other groups. Peng et al. (2005) fabricated Si nanowire solar cells by galvanic wet

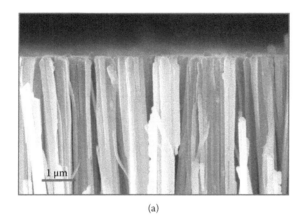

(a)

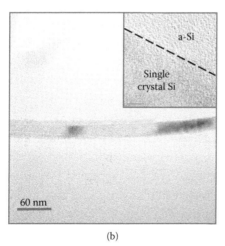

(b)

FIGURE 6.16

(a) Cross-sectional SEM image of a-Si:H-coated Si NW array fabricated by metal-assisted wet etching. (b) TEM image of such a nanowire showing partial coverage by the a-Si:H film. The inset shows a high-resolution TEM micrograph of the interface between the amorphous Si film and the crystalline Si nanowire.

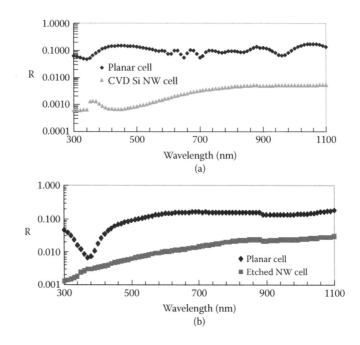

FIGURE 6.17

Specular reflectance spectra for (a) a planar Si solar cell and a CVD-grown Si NW solar cell. (b) A planar Si solar cell and a wet-etched fabricated Si NW solar cell showing significantly reduced reflectance for both types of nanowire PV devices.

etching and formed the p-n junction using a conventional $POCl_3$ diffusion process at 930°C for 30 minutes. The devices achieved a power conversion efficiency of 4.73% on polycrystalline Si substrates, and 9.31% on single-crystal Si substrates. The devices showed relatively high short-circuit current densities (~20 mA/cm²), but low fill factors and V_{oc}. It was shown that the minority carrier lifetime of the substrates, as measured by photoconductive decay, was reduced by as much as 10 times after etching of the nanowires and decreased as the nanowire lengths increased. Si NW PV cells fabricated by galvanic wet etching were also fabricated by Garnett and Yang (2008). The p-n junction was formed by depositing a polycrystalline Si layer (p-type on n-type nanowires), which yielded a maximum efficiency of 0.5%. Falk and coworker (Sivakov et al. 2009) demonstrated a wet-etched polycrystalline p+nn+ nanowire device on glass with a current density as high as 40 mA/cm² (metal point contact) and an overall power conversion efficiency of 4.4%. The same group fabricated CVD-grown Si NW solar cells on Si substrates and polycrystalline Si-coated glass substrates, demonstrating a V_{oc} of up to 280 mV and J_{sc} of ~2 mA/cm² (Stelzner et al. 2008). Finally, it is noted that several works have targeted the use of Si NWs in photoelectrochemical cells, most notably those at the California Institute of Technology (Maiolo et al. 2007) and Pennsylvania State University (Goodey et al. 2007).

In addition to silicon-based nanowire PV devices, research in III-V-based nanowire devices is emerging. LaPierre and co-workers (Czaban et al. 2009) demonstrated GaAs nanowire solar cells in which the nanostructures where synthesized by molecular beam epitaxy (MBE). Longitudinal and radial p-n junctions were fashioned. It was found that radial junctions yielded the best device performance, but the shadowing effects (due to the directional nature of MBE) degraded performance. The best devices provided an efficiency of ~0.8% for low Te-doped growth times, and it was argued that longer growth times degraded performance due to dopant outdiffusion and the aforementioned shading effects.

InP nanowire devices were reported by Goto et al. (2009). The nanowires were grown on a single-crystal substrate using a catalyst-free epitaxial growth method in which a nanotemplate assists in preferentially axial growth (Motohisa et al. 2004). The nanowires were p-type with the n-type shell layer formed by growth at a lower temperature of 600°C (Figure 6.18). The solar cells provided a power conversion efficiency of 3.37% under AM1.5G illumination with $J_{sc} = 13.7$ mA/cm^2, $V_{oc} = 0.43$ V, and FF = 0.57.

6.3.1.3 Nanowire Dye-Sensitized Solar Cells

Inorganic nanowires have also been applied to dye-sensitized solar cells (DSCs). The first DSCs were based on TiO$_x$ nanoparticles (O'Regan and Grätzel 1991), and indeed most DSCs studied in academic laboratories or under consideration for commercialization are based on nanoparticles. Nanowire-based DSCs have been demonstrated by both Aydil and coworkers (Baxter and Aydil 2005) and Yang and coworkers (Law et al. 2005). It was argued that the use of nanowires helps in charge transport relative to a nanoparticle electron-transporting layer that has significantly more interfaces/surfaces. The roughness of the nanowires is also expected to assist in providing enhanced absorption (and hence J_{sc}) relative to nanoparticle assemblies. Law et al. (2005) demonstrated a twofold increase in the short-circuit current density of titania nanowire-based DSCs compared to a nanocrystalline titania-based DSCs. The maximum power conversion efficiency under AM1.5 light was found to be ~1.5%. On the other hand, Baxter and Aydil fabricated dendritic ZnO nanowire-based dye-sensitized cells and showed an improvement of approximately

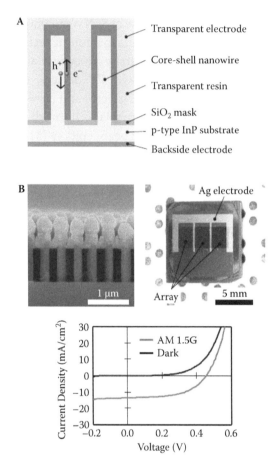

FIGURE 6.18

(a) Structure of an InP nanowire device grown by catalyst-free MOCVD in an ordered silicon oxide template. (b) Representative current-voltage data. (With permission from Goto, H., Nosaki, K., and Tomioka, K., "Growth of Core-Shell InP Nanowires for Photovoltaic Application by Selective-Area Metal Organic Vapor Phase Epitaxy," *Appl. Phys. Express* 2 [2009]: 035004. Copyright of the Japanese Society of Applied Physics.)

100 times in photocurrent and efficiency compared to smooth nanowires, with a power conversion efficiency of 0.5%. Titania nanotubes have also been fabricated and applied to DSCs (Paulose et al. 2006).

6.3.2 Carbon Nanotube–Based Solar Cells

Carbon nanotubes, another prototypical one-dimensional nanostructured materials system, are well known to possess unique optical and electrical properties, which may be of great interest to PV applications. One of the first demonstrations of a CNT p-n diode was by J. U. Lee et al. (2004), and they were found to be ideal junctions from the perspective of the ideality factor being equal to 1 under light illumination (Lee 2005). Due to various technological challenges (discussed below), it is difficult to configure large area arrays of CNTs for PV devices that take advantage of the highly ideal junctions within an undoped, electrostatically controlled CNT junction. Nevertheless, other configurations are possible; for example, Wei et al. (2007) fabricated a hybrid device in which a parallel in-plane array of double-walled CNTs formed a charge-separating junction with n-type Si (Figure 6.19). The exact nature of this junction requires further understanding; however, it was possible to obtain relatively high photocurrents (13.8 mA/cm^2) and an efficiency of 1.31%. The fill factor was found to be particularly low, ca. 19%.

CNTs have also been studied as fillers for improving the performance of organic photovoltaics (OPV). Kymakis and Amaratunga (2002) fabricated poly(3-octylthiophene) (P3OT)–single-wall carbon nanotube composite solar cells on ITO/glass substrates and found the presence of SWNTs to increase the J_{sc} by two to three orders of magnitude and the FF from 0.3 to 0.4. The V_{oc} was in the range of 0.7 to 0.9 V, and though the power conversion efficiency was low, it was concluded that indeed the CNTs increase electron transfer between the polymer and nanotube, leading to the higher observed J_{sc}. Similarly, Landi et al. (2005) fabricated composite P3OT-SWNT solar cells in a bilayer design in which a pristine P3OT layer was first formed on ITO followed by a 1 wt% loaded CNT-P3OT composite layer. These devices showed a V_{oc} of 0.97 V and a J_{sc} of 0.12 mA/cm^2. The presence of CNTs increases the absorption of the polymer layer, which had a positive effect on the J_{sc} and also increased the V_{oc}.

6.3.3 Nanowire and Nanotube Transparent Conductors

One-dimensional nanostructures also hold potential in improving the performance of conventional solar cells by improving charge collection and absorption. CNTs have the potential to replace transparent conductors since one can form a thin CNT network layer with relatively low sheet resistance and high transparency (Du Pasquier et al. 2005). Sheet resistance on the order of 200 $\Omega/\square$ (typical ITO sheet resistance is < 100 $\Omega/\square$) was obtained and the transmission was ~85% at 550 nm. At present these coatings are most relevant to organic photovoltaics, and a critical issue is the nonuniform charge extraction by such layers. Metal nanowires have

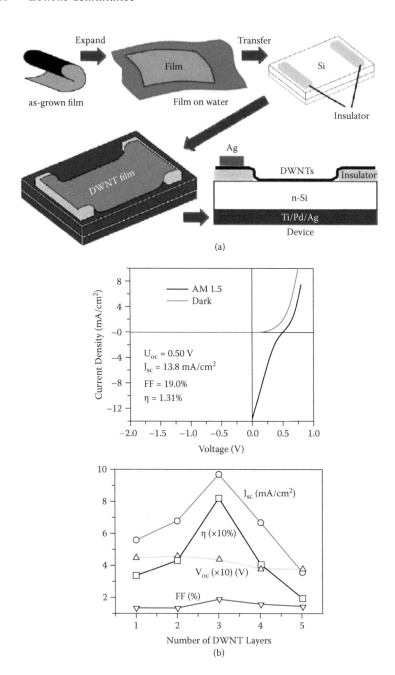

FIGURE 6.19

(a) Structure and fabrication process of a hybrid double-wall carbon nanotube film-silicon PV device grown. (b) Representative electrical data. (With permission from Wei, J., Jia, Y., Shu, Q., et al., "Double-Walled Carbon Nanotube Solar Cells," *Nano Lett.* 7 [2007]: 2317–21. Copyright of the American Chemical Society.)

also been studied as replacements for ITO. Lee et al. (2008) fabricated silver nanowire meshes and applied them to an OPV device. The Ag nanowire mesh layers achieved sheet resistances as low as ~10 $\Omega/\square$ with optical transmission of ~75% and were found to obey the general transmission sheet resistance trade-off behavior as observed with indium tin oxide thin films. Finally, similar to Si nanowires showing enhanced absorption relative to thin films (Tsakalakos et al. 2007a), transparent nanowire/rod arrays have also been shown to act as excellent omnidirectional and broadband antireflective layers with high optical transmission (Xi et al. 2007; Chang et al. 2009; Kennedy and Brett 2003).

6.4 BASIC SCIENTIFIC AND TECHNOLOGICAL CHALLENGES

The use of nanowires in PV device poses numerous scientific and technological challenges for future development of novel photovoltaic modules with high performance. It is critical that the design of nanowire PV devices with respect to the major geometrical parameters of the cell be considered in light of the material properties of the nanowire material. The nanowire length and diameter should be optimized based on the optical absorption of the array as well as the diffusion length of the material. Indeed, techniques for characterizing the relevant material properties of nanowire materials in a repeatable manner are in their infancy; there are few measurements of minority carrier lifetime/diffusion length or surface recombination velocity in semiconducting nanowires in the literature. These measurements suggest that in large-diameter nanowires the diffusion length is quite large (Kelzenberg et al. 2008), i.e., on the order of a few microns, whereas in smaller-diameter wires (ca. 80 to 200 nm) the diffusion length is smaller and is dominated by surface recombination (Allen et al. 2008). While at present such data may be somewhat dependent on the synthesis method and conditions used, these data suggest relatively large-diameter wires are required in the case of Si nanowire PV cells.

It is also evident that nanowire solar cells possess reasonable short-circuit current densities, yet their open-circuit voltages and fill factors are generally low. Among the factors that may be attributed to this observation is the potential for areas of local shunting in the devices. Furthermore, there may be higher series resistance losses in the thin layers used to form the p-n junction, due to contact resistance (nonohmic contacts), or in the TCO used. The inherent quality of the p-n junction in nanowire devices is also unexplored, since other array-level features often mask this. Inadequate passivation and a high density of recombination centers due to unintended impurities or line defects are also features that must be considered as potentially limiting performance of nanowire devices. Of particular significance is that the few single-nanowire PV measurements performed to date (Tian et al. 2007; Kelzenberg et al. 2008) yield higher power conversion efficiency values, and the efficiencies extracted from well-ordered arrays (Goto et al. 2009) are higher than the array values (~12% estimated for a single nanowire compared to ~3.4%). This points to the device-level issues described above as playing a prominent role in limiting the performance of nanowire solar cells.

Use of carbon nanotube–based solar cells also requires further research. A critical challenge is the ability to create CNTs with a well-defined bandgap. At present this is not possible, though efforts to separate CNTs based on their semiconducting or metallic nature have met with some success (Campidelli et al. 2007). Further research is necessary to provide full separation based on bandgap (not merely diameter, since similar diameter CNTs can have a different band structure). Indeed, the bandgap of most CNTs is relatively small (<0.8 eV), and based on the established empirical equations of the inverse proportionality of CNT bandgap to diameter, the diameter of CNTs with a bandgap of 1.1 eV or greater is required to be less than 0.8 nm (Weisman et al. 2003). The abundance of CNTs with such small diameters obtained by conventional methods at present is relatively low; hence, methods to enhance such diameters or select the correct chiralities are needed. Another basic materials challenge for CNTs is control of doping. Work on p-n junctions has focused on electrostatic doping, though a more likely practical device would require chemical doping. Control of chemical doping in CNTs has proven to be challenging; dopants such as potassium that have typically been used are generally unstable (Zhou et al. 2000).

6.5 SUMMARY

One-dimensional nanostructures such as nanowires, nanorods, and nanotubes have recently been shown to be of great interest for future solar cells and modules. Nanowires device form a new class of PV devices that are gaining in interest by the solar energy and broader applied physics/materials communities. Nanowire devices offer several advantages, including enhanced light trapping, potential for low-cost PV cells on flexible substrates, and the ability to separate optical absorption from charge transport in indirect bandgap materials by allowing for lateral extraction of photogenerated charge carriers. Nanowires have been used in all-inorganic solar cells based on silicon and III-V materials, as well as in organic and dye-sensitized solar cells. Similarly, carbon nanotubes have been applied to organic photovoltaics and shown to improve the short-circuit current density by two to three orders of magnitude. Finally, one-dimensional nanostructures are being applied to conventional solar cells as novel transparent conductors or antireflective layers.

6.6 FUTURE OUTLOOK

Nanowire- and nanotube-based solar cells are in the early stages of development. There are numerous fundamental materials and device challenges that must be solved in order to improve the performance of these novel devices. Indeed, these new device structures were only demonstrated in the last few years, yet there is growing interest in the research community, including from the broader photovoltaic community (Tsakalakos et al. 2008). It is noted that thin-film solar cells have been known for over thirty years, but it has only been in the last five or so years that thin-film solar cells, and in particular those based on CdTe, have become a significant part of the PV market. Nanowire/tube solar cells should have a shorter time to market than thin-film cells owing to the concurrent growth of the PV industry, though concerted efforts are required to address the basic materials, devices, and manufacturing challenges of solar cells based on one-dimensional nanostructures.

ACKNOWLEDGMENTS

The author expresses his deepest gratitude to the following for their technical contributions, collaboration, and useful conversations with respect to the data presented: J. Balch, Dr. P. Codella, Dr. R. R. Corderman, J. Fronheiser, Dr. B. A. Korevaar, M. Larsen, M. Pietryczowski, Dr. J. Rand,

Dr. R. Rodrigues, Dr. F. Sharifi, Dr. V. Smentkowski, Dr. O. Sulima, Dr. F. Vitsee, and R. Wortman. The support of this work by M. Beck, Dr. M. L. Blohm, E. Butterfield, Dr. T. Feist, K. Fletcher, M. Idelchik, Dr. C. Korman, Dr. C. Lavan, Dr. J. LeBlanc, J. Likar, Dr. D. Merfeld, B. Norman, Dr. S. Rawal, P. Rosecrans, and G. Trant is greatly appreciated.

REFERENCES

Alet, P.-J., Palacin, S., Roca, P., et al. 2006. Hybrid solar cells based on thin-film silicon and P3HT. *Eur. Phys. J. Appl. Phys.* 36:231–34.

Alizadeh, A., Sharma, P., Ganti, S. P., LeBoeuf, S. F., and Tsakalakos, L. 2004. Templated wide bandgap nanostructures. *J. Appl. Phys.* 95:8199–206.

Allen, J. E., Hemesath, E. R., Perea, D. E., et al. 2008. High-resolution detection of Au catalyst atoms in Si nanowires. *Nature Nanotechnol.* 3:168–73.

Ashfold, M. N. R., Claeyssens, F., Fuge, G. M., and Henley, S. J. 2004. Pulsed laser ablation and deposition of thin films. *Chem. Soc. Rev.* 33:23–31.

Bae, J., Kulkarni, N. N., Zhou, J. P., Ekerdt, J. G., and Shih, C.-K. 2008. VLS growth of Si nanocones using Ga and Al catalysts. *J. Crystal Growth* 310:4407–11.

Bao, X.-Y., Soci, C., Susac, D., et al. 2008. Heteroepitaxial growth of vertical GaAs nanowires on Si (111) substrates by metal-organic chemical vapor deposition. *Nano Lett.* 8:3755–60.

Baxter, J. B., and Aydil, E. S. 2005. Nanowire-based dye-sensitized solar cells. *Appl. Phys. Lett.* 86:053114.

Bemski, G. 1958. Recombination properties of gold in silicon. *Phys. Rev.* 111:1515–18.

Benoit, C., Bastide, S., and Lévy-Clément, C. 2008. Formation of Si nanowire arrays by metal-assisted chemical etching. In *Proceedings of the 23rd European Photovoltaic Solar Energy Conference*, 1DV2.13. Valencia, Spain.

Bertnessa, K. A., Roshkoa, A., Sanforda, N. A., Barkera, J. M., and Davydov, A. V. 2006. Spontaneously grown GaN and AlGaN nanowires. *J. Crystal Growth* 287:522–27.

Bolton, K., Ding, F., and Rosén, A. 2006. Atomistic simulations of catalyzed carbon nanotube growth. *J. Nanosci. Nanotechnol.* 6:1211–24.

Bonard, J.-M., Salvetat, J.-P., Stockli, T., de Heer, W. A., Forro, L., and Chatelain, A. 1998. Field emission from single-wall carbon nanotube films. *Appl. Phys. Lett.* 73:918.

Campidelli, S., Meneghetti, M., and Prato, M. 2007. Separation of metallic and semiconducting single-walled carbon nanotubes via covalent functionalization. *Small* 3:1672.

Chan, C. K., Peng, H., and Liu, G. 2008. High-performance lithium battery anodes using silicon nanowires. *Nature Nanotechnol.* 3:31–35.

Chang, C. H., Yu, P., and Yang, C. S. 2009. Broadband and omnidirectional antireflection from conductive indium-tin-oxide nanocolumns prepared by glancing-angle deposition with nitrogen. *Appl. Phys. Lett.* 94:051114.

Chang, Y.-F., Chou, Q.-R., Lin, J.-Y., and Lee, C.-H. 2007. Fabrication of high-aspect-ratio silicon nanopillar arrays with the conventional reactive ion etching technique. *Appl. Phys. A* 86:193–96.

Cheung, C. L., Nikoli, R. J., Reinhardt, C. E., and Wang, T. F. 2006. Fabrication of nanopillars by nanosphere lithography. *Nanotechnology* 17:1339–43.

Chhowalla, M., Teo, K. B. K., Ducati, C., et al. 2001. Growth process conditions of vertically aligned carbon nanotubes using plasma enhanced chemical vapor deposition. *J. Appl. Phys.* 90:5308.

Chrisey, D. B., and Hubler, G. K., ed. 2003. *Pulsed laser deposition of thin films*, 648. New York: Wiley-VCH.

Chung, S. W., Yu, J.-Y., and Heath, J. R. 2000. Silicon nanowire devices. *Appl. Phys. Lett.* 76:2068.

Cui, Y., Duan, X., Hu, J., and Lieber, C. M. 2000. Doping and electrical transport in silicon nanowires. *J. Phys. Chem. B* 104:5213.

Cui, Y., Lauhon, L. J., Gudiksen, M. S., Wang, J., and Lieber, C. M. 2001. Diameter-controlled synthesis of single-crystal silicon nanowires. *Appl. Phys. Lett.* 78:2214.

Czaban, J. A., Thompson, D. A., and LaPierre, R. R. 2009. GaAs core-shell nanowires for photovoltaic applications. *Nano Lett.* 9:148–54.

Debnath, R. K., Meijers, R., Richter, T., Stoica, T., Calarco, T. R., and Lüth, H. 2007. Mechanism of molecular beam epitaxy growth of GaN nanowires on Si(111). *Appl. Phys. Lett.* 90:123117.

Dick, K. A., Kodambaka, S., and Reuter, M. C., et al. 2007. The morphology of axial and branched nanowire heterostructures. *Nano Lett.* 7:1817–22.

Duan, X., Huang, Y., Agarwal, R., and Lieber, C. M. 2003. Single-nanowire electrically driven lasers. *Nature* 421:241–45.

Duan, X., and Lieber, C. M. 2000. Laser-assisted catalytic growth of single crystal GaN nanowires. *J. Am. Chem. Soc.* 122:188–89.

Du Pasquier, A., Unalan, H. E., Kanwal, A., Miller, S., and Chhowalla, M. 2005. Conducting and transparent single-wall carbon nanotube electrodes for polymer-fullerene solar cells. *Appl. Phys. Lett.* 87:203511.

Ertekin, E., Greaney, P. A., Chrzan, D. C., and Sands, T. D. 2005. Equilibrium limits of coherency in strained nanowire heterostructures. *J. Appl. Phys.* 97:114325.

Fang, X.-S., Ye, C.-H., Xu, X.-X., Xie, T., Wu, Y.-C., and Zhang, L.-D. 2004. Synthesis and photoluminescence of α-Al_2O_3 nanowires. *J. Phys. Condens. Matter* 16:4157–63.

Frederick, C. K., Au, K. W., Wong, Y. H., et al. 1999. Electron field emission from silicon nanowires. *Appl. Phys. Lett.* 75:1700.

Garnett, E. C., and Yang, P. 2008. Silicon nanowire radial p-n junction solar cells. *J. Am. Chem. Soc.* 130:9224–25.

Greene, L. E., Law, M., Goldberger, J., et al. 2003. Low-temperature wafer-scale production of ZnO nanowire arrays. *Angew. Chem. Int. Ed.* 42:3031–34.

Goncher, G., and Solanki, R. 2008. Semiconductor nanowire photovoltaics. *Proc. SPIE* 7047:70470L.

Goncher, G., Solanki, R., Carruthers, J. R., Conley, J., and Ono, Y. 2006. p-n junctions in silicon nanowires. *J. Electr. Mater.* 35:1509–12.

Goodey, A. P., Eichfeld, S. M., Lew, K.-K., Redwing, J. M., and Mallouk, T. E. 2007. Silicon nanowire array photoelectrochemical cells. *J. Am. Chem. Soc.* 129:12344–45.

Goto, H., Nosaki, K., and Tomioka, K. 2009. Growth of core-shell InP nanowires for photovoltaic application by selective-area metal organic vapor phase epitaxy. *Appl. Phys. Express* 2:035004.

Griffiths, H., Xu, C., Barrass, T., et al. 2007. Plasma assisted growth of nanotubes and nanowires. *Surface Coatings Technol.* 201:9215–20.

Guo, T., Nikolaev, P., Rinzler, A. G., Tomanek, D., Colbert, D. T., and Smalley, R. E. 1995. Self-assembly of tubular fullerenes. *J. Phys. Chem.* 99:10694.

Hannon, J. B., Kodambaka, S., Ross, F. M., and Tromp, R. M. 2006. The influence of the surface migration of gold on the growth of silicon nanowires. *Nature* 440:69–71.

Hanrath, T., and Korgel, B. A. 2003. Supercritical fluid-liquid-solid (SFLS) synthesis of Si and Ge nanowires seeded by colloidal metal nanocrystals. *Adv. Mater.* 15:437.

Haraguchi, K., Katsuyama, T., Hiruma, K., and Ogawa, K. 1992. GaAs p-n junction formed in quantum wire crystals. *Appl. Phys. Lett.* 60:745.

Heath, J. R., and LeGoues, F. K. 1993. A liquid solution synthesis of single crystal germanium quantum wires. *Chem. Phys. Lett.* 208:263.

Hofmann, S., Ducati, C., Neill, J., et al. 2003. Gold catalyzed growth of silicon nanowires by plasma enhanced chemical vapor deposition. *J. Appl. Phys.* 94:6005.

Hofmann, S., Sharma, R., Ducati, C., et al. 2007. In situ observations of catalyst dynamics during surface-bound carbon nanotube nucleation. *Nano Lett.* 7:602–8.

Holmes, J. D., Johnston, K. P., Doty, R. C., and Korgel, B. A. 2000. Control of thickness and orientation of solution-grown silicon nanowires. *Science* 287:1471–73.

Huang, M. H., Mao, S., Feick, H., et al. 2001a. Room-temperature ultraviolet nanowire nanolasers. *Science* 292:1897.

Huang, M. H., Wu, Y., Feick, H., Tran, N., Weber, E., and Yang, P. 2001b. Catalytic growth of zinc oxide nanowires by vapor transport. *Adv. Mater.* 13:113–16.

Huang, Y., Duan, X., and Lieber, C. M. 2005. Nanowires for integrated multicolor nanophotonics. *Small* 1:142–47.

Huang, Y. H., Okumura, H., Hadjipanayis, G. C., and Weller, D. 2002. CoPt and FePt nanowires by electrodeposition. *J. Appl. Phys.* 91:6869.

Hulteen, J. C., and Van Duyne, R. P. 1995. Nanosphere lithography: A materials general fabrication process for periodic particle array surfaces. *J. Vac. Sci. Technol. A* 13:1553–58.

Huynh, W. U., Dittmer, J. J., and Alivisatos, A. P. 2002. Hybrid nanorod-polymer solar cells. *Science* 295:2425–27.

Iacopi, F., Vereecken, P. M., Schaekers, M., et al. 2007. Plasma enhanced CVD growth of Si nanowires with low melting point metal catalysts: An effective alternative to Au-mediated growth. *Nanotechnology* 18:505307.

Iijima, S., and Ichihashi, T. 1993. Single-shell carbon nanotubes of 1-nm diameter. *Nature* 363:603–5.

Jensen, L. E., Bjrk, M. T., Jeppesen, S., Persson, A. I., Ohlsson, J., and Samuelson, L. 2004. Role of surface diffusion in chemical beam epitaxy of InAs nanowires. *Nano Lett.* 4:1961–64.

Joyce, H. J., Qiang Gao, Q., Tan, H. H., et al. 2009. Unexpected benefits of rapid growth rate for III-V nanowires. *Nano Lett.* 9:695–701.

Jung, G. Y., Ganapathiappan, S., Li, X., et al. 2004. Fabrication of molecular-electronic circuits by nanoimprint lithography at low temperatures and pressures. *Appl. Phys. A* 78:1169–73.

Kamins, T. I., Williams, S. R., Basile, D. P., Hesjedal, T., and Harris, J. S. 2001. Ti-catalyzed Si nanowires by chemical vapor deposition: Microscopy and growth mechanisms. *J. Appl. Phys.* 89:1008–16.

Kang, W. P., Wong, Y. M., Davidson, J. L., et al. 2006. Carbon nanotubes vacuum field emission differential amplifier integrated circuits. *Electron. Lett.* 42:210–11.

Kayes, B. M., Lewis, N. S., and Atwater, H. A. 2005. Comparison of the device physics principles of planar and radial p-n junction nanorod solar cells. *J. Appl. Phys.* 97:114302.

Kayes, B. M., Spurgeon, J. M., Sadler, T. C., Lewis, N. S., and Atwater, H. A. 2006. Growth and characterization of silicon nanorod arrays for solar cell applications. In *Proceedings of the IEEE 4th World Conference on Photovoltaic Energy Conversion*, Waikoloa, HI, May 7–12, vol. 1, pp. 221–24.

Kelzenberg, M. D., Turner-Evans, D. B., Kayes, B. M., et al. 2008. Photovoltaic measurements in single-nanowire silicon solar cells. *Nano Lett.* 8:710–14.

Kemp, N. T., McGrouther, D., Cochrane, J. W., and Newbury, R. 2007. Bridging the gap: Polymer nanowire devices. *Adv. Mater.* 19:2634–38.

Kennedy, S. R., and Brett, M. J. 2003. Porous broadband antireflection coating by glancing angle deposition. *Appl. Opt.* 42:4573–79.

Kong, J., Soh, H. T., Cassell, A. M., Quate, C. F., and Dai, H. J. 1998. Synthesis of individual singlewalled carbon nanotubes on patterned silicon wafers. *Nature* 395:878.

Kymakis, E., and Amaratunga, G. A. J. 2002. Single-wall carbon nanotube/conjugated polymer photovoltaic devices. *Appl. Phys. Lett.* 80:112.

Landi, B. J., Raffaelle, R. P., Castro, S. L., and Bailey, S. G. 2005. Single-wall carbon nanotube-polymer solar cells. *Progr. Photovolt. Res. Appl.* 13:165.

Law, M., Greene, L. E., Johnson, J. C., Saykally, R., and Yang, P. 2005. Nanowire dye-sensitized solar cells. *Nature Mater.* 4:455–59.

Lee, J. U. 2005. Photovoltaic effect in ideal carbon nanotube diodes. *Appl. Phys. Lett.* 87:073101.

Lee, J. U., Gipp, P. P., and Heller, C. M. 2004. Carbon nanotube p-n junction diodes. *Appl. Phys. Lett.* 85:145.

Lee, J.-Y., Connor, S. T., Cui, Y., and Peumans, P. 2008. Solution-processed metal nanowire mesh transparent electrodes. *Nano Lett.* 8:689–92.

Lewis, P. A., Ahmed, H., and Sato, T. 1998. Silicon nanopillars formed with gold colloidal particle masking. *J. Vac. Sci. Technol. B* 16:2938–41.

Li, D., Huxtable, S. T., Abramson, A. R., and Majumdar, A. 2005. Thermal transport in nanostructured solid-state cooling devices. *J. Heat Transfer* 127:108–14.

Li, Y., Mann, D. Rolandi, M., et al. 2004. Preferential growth of semiconducting single-walled carbon nanotubes by a plasma enhanced CVD method. *Nano Lett.* 4:317–21.

Li, Z., Rajendran, B., Kamins, T. I., Li, X., Chen, Y., and Williams, R. S. 2005. Silicon nanowires for sequence-specific DNA sensing: Device fabrication and simulation. *Appl. Phys. A* 80:1257–63.

Liu, J., Duan, J. L., Toimil-Molares, M. E., et al. 2006. Electrochemical fabrication of single-crystalline and polycrystalline Au nanowires: The influence of deposition parameters. *Nanotechnology* 17:1922–26.

Maiolo III, J. R., Kayes, B. M., Filler, M. A., et al. 2007. High aspect ratio silicon wire array photoelectrochemical cells. *J. Am. Chem. Soc.* 129:12346–47.

Marchi, A., Gnani, E., Reggiani, S., Rudan, M., and Baccarani, G. 2006. Investigating the performance limits of silicon-nanowire and carbon-nanotube FETs. *Solid-State Electronics* 50:78–85.

Massalski, T. B., ed. 1990. *Binary alloy phase diagrams*, 428. Materials Park, OH: ASM International.

Masuda, H., and Fukuda, K. 1995. Ordered metal nanohole arrays made by a two-step replication of honeycomb structures of anodic alumina. *Science* 268:1466.

Melosh, N. A., Boukai, A., Diana, F., et al. 2003. Ultrahigh-density nanowire lattices and circuits. *Science* 300:112–15.

Michotte, S., Matefi-Tempfli, S., and Piraux, L. 2003. Current–voltage characteristics of Pb and Sn granular superconducting nanowires. *Appl. Phys. Lett.* 82:4119.

Misewich, J. A., Martel, R., Avouris, Ph., Tsang, J. C., Heinze, S., and Tersoff, J. 2003. Electrically induced optical emission from a carbon nanotube FET. *Science* 300:783–86.

Mohan, P., Motohisa, J., and Fukui, T. 2006. Fabrication of InP/InAs/InP core-multishell heterostructure nanowires by selective area metalorganic vapor phase epitaxy. *Appl. Phys. Lett.* 88:133105.

Morales, A. M., and Lieber, C. M. 1998. A laser ablation method for the synthesis of crystalline semiconductor nanowires. *Science* 279:208–11.

Motohisa, J., Noborisaka, J., Takeda, J., Inari, M., and Fukui, T. 2004. Catalyst-free selective-area MOVPE of semiconductor nanowires on (111)B oriented substrates. *J. Crystal Growth* 272:180–85.

Muskens, O. L., Diedenhofen, S. L., Kaas, B. C., et al. 2009. Large photonic strength of highly tunable resonant nanowire materials. *Nano Lett.* 9:930–34.

Nam, K. T., Kim, D.-W., Yoo, P. J., et al. 2006. Virus-enabled synthesis and assembly of nanowires for lithium ion battery electrodes. *Science* 312:885–88.

Navas, D., Pirota, K. R., Mendoza Zelis, P., Velazquez, D., Ross, C. A., and Vazquez, M. 2008. Effects of the magnetoelastic anisotropy in Ni nanowire arrays. *J. Appl. Phys.* 103:07D523.

Novotny, C. J., Yu, E. T., Yu, P. K. L. 2008. In P nanowire/polymer hybrid photodiode. *Nano Lett.* 8:775–9.

O'Regan, B., and Grätzel, M. 1991. A low-cost, high-efficiency solar cell based on dye-sensitized colloidal TiO2 films. *Nature* 353:737–40.

Paladugu, M., Zou, J., Guo, Y.-N., et al. 2008. Nature of heterointerfaces in GaAs/InAs and InAs/GaAs axial nanowire heterostructures. *Appl. Phys. Lett.* 93:101911.

Panish, M. B., and Temkin, H. 1989. Gas-source molecular beam epitaxy. *Annu. Rev. Mater. Sci.* 19:209–29.

Paulose, M., Shankar, K., Varghese, O. K, Mor, G. K., and Grimes, C. A. 2006. Application of highly-ordered TiO nanotube-arrays in heterojunction dye-sensitized solar cells. *J. Phys. D Appl. Phys.* 39:2498–503.

Park, B. T., and Yong, K. 2004. Controlled growth of core-shell Si–SiO$_x$ and amorphous SiO$_2$ nanowires directly from NiO/Si. *Nanotechnology* 15:S365–70.

Peng, K. Q., Huang, Z. P., and Zhu, J. 2004. Fabrication of large-area silicon nanowire p-n junction diode arrays. *Adv. Mater.* 16:73–76.

Peng, K., Xu, Y., Wu, Y., Yan, Y., Lee, S.-T., and Zhu, J. 2005. Aligned single-crystalline Si nanowire arrays for photovoltaic applications. *Small* 1:1062.

Peng, X., Manna, L., Yang, W., et al. 2000. Shape control of CdSe nanocrystals. *Nature* 404:59.

Persson, A. I., Larsson, M. W., Stenström, S., Ohlsson, B. J., Samuelson, L., and Wallenberg, L. E. 2004. Solid-phase diffusion mechanism for GaAs nanowire growth. *Nature Mater.* 3:677–81.

Prosa, T., Alvis, R., Tsakalakos, L., and Smentkowski, V. 2008. Analysis of silicon nanowires by laser atom probe tomography prepared by a protected lift-out processing technique. *Microscopy Microanal.* 14(Suppl. 2):456–57.

Puretzky, A. A., Geohegan, D. B., Fan, X., and Pennycook, S. J. 2000. Dynamics of single-wall carbon nanotube synthesis by laser vaporization. *Appl. Phys. A* 70:153–60.

Putnam, M. C., Filler, M. A., Kayes, B. M., et al. 2008. Secondary ion mass spectrometry of vapor-liquid-solid grown, Au-catalyzed, Si wires. *Nano Lett.* 8:3109–13.

Rabin, O., Herz, P. R., Lin, Y. M., Akinwande, A. I., Cronin, S. B., and Dresselhaus, M. S. 2003. Formation of thick porous anodic alumina films and nanowire arrays on silicon wafers and glass. *Adv. Funct. Mater.* 13:631–38.

Rappich, J., Lust, S., Sieber, I., Henrion, W., Dohrmann, J. K., and Fuhs, W. 2000. Light trapping by formation of nanometer diameter wire-like structures on μc-Si thin films. *J. Noncryst. Sol.* 266–69:284–89.

Reich, S., Li, L., and Robertson, J. 2005. Structure and formation energy of carbon nanotube caps. *Phys. Rev. B* 72:165423.

Ross, F. M., Tersoff, J., and Reuter, M. C. 2005. Sawtooth faceting in silicon nanowires. *Phys. Rev. Lett.* 95:146104.

Routkevitch, D., Tager, A. A., Haruyama, J., Almawlawi, D., Moskovits, M., and Xu, J. M. 1996. Nonlithographic nano-wire arrays: Fabrication, physics, and device applications. *IEEE Trans. Electr. Dev.* 43:1646–58.

Sander, M. S., Prieto, A. L., Gronsky, R., Sands, T., and Stacy, A. M. 2002. Fabrication of high-density, high aspect ratio, large-area bismuth telluride nanowire arrays by electrodeposition into porous anodic alumina templates. *Adv. Mater.* 14:665–67.

Schmid, W., and Reiner, J. 1982. Minority-carrier lifetime in gold-diffused silicon at high carrier concentrations. *J. Appl. Phys.* 53:6250–52.

Schonenberger, C., van der Zande, B. M. I., Fokkink, L. G. J., et al. 1997. Template synthesis of nanowires in porous polycarbonate membranes: Electrochemistry and morphology. *J. Phys. Chem. B* 101:5497–505.

Schricker, A. D., Davidson, F. M., Wiacek, R. J., and Korgel, B. A. 2006. Space charge limited currents and trap concentrations in GaAs nanowires. *Nanotechnology* 17:2681–88.

Schubert, L., Werner, P., Zakharov, N. D., et al. 2004. Silicon nanowhiskers grown on <111> Si substrates by molecular-beam epitaxy. *Appl. Phys. Lett.* 84:4968.

Singh, R. K., and Kumar, D. 1998. Pulsed laser deposition and characterization of high-T_c $YBa_2Cu_3O_{7-x}$ superconducting thin films. *Mater. Sci. Eng. Rep.* 22:113–85.

Sivakov, V., Andrä, G., Gawlik, A., et al. 2009. Silicon nanowire-based solar cells on glass: Synthesis, optical properties, and cell parameters. *Nano Lett.*, 9:1549–54.

Skinner, B. J. 1979. Earth resources. *Proc. Natl. Acad. Sci. USA* 76:4212–17.

Stelzner, Th., Pietsch, M., Andrä, G., Falk, F., Ose, E., and Christiansen, S. 2008. Silicon nanowire-based solar cells. *Nanotechnology* 19:295203.

Street, R. A., Qi, P., Lujan, R., and Wong, W. S. 2008. Reflectivity of disordered silicon nanowires. *Appl. Phys. Lett.* 93:163109.

Sunkara, M. K., Sharma, S., Miranda, R., Lian, G., and Dickey, E. C. 2001. Bulk synthesis of silicon nanowires using a low-temperature vapor-liquid-solid method. *Appl. Phys. Lett.* 79:1546.

Sze, S. M. 1969. *Physics of semiconductor devices*, 21, 37, 850. New York: Wiley-Interscience.

Tambe, M. J., Lim, S. K., Smith, M. J., Allard, L. F., and Gradečak, S. 2008. Realization of defect-free epitaxial core-shell GaAs/AlGaAs nanowire heterostructures. *Appl. Phys. Lett.* 93:151917.

Tan, T. H., Li, N., and Gosele, U. 2003. Is there a thermodynamic size limit of nanowires grown by the vapor-liquid-solid process? *Appl. Phys. Lett.* 83:1199–201.

Teo, K. B. K., Lacerda, R. G., Yang, M. H., et al. 2004. Carbon nanotube technology for solid state and vacuum electronics. *IEE Proc. Circuits Devices Syst.* 151:443–51.

Thurn-Albrecht, T., Schotter, J., and Kastle, G. A. 2000. Ultrahigh-density nanowire arrays grown in self-assembled diblock copolymer templates. *Science* 290:2126.

Tian, B., Zheng, X., Kempa, T. J., et al. 2004. Coaxial silicon nanowires as solar cells and nanoelectronic power sources. *Nature* 449:885.

Toimil Morales, M. E., Buschmann, V., and Dobrev, D. 2001. Single-crystalline copper nanowires produced by electrochemical deposition in polymeric ion track membranes. *Adv. Mater.* 13:62.

Tong, H. D., Chen, S., van der Wiel, W. G., Carlen, E. T., and van den Berg, A. 2009. Novel top-down wafer-scale fabrication of single crystal silicon nanowires. *Nano Lett.* 9:1015–22.

Tsakalakos, L., Balch, J., Fronheiser, J., Korevaar, B. A., Sulima, O., and Rand, J. 2008. Silicon nanowire solar cells: Device physics, fabrication, and optoelectronic properties. In *Proceedings of the 23rd European Photovoltaic Solar Energy Conference*, pp. 11–16, Valencia, Spain.

Tsakalakos, L., Balch, J. Fronheiser, J., et al. 2007a. Strong broadband optical absorption in silicon nanowire films. *J. Nanophoton* 01:013552.

Tsakalakos, L., Fronheiser, J., Rowland, L., Rahmane, M. Larsen, M., and Gao, Y. 2007b. SiC nanowires by silicon carburization. *Mater. Res. Soc. Symp. Proc.* 963:0963-Q11-03.

Tsakalakos, L., Rahmane, M. Larsen, M., et al. 2005. Mo_2C nanowires and nanoribbons on Si by two-step vapor-phase growth. *J. Appl. Phys.* 98:044317.

Urban, J. J., Yun, W. S., Gu, Q., and Park, H. 2002. Synthesis of single-crystalline perovskite nanorods composed of barium titanate and strontium titanate. *J. Am. Chem. Soc.* 124:1186–87.

Valizadeh, S., Abid, M., Hernandez-Ramırez, F., Romano Rodrıguez, A., Hjort, K., and Schweitz, J. A. 2006. Template synthesis and forming electrical contacts to single Au nanowires by focused ion beam techniques. *Nanotechnology* 17:1134–39.

Vieu, C., Carcenac, F., Pepin, A., et al. 2000. Electron beam lithography: Resolution limits and applications. *Appl. Surface Sci.* 164:111–17.

Wagner, R. S., and Ellis, W. C. 1964. Vapor-liquid-solid mechanism of single crystal growth. *Appl. Phys. Lett.* 4:89.

Wang, W., Jia, F., Huang, Q., and Zhang, J. 2005. A new type of low power thermoelectric micro-generator fabricated by nanowire array thermoelectric material. *Microelectronics Eng.* 77:223–29.

Wang, Y., Schmidt, V., Senz, S., and Gösele, U. 2006. Epitaxial growth of silicon nanowires using an aluminum catalyst. *Nature Nanotechnol.* 1:186–89.

Wei, J., Jia, Y., Shu, Q., et al. 2007. Double-walled carbon nanotube solar cells. *Nano Lett.* 7:2317–21.

Weisman, R. B., and Bachilo, S. M. 2003. Dependence of optical transition energies on structure for single-walled carbon nanotubes in aqueous suspension: An empirical Kataura plot. *Nano Lett.* 3:1235–38.

Whitney, T. M., Jiang, J. S., Searson, P. C., and Chien, C. L. 1993. Fabrication and magnetic properties of arrays of metallic nanowires. *Science* 261:1316–19.

Wu, Y., Cui, Y., Huynh, L., Barrelet, C. J., Bell, D. C., and Lieber, C. M. 2004. Controlled growth and structures of molecular-scale silicon nanowires. *Nano Lett.* 4:433–36.

Wu, Y., Fan, R., and Yang, P. 2002a. Block-by-block growth of single-crystalline Si/SiGe superlattice nanowires. *Nano Lett.* 2:83–86.

Wu, Z. H., Mei, X. Y., Kim, D., Blumin, M., and Ruda, H. E. 2002b. Growth of Au-catalyzed ordered GaAs nanowire arrays by molecular-beam epitaxy. *Appl. Phys. Lett.* 81:5177.

Xi, J.-Q., Schubert, M. F., Kim, J. K., et al. 2007. Optical thin-film materials with low refractive index for broadband elimination of Fresnel reflection. *Nature Photonics* 1:176–79.

Xu, D., Xu, Y., Chen, D., Guo, G., Gui, L., and Tang, Y. 2000. Preparation of CdS single-crystal nanowires by electrochemically induced deposition. *Adv. Mater.* 12:520–22.

Yin, A. J., Li, J., Jian, W., Bennett, A. J., and Xu, J. M. 2001. Fabrication of highly ordered metallic nanowire arrays by electrodeposition. *Appl. Phys. Lett.* 79:1039.

Zhang, R.-Q., Lifshitz, Y., and Lee, S.-T. 2003. Oxide-assisted growth of semiconducting nanowires. *Adv. Mater.* 15:635–40.

Zhao, Q., Zhang, H., Xu, X., et al. 2005. Optical properties of highly ordered AlN nanowire arrays grown on sapphire substrate. *Appl. Phys. Lett.* 86:19310.

Zhao, Y.-P., Ye, D.-X, Wang, G.-C., and Lu, T.-M. 2003. Designing nanostructures by glancing angle deposition. *Proc. SPIE* 5219:59–73.

Zhou, C., Kong, J., Yenilmez, E., and Dai, H. 2000. Modulated chemical doping of individual carbon nanotubes. *Science* 290:1552–55.

7

Semiconductor Nanowires: Contacts and Electronic Properties

Emanuel Tutuc and En-Shao Liu

7.1 INTRODUCTION

Semiconductor nanowires, namely, highly anisotropic crystals with diameters of the order of a few tens of nanometers and aspect ratio of 1:100 to 1:1,000, have gained increased scrutiny recently, motivated in part by issues associated with complementary metal-oxide-semiconductor device scaling. Depending on device design and host semiconductor, semiconductor nanowires can potentially serve as high-speed and low-power transistors for high-performance electronics, flexible electronics, and sensors, and can also serve as a test bed to study the electron physics at the nanoscale.

Semiconductor nanowire array-based photovoltaic devices have also attracted interest recently. Such devices may enable more efficient photogenerated carrier collection, which is particularly relevant for thin-film materials with low minority-carrier diffusion length.

The fabrication of semiconductor nanowires is based on two main approaches: bottom up using self-assembled growth, or top down using a combination of lithography and etching. The bottom-up approach employs the vapor-liquid-solid (VLS) growth mechanism (Wagner and Ellis 1964). Here, metal nanoparticles are used as catalysts to decompose a precursor gas (e.g., SiH_4). The semiconductor atoms then form a eutectic alloy with the metal catalysts, which is generally liquid at the growth temperatures. The semiconductor atoms condense at the metal-substrate interface, resulting in acicular growth. Gold is most commonly used as metal catalyst, although VLS growth using other metal catalysts, e.g., Ni, Cu, Pt, and Co, is also possible (Givargizov 1987). The chemical precursor varies depending on the desired semiconductor nanowire growth. Examples of precursors used for group IV nanowire growth are SiH_4, Si_2H_4, and $SiCl_4$ for silicon, and GeH_4 and Ge_2H_6 for germanium nanowire growth. For III-V compound semiconductor nanowire growth, mixtures of metalorganic and hydride precursors are used, such as trimethylindium (TMI) and arsine (AH_3) for InAs growth, TMI and phosphine for InP growth, trimethylgallium and AsH_3 for GaAs nanowires, etc.

While the VLS growth mechanism provides a very versatile and relatively inexpensive method for semiconductor device fabrication at the nanoscale, there are also drawbacks associated with uncertainty in nanowire placement, metal incorporation into the nanowire (Hannon et al. 2006), and potential metal cross-contamination. In particular, Au impurities are deep traps in Si, and reduce the minority carrier recombination time (Bemski 1958; Abbas 1984). This in turn increases the ideality factor and reverse current in a p-n junction, and can adversely affect the efficiency of a nanowire-based photovoltaic device.

The second, top-down approach relies primarily on advanced lithography and etching to define nanowires from a bulk substrate. Vertical nanowire arrays can be defined using deep reactive ion etching out of bulk substrates, and horizontal nanowires can be fabricated using silicon-on-insulator substrates (Wang et al. 2006), and potentially other semiconductor-on-insulator substrates. Figure 7.1 data illustrate two examples of semiconductor nanowires fabricated using bottom-up (a) and top-down (b) approaches.

Semiconductor nanowires represent an interesting platform for photovoltaic applications, as they can enable nonplanar device geometries where the photon absorption and minority carrier diffusion length are decoupled. Figure 7.2 illustrates an example, consisting of an array of radial p-n junction nanorods. Kayes et al. (2005) have theoretically examined the performance of such a photovoltaic device and shown that in certain conditions this device geometry can outperform a *planar* solar cell fabricated using the same host semiconductor. In light of the potential applications of semiconductor nanowires for electronic and photovoltaics devices, it is therefore relevant to ask: How do we make electrical contacts to a semiconductor nanowire? What is the underlying physics governing the carrier (electron or hole) injection into a semiconductor nanowire? How do we realize low-resistance electrical contacts to semiconductor nanowires? Indeed, despite the relatively mundane aspect of the topic, the performance of nanowire devices has been, with a few exceptions, largely determined by electrical contacts rather than their intrinsic electronic properties. The

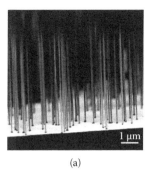

(a) (b)

FIGURE 7.1

(a) Scanning electron micrograph of Si nanowire grown epitaxially on a Si substrate using the vapor-liquid-solid growth mechanism. The nanowires are grown using $SiCl_4$ as precursor and Au nanoparticles as metal catalyst. (From Goldberger, J., Hochbaum, A. I., Fan, R., et al., "Silicon Vertically Integrated Nanowire Field Effect Transistors," *Nano Letters* 6 [2006]: 973–77.) (b) Scanning electron micrograph of silicon nanowires with different lengths defined using reactive ion etching along with a self-assembled layer of polystyrene spheres, which serves as mask. (From Huang, Z., Fang, H., and Zhu, J., "Fabrication of Silicon Nanowire Arrays with Controlled Diameter, Length and Density," *Advanced Materials* 19 [2007]: 744–48. With permission.)

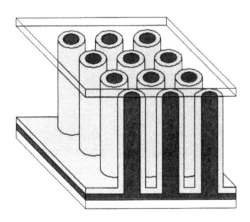

FIGURE 7.2

Schematic cross-section of a radial p-n junction solar cell. Light is incident on the top surface. The light (dark) grey area represents the n-type (p-type) doped semiconductor. The photogenerated carriers are collected radially, while the incident light is absorbed along the nanorod. The device geometry allows the photogenerated carriers to be collected on a distance much shorter than the absorption length, which in turn provides an efficiency enhancement by comparison to a planar cell. (Reproduced from Kayes, B. M., Atwater, H. S., and Lewis, N. S., "Comparison of the Device Physics Principles of Planar and Radial p-n Junction Nanorod Solar Cells," *Journal of Applied Physics* 97 [2005]: 114302. With permission.)

purpose of this chapter is to shed light on some of the questions regarding contact engineering in semiconductor nanowires.

Section 7.2 describes the metal-nanowire contact resistance measurement and a simple transmission line model that allows the extraction of the specific contact resistance. Section 7.3 is focused on the impact of the metal-semiconductor Schottky barrier on carrier injection in semiconductor nanowires, and its dependence on nanowire doping and metal profile into the nanowire. Section 7.4 presents a simple recipe for achieving low contact resistance to nanowires by doping, and existing experimental approaches to electrically dope nanowires. Sections 7.3 and 7.4 will be focused primarily on the germanium nanowire, a system the authors are most familiar with, but the conclusions are generally applicable to most semiconductor nanowire systems. Lastly, Section 7.5 summarizes the specific contact resistances reported thus far for semiconductor nanowires.

7.2 NANOWIRE-SPECIFIC CONTACT RESISTANCE: MEASUREMENT AND MODEL

A relatively simple method to investigate the electron transport through a semiconductor nanowire is to fabricate a back-gated nanowire field effect transistor (FET) with metal contacts. The nanowires are typically "harvested" in a solvent solution, which is subsequently dispersed onto a substrate. To serve as a gate, the substrate consists of a dielectric layer grown on a conducting substrate, e.g., SiO_2 thermally grown on a doped Si substrate. Once dispersed, the contacts are defined using aligned electron beam lithography, metal deposition, and lift-off. An example of a single nanowire back-gated FET with metal contacts is shown in Figure 7.3a.

Once fabricated, two types of measurement can be performed on such devices. In a two-terminal, gate-dependent measurement, the current is measured across two adjacent terminals as a function of the applied voltage. This probes the nanowire intrinsic resistance (R_{ch}) *in series* with the contact resistances, R_{c1} and R_{c2} of Figure 7.3b:

$$R_{ch} + R_{c1} + R_{c2} = \frac{\Delta V_{2p}}{I_{2p}}$$

(7.1)

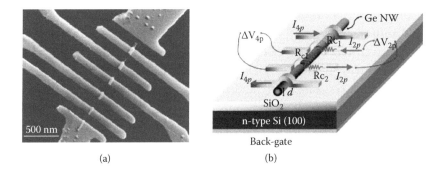

FIGURE 7.3

(a) Scanning electron micrograph (SEM) of a Ge nanowire multiterminal device with Ni contacts. (b) Schematic representation of the two-point and four-point current-voltage measurements on a nanowire.

The nanowire intrinsic resistance (R_{ch}) is determined by the carrier concentration and mobility:

$$R_{ch} = \frac{L}{n_L e \mu} \tag{7.2}$$

where L is the nanowire length, n_L is the electron (or hole) concentration per unit length, μ is the electron (hole) mobility, and e is the electron charge. The carrier concentration dependence on the back-gated bias (V_{bg}) is given by

$$n_L = e^{-1} C_L \cdot \left(V_{bg} - V_T \right) \tag{7.3}$$

where C_L is the back-gated nanowire capacitance per unit length, and V_T represents the threshold voltage, at which the carrier concentration in the nanowire is zero. A measurement of the intrinsic nanowire resistance R vs. V_{bg} allows the extraction of the carrier mobility via

$$\mu = L C_L^{-1} \cdot \frac{dR_{ch}^{-1}}{dV_{bg}} \tag{7.4}$$

We note that the mobility extraction requires the C_L values to be known. While individual nanowire capacitance measurements have been reported (Tu et al. 2007), C_L typical values for back-gated NW devices are ~100 aF/μm, and accurate measurements of this capacitance are difficult. Alternatively, C_L can be calculated using commercial device simulators (e.g., Sentaurus© Synopsis, Q3D Extractor© Ansoft).

The contact resistances, R_{c1} and R_{c2} of Figure 7.3b, consist of the metal lead resistance in series with the metal-nanowire junction resistance. The latter is characterized by a Schottky barrier and depends on multiple factors, such as metal-semiconductor Schottky barrier height, applied back-gated bias, metal protrusion into the nanowire, nanowire doping, etc. Consequently, it is desirable to eliminate the contact resistance contributions, which can be accomplished by performing a four-point resistance measurement. A current (I_{4p}) is forced between the outer terminals of the Figure 7.3b device, and the voltage drop across the inner terminals is measured (ΔV_{4p}). Because no current flows through the inner contacts, the four-point measurement probes the intrinsic nanowire resistance, without contact resistance contributions:

$$R_{ch} = \frac{\Delta V_{4p}}{I_{4p}} \qquad (7.5)$$

By performing a two-point and a four-point measurement, both the nanowire and contact resistances can be determined. In the following we describe a simple transmission line model (Mohney et al. 2005), which relates the measured contact resistance to the specific contact resistance, and in the next section we examine in more detail the underlying physics of electron injection at the metal-nanowire junction. Let ρ be the nanowire resistivity, ρ_c the specific contact resistance, L_c the nanowire length covered by the metal contact, and d the nanowire diameter. ρ is related to the nanowire resistance via $R_{ch} = \rho L / (\pi d^2 / 4)$ and can be determined from a four-point measurement and using the nanowire dimensions (Figure 7.4a).

The nanowire section under the metal contact can be decomposed into infinitesimal longitudinal segments (Figure 7.4b). A segment of length dx has a resistance $\rho dx / (\pi d^2 / 4)$ and is connected to the metal contact by a resistance $2\rho_c / (\theta \cdot d \cdot dx)$; θ is the angle covered by the metal contact on the nanowire circumference, and to a good approximation $\theta \approx \pi$ (see Figure 4a). When a voltage V_0 is applied on the metal contact, current will flow through the resistor network of Figure 7.4b. The voltage drop and the

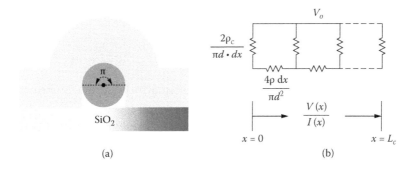

FIGURE 7.4

(a) Cross-sectional representation of a metal-nanowire contact. The transmission line model employed here assumes the metal covers roughly half the nanowire circumference. (b) Resistor network model and current flow along a metal-nanowire contact. The V_0 potential line represents the metal contact. The vertical resistances are determined by the specific contact resistance, and the horizontal resistances are determined by the nanowire resistivity. The metal contact covers a nanowire section of length L_c. (Reproduced from Mohney, S. E., Wang, Y., Cabassi, M. A., et al., "Measuring the Specific Contact Resistance to Semiconductor Nanowires," *Solid-State Electronics* 49 [2005]: 227–32. With permission.)

current flow at position x along the nanowire section under the contact can be related to the nanowire resistivity and contact resistance as follows:

$$\frac{dV}{dx} = -\frac{4\rho}{\pi d^2} I(x) \tag{7.6}$$

$$V_0 - V(x) = \frac{2\rho_c}{\pi d}\frac{dI}{dx} \tag{7.7}$$

Equations 7.6 and 7.7 along with the boundary condition $I(x=0)=0$ provide the following solution for the current:

$$I(x) = I_0 \cdot \frac{\sinh(x/L_T)}{\sinh(L_c/L_T)} \tag{7.8}$$

where the transmission length L_T is given by $L_T = \sqrt{d \cdot \rho_c / 2\rho}$. Using Equations 7.7 and 7.8, along with the boundary condition $V(x=L_c)=0$, the relation between the contact resistance (R_c) and the specific contact resistance (ρ_c) and nanowire resistivity (ρ) is readily obtained:

$$R_c = \frac{V_0}{I_0} = \frac{2\sqrt{2\rho_c\rho}}{\pi d^{3/2}} \coth\left(\frac{L_c}{L_T}\right) \tag{7.9}$$

Since R_c and ρ can be measured experimentally, Equation 7.9 provides a method for extracting ρ_c.

One note on the applicability of the model presented here: in deriving Equation 7.9, we assumed that the contact can be partitioned into infinitesimal sections, and Ohm's law applied along each of these sections. Consequently, the carrier mean free path in the nanowire should be much smaller than L_c. Furthermore, the contact resistance R_c should be much larger than the quantum limit, $(e^2/2h)/N = 12.9/N$ kΩ, where N is the number of subbands occupied by the electrons in the nanowire, and h is Planck's constant (Datta 1989).

7.3 ROLE OF SCHOTTKY BARRIER IN METAL-TO-NANOWIRE CONTACTS

In this section we will examine in more detail the physics of the metal-nanowire contact. The metal-nanowire junction is characterized by a Schottky barrier, common for any metal-semiconductor interface, and the carriers are injected primarily by tunneling through this barrier. Unlike a bulk sample, however, an applied gate bias can also change the width of the Schottky barrier, which results in a gate-dependent metal-nanowire contact resistance.

Let us first examine the device characteristics of a back-gated Ge nanowire with metal (Ni) contacts. The device, shown schematically in Figure 7.5, consists of an undoped Ge nanowire on a 10-nm thick SiO$_2$ dielectric grown on an n-type Si substrate. The source and drain Ni contacts are fabricated using e-beam lithography and lift-off. The nanowire diameter is $d = 35$ nm and the channel length is $L = 600$ nm. Two types of device characteristics are measured: (1) the drain current (I_d) vs. the applied drain-source bias (V_{ds}) measured at fixed values of the back-gated bias (V_{gs}) (output characteristics), and (2) I_d vs. V_{gs} measured at fixed values of V_{ds} (transfer characteristics). The data are shown in Figure 7.5a. The transfer characteristics of Figure 7.5a show ambipolar behavior: as V_{gs} is changed at constant V_{ds}, I_d decreases, goes through a minimum, and

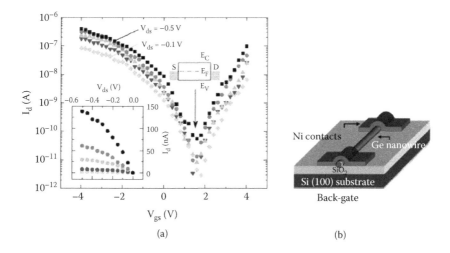

FIGURE 7.5

(a) I_d-V_{gs} of an undoped, back-gated Ge nanowire FET. Different symbols represent data measured at V_{ds} from −0.1 V to −0.5 V in steps of −0.1 V. The inset shows the output characteristics of the hole injection branch, measured at V_{gs} from −4 V to 0 V in steps of 1 V. (Data reproduced from Tutuc, E., Chu, J. O., Ott, J. A., et al., "Doping of Germanium Nanowire Grown in Presence of PH₃," *Applied Physics Letters* 89 [2006]: 263101.) (b) Schematic representation of the back-gated Ge nanowire FET with Ni contacts. (With permission.)

increases for both large negative or positive V_{gs} values. The data can be explained as follows: at large negative V_{gs} values, holes are induced in the nanowire, and the Fermi level (E_F) is close to the valence band (E_V). For large positive values of V_{gs}, electrons are the majority carriers in the nanowire. In both cases, the nanowire conductance is large. For a gate bias of $V_{gs} \sim 2V$ the nanowire conductance reaches a minimum, the Fermi level in the nanowire is close to mid-gap, neither type of carrier (electrons or holes) dominates, and their concentration is close to the intrinsic carrier concentration in the host semiconductor ($\sim 2 \times 10^{13}$ cm⁻³ for Ge).

While the data of Figure 7.5 can be understood in this simple picture, quantitatively the nanowire conductance is more than one order of magnitude lower than expected. Indeed, the measured resistance at $V_{gs} = -4$ V is ~ 3 MΩ. Using Equations 7.2 and 7.3, and assuming a typical mobility of ~ 100 cm²/Vs, and a dielectric capacitance per unit length of $C_L \sim 0.1$nF/m, the estimated nanowire resistance should be 0.2 MΩ. Clearly, the contact resistance of the metal-nanowire Schottky barrier is much larger than the intrinsic nanowire resistance.

To better understand the transport in back-gated nanowire FETs, in Figure 7.6 we present self-consistent device simulations of back-gated Ge field effect transistors with Ni contacts, using the numerical simulator Sentaurus (copyright Synopsis, Version A-2007.12). The device structure consists of a planar back-gated Ge FET with metal (Ni) contacts (Figure 7.6 inset), with the back-gate consisting of a highly doped Si substrate separated from the Ge channel by a 10-nm thick SiO_2 dielectric layer. The device characteristics are simulated for different metal contact depths (l_{diff}), while keeping other device structure parameters constant. The carrier transport model used is drift diffusion assuming bulk mobilities for Ge, and the simulator solves the Poisson and carrier continuity equations self-consistently. The carrier injection is modeled using nonlocal tunneling at metal-semiconductor Schottky contacts (Ieong et al. 1998). The Schottky barrier height for the Ni to Ge valence band was chosen as 0.17 eV (Sze 1981). The thermionic emission over the barrier is negligible and thus not considered in the simulation. For simplicity, the simulations shown here are for a two-dimensional structure with the channel width and thickness similar to the NW diameter. These data, however, capture the essential device physics of a back-gated NW device and reproduce qualitatively and semiquantitatively the experimental results of Figure 7.5.

Several observations can be made about Figure 7.6 data. The device characteristics are ambipolar, consistent with electron and hole injection from the metal contact. The current (I_D) *does* depend on the metal contact protrusion (l_{diff}) into the semiconductor channel, which in turn can be controlled by thermal diffusion during annealing. Moreover, the current dependence on l_{diff} is nonmonotonic. For small l_{diff} values, the current increases with l_{diff}. This observation explains a general experimental finding, namely, that annealing reduces the nanowire contact resistance. I_D reaches a maximum for an optimum l_{diff} value, in our case $l_{diff} = 30$ nm, and then *decreases* as l_{diff} is further increased. We examine in the following the underlying physics of the I_D vs. l_{diff} nonmonotonic dependence, but a key conclusion we can readily draw is that the carrier injection is highly sensitive to the contact metal profile, which in turn translates into significant contact resistance variability (Cui et al. 2003).

Figure 7.7 data show the calculated current density (a) and energy band diagram (b) along the device of Figure 7.6. These data were calculated for $l_{diff} = 30$nm, the optimum value of Figure 7.6, and for $V_{GS} = -4$ V and $V_{DS} = -1$ V. Figure 7.7a data show that current flows primarily near the bottom oxide-semiconductor interface, where the hole inversion layer forms. The energy band profile along the device reveals an interesting finding. The applied V_{DS} bias falls primarily on the Schottky barrier at the source

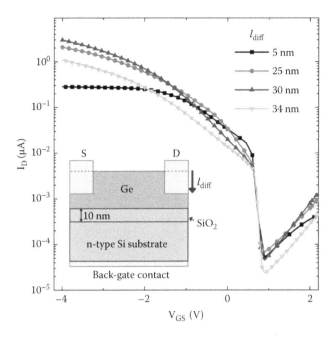

FIGURE 7.6

Simulated I_D-V_{GS} data for a planar, back-gated Ge FET with Ni contacts, at different metal contact depth values (l_{diff}), and at V_{DS} = –1 V. The channel length is 700 nm, and the channel width and thickness are both 35nm. Inset: Schematic representation of the simulated device structure. (Reproduced from Liu, E. S., Jain, N., Varahramyan, K. M., et al., "Role of Metal-Semiconductor Contact in Nanowire Field-Effect Transistors," to appear in *IEEE Transactions on Nanotechnology*, 2010. With permission.)

contact, which controls the hole injection in the channel. This in turn makes the device characteristics highly dependent on the electric field along the source contact Schottky barrier, controlled by V_{GS} and l_{diff}, and relatively insensitive to the carrier mobility in the channel. In Figure 7.8 we show the calculated energy band vs. depth profile at the metal-semiconductor source contact for the device of Figures 7.6 and 7.7, for different V_{GS} values, and at V_{DS} = –1 V and l_{diff} = 30 nm. These data clearly show that the electric field along the source contact Schottky barrier increases at higher gate biases, which in turn increases the tunneling rate through this barrier, reducing the contact resistance.

Lastly, let us examine the impact of contact metal depth on carrier injection. In Figure 7.9b we show the valence band profiles at the source contact in a back-gated Ge FET with Ni contacts, for different metal contact depth values, l_{diff} = 5, 30, and 34 nm. At small l_{diff} values (Figure 7.9b, lower left panel), holes tunnel through a relatively wide Schottky barrier

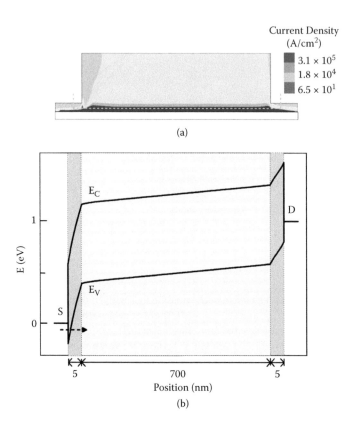

(a)

(b)

FIGURE 7.7

(a) Cross-sectional contour plot showing the current density in a back-gated Ge FET with Ni contacts, and with $l_{diff} = 30$ nm, $V_{GS} = -4$ V, and $V_{DS} = -1$ V. The conduction in the Ge channel takes place primarily near the bottom oxide interface, where the holes accumulate. (b) Conduction (E_C) and valence (E_V) energy band profile along the dashed path shown in the upper panel. Note that the applied $V_{DS} = -1$ V bias falls primarily across the metal-semiconductor barrier at the source contact, which in turn renders the device current highly dependent on the tunneling through the Schottky barrier at the source contact, and relatively insensitive to the carrier mobility in the channel. (Reproduced from Liu, E. S., Jain, N., Varahramyan, K. M., et al., "Role of Metal-Semiconductor Contact in Nanowire Field-Effect Transistors," to appear in *IEEE Transactions on Nanotechnology*, 2010. With permission.) (See color insert following page 206.)

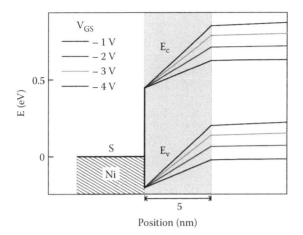

FIGURE 7.8

Conduction (E_C) and valence (E_V) energy band profile along the metal-nanowire Schottky barrier at the source contact, for different back-gated biases. The profile is taken along the dashed line of Figure 7.7, in the vicinity of the source contact. The drain-source bias and metal contact depth values are V_{DS} = –1 V and l_{diff} = 30 nm, respectively, consistent with Figure 7.7 data. As V_{GS} becomes more negative, the electric field along the Schottky barrier increases. This in turn increases the carrier tunneling through the Schottky barrier and reduces the contact resistance. The device characteristics are dominated by the carrier injection at the source contact, which in a simple picture acts as a gate-tunable contact resistance. (Reproduced from Liu, E. S., Jain, N., Varahramyan, K. M., et al., "Role of Metal-Semiconductor Contact in Nanowire Field-Effect Transistors," to appear in *IEEE Transactions on Nanotechnology*, 2010. With permission.)

before entering the channel. This translates in reduced carrier injection efficiency, and a low drive current. This situation is labeled as the *injection-limited* regime in Figure 7.9a, where we summarize the I_D vs. l_{diff} dependence for two types of nanowire FETs, using Ge and Si as channel materials. As the contact metal protrudes into the semiconductor, at the same V_{DS} and V_{GS} values, the electric field at the source contact metal-NW interface increases. This reduces the Schottky barrier width, which leads to an increase of carrier tunneling rate and I_D. At an optimized l_{diff} value the electric field across the junction is high and there is also a large density of states (DOS) in the semiconductor aligned with the metal Fermi level. In this regime, labeled as *optimum* in Figure 7.9, the hole injection efficiency and I_D reach a maximum. For larger l_{diff}, however, as illustrated in the lower-right panel in Figure 7.9b, despite a high electric field at the metal-semiconductor Schottky barrier, the injection is restricted only to the high-energy holes in contact metal, because the DOS in the nanowire

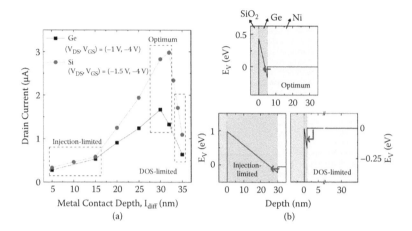

FIGURE 7.9

(a) Simulated I_D values at fixed V_{DS} and V_{GS} values for back-gated Ge and Si FET as a function of metal contact depth (l_{diff}). The simulated structure has a channel width and thickness of 35 nm for both the Ge and Si FET. $V_{GS} = -4$ V in both cases, and the V_{DS} biases are –1 and –1.5 V for Ge and Si FETs, respectively. In both cases I_D shows a non-monotonic dependence on l_{diff}. Depending on the l_{diff} value, three different regimes are defined: injection-limited, optimum, and DOS-limited regime. (b) Valence band vs. depth profile at the metal-semiconductor source contact for a Ge FET, corresponding to the three regimes of panel (a). The arrows indicate the carrier tunneling through the Schottky barrier. (Reproduced from Liu, E. S., Jain, N., Varahramyan, K. M., et al., "Role of Metal-Semiconductor Contact in Nanowire Field-Effect Transistors," to appear in *IEEE Transactions on Nanotechnology*, 2010. With permission.)

aligned with the Fermi level of the metal contact is zero. Consequently, the tunneling rate decreases, and I_D is reduced (*DOS-limited* regime of Figure 7.9a).

To conclude this section, the carrier injection through the metal-semiconductor Schottky barrier is the primary factor controlling the electrical characteristics of nanowire FETs with metal contacts. The electric field at the source contact Schottky barrier depends on gate bias and metal contact depth, which in turn is highly sensitive to anneal conditions. In particular, the drain current displays a nonmonotonic trend as a function on metal contact depth. This behavior can be explained by considering the interplay between carrier injection efficiency through the metal-NW Schottky barrier, which increases with the metal contact depth, and the number of available states in the NW between contact and the bottom oxide, which decreases with the metal contact depth. At small contact depth the current is limited by a low carrier tunneling rate through the metal-semiconductor junction, and at large contact depth by the reduced density of states in the

NW. The device current reaches a maximum at an optimum metal contact depth characterized by a high enough tunneling rate and a sufficient number of states in the NW. These observations imply that small variations in the metal contact profile or device geometry induce a significant variability in the observed device characteristics. These conclusions mirror similar findings related to the device physics of carbon nanotube field effect transistors, where the metal contact geometry and metal-nanotube Schottky barrier critically impact the carrier injection (Heinze et al. 2002; Chen et al. 2005).

Finally, we note that although in our study we use Ge and Si NWs as channel materials, our conclusions apply broadly to back-gated FETs with metal contacts using either two-dimensional or one-dimensional device geometries, and are qualitatively independent of the host semiconductor.

7.4 NANOWIRE DOPING: A RECIPE FOR LOW CONTACT RESISTANCE

In light of the conclusions reached in the previous section, namely, reduced carrier injection at the metal-nanowire junction because of the metal-semiconductor Schottky barrier, and also significant variability because of the Schottky barrier width dependence on gate bias and contact metal profile, it becomes highly desirable to immunize nanowire-based electrical or electro-optical devices against these issues. As we show here, one straightforward solution is to electrically dope the nanowires. The width of a metal-semiconductor Schottky barrier is proportional to $1/\sqrt{N_{D,A}}$, where N_D and N_A represent the donor and acceptor concentrations, respectively (Sze 1981, Chapter 5). For very high doping levels, this width is reduced and the barrier becomes transparent, i.e., allows high tunneling current and low contact resistance. Indeed, the transmission coefficient through the barrier increases exponentially with $\sqrt{N_{D,A}}$ (Sze 1981, Chapter 5).

In situ doping of self-assembled Si and Ge NWs grown using the VLS mechanism can be achieved by adding a dopant agent, e.g., B_2H_6 or PH_3, in the growth chamber along with the growth precursor (Greytak et al. 2004; Wang et al. 2005; Tutuc et al. 2006). High doping levels, up to 1 to 4×10^{19} cm^{-3}, can be achieved using this approach. In the case of phosphorous-doped Ge NWs with Ni contacts, the metal-NW contact resistance determined using two-point and four-point measurements becomes as low as ~500 Ω, for doping levels of ~1 to 4×10^{19} cm^{-3}. However, the dopant atoms incorporate primarily via conformal growth during *in situ* doping, with a doping concentration proportional to the exposure to the doping agent (Tutuc et al. 2006). This, in turn, prevents axial doping modulation or the realization of uniform *and* high doping concentrations using *in situ* doping.

Ho et al. (2008) demonstrated controlled nanoscale doping of Si nanoribbons using high temperature diffusion from molecular monolayers containing dopant (phosphorous or boron) atoms. Using this approach, doping concentrations as high as 2×10^{20} cm^{-3} have been demonstrated in Si, for both phosphrous and boron dopants. The doping is controlled by thermal diffusion, which requires a relatively high (~1,000°C) thermal budget, and concentration-dependent dopant diffusivities can affect the doping profile.

Ion implantation, a doping technique widely used in the microelectronics industry, allows for precise control of the doping concentration and depth. The main drawback associated with ion implantation is the ion beam-induced damage, which can cause nanowire amorphization. For bulk samples the amorphized region can be recrystallized using solid phase epitaxial regrowth, with the substrate acting as template (Narayan 1982). However, if the nanowires are fully amorphized during ion implantation, the solid phase regrowth is no longer possible (Greene et al. 2000). This sets upper boundaries for the dose and implant energy. Recent studies (Cohen et al. 2007; Colli et al. 2008) showed successful doping of Si NWs by ion implantation.

Using low-energy ion implantation, Nah et al. (2008) demonstrated Ge NW doping with levels up to 2×10^{20} cm^{-3}. In this study the nanowires are dispersed on a HfO$_2$ substrate, and implanted with boron at an energy of 3 keV and at different doses. Following activation at relatively low temperatures (400 to 600°C), multiterminal back-gated devices are fabricated in order to determine the doping concentration (Figure 7.10). Figure 7.11 data show the four-point current-voltage characteristics measured for B-implanted Ge nanowires. Figure 7.11a data indicate the nanowires have a high, up to ~10^{-4} S, conductance, which is roughly proportional to the implant dose. The gate dependence of I_d vs. ΔV_{4p} data of Figure 7.11b allows the extraction of the mobility and carrier concentration, using Equations 7.2 to 7.4. This study showed that doping concentrations up to 2×10^{20} cm^{-3} can be achieved using low energy ion implantation, and that the ion-beam-induced damage can be removed using a reduced, as low as 400°C,

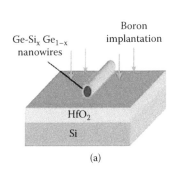
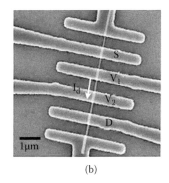

(a)   (b)

FIGURE 7.10

(a) Schematic representation of a Ge-Si$_x$Ge$_{1-x}$ core-shell nanowire dispersed on a HfO$_2$/Si substrate for ion implantation. (b) SEM of a multiterminal NW device. The outer terminals (S, D) are used to flow current, and the inner terminals (V_1, V_2) probe the voltage drop ($\Delta V_{4p} = V_2 - V_1$) along the nanowire. (Reproduced from Nah, J., Varahramyan, K., Liu, E.-S., et al., "Doping of Ge-Si$_x$Ge$_{1-x}$ Core-Shell Nanowires Using Low Energy Ion Implantation," *Applied Physics Letters* 93 [2008]: 203108. With permission.)

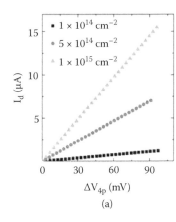
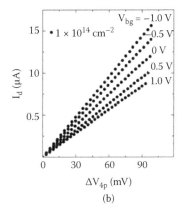

(a)   (b)

FIGURE 7.11

(a) I_d vs. ΔV_{4p} measured for B-implanted Ge-SiGe core-shell nanowires, at different implant doses. These data indicate the nanowire conductivity increases proportional to implant dose. (b) I_d vs. ΔV_{4p} measured at different back-gated bias (V_{bg}), as indicated in the figure. The I_d vs. ΔV_{4p} dependence on V_{bg} can be used to extract the mobility and doping concentrations according to Equations 7.2 to 7.4. (Reproduced from Nah, J., Varahramyan, K., Liu, E.-S., et al., "Doping of Ge-Si$_x$Ge$_{1-x}$ Core-Shell Nanowires Using Low Energy Ion Implantation," *Applied Physics Letters* 93 [2008]: 203108. With permission.)

thermal budget. By combining two-point and four-point measurements, the contact resistance (R_c) was determined. The R_c values for the Ni-Ge NW contact were as low as 300 Ω ± 200 Ω for doping levels above 10^{19} cm^{-3}, and ~2,200 Ω ± 2,000 Ω for doping levels below 10^{19} cm^{-3}. The corresponding specific contact resistance is 1.1×10^{-9} ± 2.2×10^{-10} $\Omega \cdot$cm^2 for doping levels above 10^{19} cm^{-3}, and 3.4×10^{-8} ± 2.2×10^{-8} $\Omega \cdot$cm^2 for doping levels below 10^{19} cm^{-3}.

One key benefit of achieving high doping levels in nanowires is that it enables the realization of high-performance, namely, high-current and high-on-off ratio, nanowire transistors with highly doped source (S) and drain (D). The highly doped S/D terminals allow for efficient, unipolar carrier injection into the NW and also for a low contact resistance. Figure 7.12 outlines the process flow for the fabrication of Ge-Si$_x$Ge$_{1-x}$ core-shell NW FETs with highly doped source and drain (Nah et al. 2009). The nanowires are first dispersed onto a HfO$_2$/Si substrate (Figure 7.12a). Next an 8.5-nm thick HfO$_2$ film is grown by atomic layer deposition (Figure 7.12b). The top gate is patterned using e-beam lithography, metal deposition (TaN), and lift-off, followed by top-gated dielectric etching using hydrofluoric acid (Figure 7.12c). The device then undergoes a low energy boron implant, with a dose of 1×10^{15} cm^{-2}. This step results in highly doped nanowire regions *outside* the TaN gate, which can subsequently serve as source/drain, while the relatively thick (~100 nm) TaN metal gate prevents the nanowire doping, and maintains the FET channel undoped. After dopant activation, the metal (Ni) terminals are defined using e-beam lithography, metal deposition, and lift-off (Figure 7.12d). An SEM of a Ge NW FET with highly doped S/D is shown in Figure 7.12d; the nanowire sections outside the TaN gates (darker areas) are highly doped, while the FET channel under the gates remains undoped.

To illustrate the benefit of using highly doped source and drain in Figure 7.13, we compare the device characteristics of top-gated Ge-Si$_x$Ge$_{1-x}$ core-shell NW FETs using both doped (Figure 7.13a) and undoped (Figure 7.13b) source and drain. The main panels show the output characteristics, and the insets show the transfer characteristics. Both devices have similar (720 nm) channel lengths. Two noteworthy observations can be made from Figure 7.13 data. First, the current measured at given drain and gate bias values in the NW FET with highly doped source and drain is two orders of magnitude *larger* than the device with undoped source and drain. Second, the on/off-state current ratio, defined for an operating

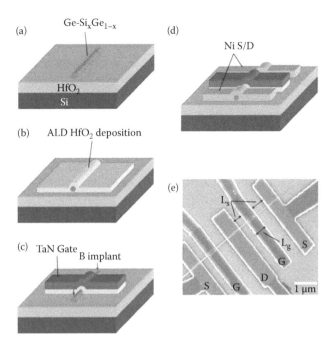

FIGURE 7.12

Schematic representation of a Ge-Si$_x$Ge$_{1-x}$ core-shell NW dispersed on a HfO$_2$/Si substrate. (b) ALD of HfO$_2$ on the NW. (c) Gate metal (TaN) deposition and HfO$_2$ layer etching, followed by B ion implantation. (d) B activation anneal and Ni deposition on the S/D regions. (e) SEM of a NW dual-gated FET with Ni contacts. The darker areas (G) represent the top gates, and the lighter areas represent the S/D. (Reproduced from Nah, J., Liu, E.-S., Shahrjerdi, D., et al., "Realization of Dual-Gated Ge-SiGe Core-Shell Nanowire Field Effect Transistors with Highly Doped Source and Drain," *Applied Physics Letters* 94 [2009]: 063117. With permission.)

widow of 1 V, is enhanced by an order of magnitude. Clearly, the device performance is greatly enhanced after S/D doping.

Using the data of Figure 7.11, the resistance of the nanowire sections outside the top gate, which represents the external contact resistance, is estimated to be $R_c \cong 22$ kΩ, which is small compared to the measured total device resistance, $R_m = 130$ kΩ at $V_g = -1$ V. This implies that the channel resistance, $R_{ch} = R_m - R_c \cong 108$ kΩ at $V_g = -1$ V, is now the dominant component of the device resistance, unlike the case of NW FET with metal contacts, where the contact resistance is the dominant component. Such NW FETs with highly doped source and drain, where the channel resistance determines the device characteristics, also allows for an accurate measurement of the intrinsic mobility (μ) using $R_{ch} = L_g [C_L (V_g - V_T) \mu]^{-1}$; C_L is the dielectric capacitance per unit length, and V_T the threshold voltage defined by linearly extrapolating the I_d vs. V_g data (Figure 7.13b) to zero I_d. For the device considered in Figure 7.13a, the capacitance value of C_L = 1.56 nF/m was calculated using a finite element method (Ansoft Q3D Extractor®); the extracted hole mobility using this C_L value was μ = 60 cm²/Vs.

7.5 SUMMARY OF REPORTED SPECIFIC CONTACT RESISTANCES

In this chapter we have attempted to shed light on a number of issues regarding electrical contacts to and intrinsic electronic properties of semiconductor nanowires. The main message can be summarized as follows. The electrical properties of semiconductor nanowire devices can be, depending on the device design and geometry, easily dominated by the contact resistance rather than the intrinsic nanowire resistance. The metal-nanowire resistance is controlled by the height and width of the Schottky barrier at the metal-nanowire junction. Unlike a bulk sample, however, where the Schottky barrier width depends only on the semiconductor doping level, the metal-nanowire Schottky barrier width depends on the metal profile in the nanowire and the electric field that can be applied using a gate. These factors can lead to a significant variability in contact resistance, depending on the device design. Consequently, the intrinsic electronic properties such as carrier mobility can no longer be extracted by assuming simple device models, unless the contact resistance becomes lower than the intrinsic nanowire resistance. A simple recipe to obtain low-contact resistances is to electrically dope the nanowires. This can be done either *in situ*, during growth, or *ex situ* using ion implantation or high-temperature diffusion. The metal-nanowire Schottky barrier becomes highly transparent when the nanowires are doped, allowing for low contact resistance.

In Table 7.1 we summarize the metal-semiconductor nanowire contact resistances reported so far in the literature for a variety of semiconductors and metal contacts. While the contact resistances vary, these data confirm that the higher the doping level for a given semiconductor nanowire, the lower the contact resistance. Lower-band-gap semiconductors tend to show lower contact resistances as well.

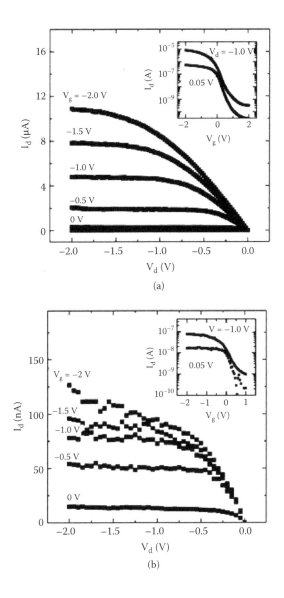

FIGURE 7.13

Drain current (I_d) vs. drain bias (V_d) data measured at different top-gated bias (V_g) values for two Ge-Si$_x$Ge$_{1-x}$ core-shell NW FETs using (a) highly doped source and drain, and (b) nominally undoped source and drain. Inset: I_d vs. V_g data measured at different V_d values for the same devices. The devices of (a) and (b) have similar channel lengths, $L_g = 720$ nm; the NW diameters are 60 and 36 nm in panels (a) and (b), respectively. These data illustrate that doping the source and drain increases the device current by two orders of magnitude. (Reproduced from Nah, J., Liu, E.-S., Shahrjerdi, D., et al., "Realization of Dual-Gated Ge-SiGe Core-Shell Nanowire Field Effect Transistors with Highly Doped Source and Drain," *Applied Physics Letters* 94 [2009]: 063117. With permission.)

TABLE 7.1

Summary of Metal-Semiconductor Nanowire Contact Resistances

Nanowire Semiconductor	Doping Type (Impurity)	Doping Level/ Resistivity	Contact Material	Contact Type	Contact Resistance (kΩ)	Specific Contact Resistance ($\Omega \cdot cm^2$) Schottky Barrier Height (eV)	Reference
Si	p (B)	1×10^{19} cm^{-3}	Ti/Au	Ohmic	260–400	4.5–5×10^{-4} $\Omega \cdot cm^2$	Mohney et al. (2005)
Si	n (P)	2×10^{19} cm^{-3}	Ti/Au	Ohmic	1.6	5×10^{-7} $\Omega \cdot cm^2$	Bjork et al. (2008)
Si	p (B)	1×10^{18} cm^{-3}	bulk Si	Ohmic	80	3.7–5×10^{-6} $\Omega \cdot cm^2$	Chaudhry et al. (2007)
Si	Undoped		Al/Au	Ohmic	2.4	2×10^{-2} $\Omega \cdot cm^2$	Richter et al. (2008)
Si	n (P)	0.01 $\Omega \cdot$cm	Ni	Schottky		0.38 eV	Woodruff et al. (2008)
Si	n (P)	0.1 $\Omega \cdot$cm	Ni	Schottky		0.49 eV	Woodruff et al. (2008)
Si	n (P)	0.3 $\Omega \cdot$cm	NiSi	Schottky		0.69 eV	Woodruff et al. (2008)
Ge/SiGe	p (B)	0.4–2×10^{20} cm^{-3}	Ni	Ohmic	0.3 ± 0.2	0.9–1.3×10^{-9} $\Omega \cdot cm^2$	Nah et al. (2008)
Ge/SiGe	p (B)	1–5×10^{18} cm^{-3}	Ni	Ohmic	2.2 ± 2	2.2–4.6×10^{-8} $\Omega \cdot cm^2$	Nah et al. (2008)
GaN	n	2×10^{17} cm^{-3}	Ti/Au	Schottky		0.019 eV	Chang et al. (2006)
GaN	n	$6.9 \times 10^{-3} \pm 3 \times 10^{-4}$ $\Omega \cdot$cm	Ti/Au	Ohmic	3.8	0.6–1.6×10^{-6} $\Omega \cdot cm^2$	Hwang et al. (2008)

Material	Type	Resistivity/Carrier concentration	Metal	Contact		Value	Reference
GaN	n	$6.9 \times 10^{-3} \pm 3 \times 10^{-4}$ Ω·cm	Al	Schottky		0.38 eV	Hwang et al. (2008)
GaN	n	$6.9 \times 10^{-3} \pm 3 \times 10^{-4}$ Ω·cm	Ti	Schottky		0.4 eV	Hwang et al. (2008)
GaN	n	$6.9 \times 10^{-3} \pm 3 \times 10^{-4}$ Ω·cm	Cr	Schottky		0.32 eV	Hwang et al. (2008)
GaN	n	$6.9 \times 10^{-3} \pm 3 \times 10^{-4}$ Ω·cm	Au	Schottky		0.42 eV	Hwang et al. (2008)
InN	n	$4.4 \times 10^{-4} \pm 1.5 \times 10^{-4}$ Ω·cm	Pd/Ti/Pt/Au	Ohmic		1.1×10^{-7} Ω·cm^2	Chang et al. (2006)
SiC	n	3×10^{19} cm^{-3}	Ni/Au	Ohmic	8.2	$5.9 \times 10^{-6} \pm 8.8 \times 10^{-6}$ Ω·cm^2	Jang et al. (2008)
SiC	n	3×10^{19} cm^{-3}	Ni/Au	Ohmic	104	$2.6 \times 10^{-6} \pm 3.4 \times 10^{-6}$ Ω·cm^2	Jang et al. (2008)
ZnSe		1 Ω·cm	Ti/Au	Ohmic	9,000	2.4×10^{-2} Ω·cm^2	Salfi et al. (2006)
SnO$_2$	n	1.8×10^{18} cm^{-3}	Pt	Schottky		0.28 ± 0.02 eV	Hernandez-Ramirez et al. (2007)
In$_2$O$_3$	Undoped	1×10^{18} cm^{-3}	Ti/Au	Ohmic	500		Jo et al. (2007)
ZnO	n(Ga)	3.9×10^{18} cm^{-3}	Ti/Au	Ohmic		$6.2\text{–}8 \times 10^{-6}$ Ω·cm^2	Yuan et al. (2008)
ZnO	n(Ga)	3.0×10^{19} cm^{-3}	Ti/Au	Ohmic		$7\text{–}9 \times 10^{-7}$ Ω·cm^2	Yuan et al. (2008)
ZnO	n(Ga)	$>1 \times 10^{20}$ cm^{-3}	Ti/Au	Ohmic		$5.3\text{–}6.8 \times 10^{-8}$ Ω·cm^2	Yuan et al. (2008)

ACKNOWLEDGMENTS

It is a pleasure to acknowledge numerous illuminating discussions with colleagues at IBM TJ Watson Research Center and the University of Texas at Austin. We thank Joerg Appenzeller, Supratik Guha, Guy Cohen, Sanjay Banerjee, L. Frank Register, Junghyo Nah, and Nitesh Jain for long-standing interactions and collaborations. We apologize for not giving adequate credit to everyone's work and for being biased toward work carried out at IBM or UT, with which we are most familiar. The work at the University of Texas was supported in part by the Defense Advanced Research Project Agency and the National Science Foundation.

REFERENCES

Abbas, C. C. 1984. A theoretical explanation of the carrier lifetime as a function of the injection level in gold-doped silicon. *IEEE Transactions on Electron Devices* 31:1428–432.

Bemski, G. 1958. Recombination properties of gold in silicon. *Physical Review* 111:1515–18.

Bjork, M. T., Knoch, J., Schmid, H., et al. 2008. Silicon nanowire tunneling field-effect transistors. *Applied Physics Letters* 92:193504.

Chang, C. Y., Chi, G. C., Wang, W. M., et al. 2006. Electrical transport properties of single GaN and InN nanowires. *Journal of Electronic Materials* 35:738.

Chaudhry, A., Ramamurthi, V., Fong, E., et al. 2007. Ultra-low contact resistance of epitaxially interfaced bridged silicon nanowires. *Nano Letters* 7:1536–41.

Chen, Z., Appenzeller, J., Knoch, J., et al. 2005. The role of metal-nanotube contact in the performance of carbon nanotube field-effect transistors. *Nano Letters* 5:1497–502.

Cohen, G. M., Rooks, M. J., Chu, J. O., et al. 2007. Nanowire metal-oxide-semiconductor field effect transistor with doped epitaxial contacts for source and drain. *Applied Physics Letters* 90:233110.

Colli, A., Fasoli, A., Ronning, C., et al. 2008. Ion beam doping of silicon nanowires. *Nano Letters* 8:2188–92.

Cui, Y., Zhong, Z., Wang, D., et al. 2003. High performance silicon nanowire field effect transistors. *Nano Letters* 3:149–52.

Datta, S. 1989. *Electronic transport in mesoscopic systems.* Cambridge: Cambridge University Press.

Givargizov, E. I. 1987. *Highly anisotropic crystals.* Dordrecht: D. Reidel Publishing Company.

Goldberger, J., Hochbaum, A. I., Fan, R., et al. 2006. Silicon vertically integrated nanowire field effect transistors. *Nano Letters* 6:973–77.

Greene, B. J., Valentino, J., Gibbons, J. F., et al. 2000. Thin single crystal silicon on oxide by lateral solid phase epitaxy of amorphous silicon and silicon germanium. *Proceedings of the 2000 MRS Spring Conference* 609:A9.3.1–6.

Greytak, A. B., Lauhon, L. J., Gudiksen, M. S., et al. 2004. Growth and transport properties of complementary germanium nanowire field-effect transistors. *Applied Physics Letters* 84:4176.

Hannon, J. B., Kodambaka, S., Ross, F. M., et al. 2006. The influence of surface migration of gold on the growth of silicon nanowires. *Nature* 440:69–71.

Heinze, S., Tersoff, J., Martel, R., et al. 2002. Carbon nanotubes as Schottky barrier transistors. *Physical Review Letters* 89:106801.

Hernandez-Ramirez, F., Tarancon, A., Casals, O., et al. 2007. Electrical properties of individual tin oxide nanowires contacted to platinum electrodes. *Physical Review B* 76:085429.

Ho, J. C., Yerulshami, R., Jacobson, Z. A., et al. 2008. Controlled nanoscale doping of semiconductors via molecular monolayers. *Nature Materials* 7:62–67.

Huang, Z., Fang, H., and Zhu, J. 2007. Fabrication of silicon nanowire arrays with controlled diameter, length and density. *Advanced Materials* 19:744–48.

Hwang, C., Hyung, J. H., Lee, S. Y., et al. 2008. The formation and characterization of electrical contacts (Schottky and ohmic) on gallium nitride nanowires. *Journal of Physics D: Applied Physics* 41:159802.

Ieong, M., Solomon, P. M., Laux, S. E., et al. 1998. Comparison of raised and Schottky source/drain MOSFETs using a novel local tunneling contact model. *Internationl Electron Device Meeting Technical Digest* 733–36.

Jang, C. O., Kim, T. H., Lee, S. Y., et al. 2008. Low resistance ohmic contacts to SiC nanowires and their applications to field-effect transistors. *Nanotechnology* 19:345203.

Jo, G., Maeng, J., Kim, T. W., et al. 2007. Effects of channel-length scaling on In_2O_3 nanowire field effect transistors studied by conducting atomic force microscopy. *Applied Physics Letters* 90:173106.

Kayes, B. M., Atwater, H. S., and Lewis, N. S. 2005. Comparison of the device physics principles of planar and radial p-n junction nanorod solar cells. *Journal of Applied Physics* 97:114302.

Liu, E. S., Jain, N., Varahramyan, K. M., et al. 2010. Role of metal-semiconductor contact in nanowire field-effect transistors. To appear in *IEEE Transactions on Nanotechnology*.

Mohney, S. E., Wang, Y., Cabassi, M. A., et al. 2005. Measuring the specific contact resistance to semiconductor nanowires. *Solid-State Electronics* 49:227–32.

Nah, J., Liu, E.-S., Shahrjerdi, D., et al. 2009. Realization of dual-gated Ge-SiGe core-shell nanowire field effect transistors with highly doped source and drain. *Applied Physics Letters* 94:063117.

Nah, J., Varahramyan, K., Liu, E.-S., et al. 2008. Doping of $Ge-Si_xGe_{1-x}$ core-shell nanowires using low energy ion implantation. *Applied Physics Letters* 93:203108.

Narayan, J. 1982. Interface structures during solid-phase-epitaxial growth in ion implanted semiconductors and a crystallization model. *Journal of Applied Physics* 53:8607–14.

Richter, C. A., Xiong, H. D., Zhu, X. X., et al. 2008. Methodology for electrical characterization of semiconductor nanowires. *IEEE Transactions on Electron Devices* 55:3086–95.

Salfi, J., Philipose, U., de Sousa, C. F., et al. 2006. Electrical properties of ohmic contacts to ZnSe nanowires and their application to nanowire-based photodetection. *Applied Physics Letters* 89:261112.

Sze, S. M. 1981. *Physical of semiconductor devices.* New York: Wiley-Interscience.

Tu, R., Zhang, L., Nishi, Y., et al. 2007. Measuring the capacitance of individual semiconductor nanowires for carrier mobility assessment. *Nano Letters* 7:1561–65.

Tutuc, E., Chu, J. O., Ott, J. A., et al. 2006. Doping of germanium nanowire grown in presence of PH_3. *Applied Physics Letters* 89:263101.

Wagner, R. S., and Ellis, W. C. 1964. Vapor-liquid-solid mechanism of single crystal growth. *Applied Physics Letters* 4:89–90.

Wang, D., Sheriff, B. A., and Heath, J. R. 2006. Silicon p-FETs from ultrahigh density nano-wire arrays. *Nano Letters* 6:1096–100.

Wang, Y., Lew, K., Ho, T., et al. 2005. Use of phosphine as an n-type dopant source for capor-liquid-solid growth of silicon nanowires. *Nano Letters* 5:2139–43.

Woodruff, S. M., Dellas, N. S., Liu, B. Z., et al. 2008. Nickel and nickel silicide Schottky bar-rier contacts to n-type silicon nanowires. *Journal of Vacuum Science and Technology B* 26:1592–96.

Yuan, G. D., Zhang, W. J., Jie, J. S., et al. 2008. Tunable n-type conductivity and transport properties of Ga-doped ZnO nanowire arrays. *Advanced Materials* 20:168–73.

8

Quantum Dot Solar Cells

Seth M. Hubbard, Ryne Raffaelle, and Sheila Bailey

8.1 INTRODUCTION

The first rigorous treatment of the limiting efficiencies of solar photovoltaic energy converters was given by Shockley and Queisser (1961). The authors made use of the fact that radiative recombination within the solar absorber will place the ultimate limit on efficiency. Their initial limit was near 44% for a single material solar absorber operating in an absolute zero ambient and illuminated with a 6,000K blackbody spectrum. This number has since been revised to 33% for an AM1.5 spectrum and the cell operating near room temperature (De Vos and Pauwels, 1981; Wuerfel and Ruppel, 1980).

Since the initial inception of the Shockley–Queisser (SQ) limit in 1961, various techniques have been proposed to increase it. These range in scope from simply more efficiently splitting the solar spectrum among multiple energy gap absorbers (e.g., tandem or multijunction cells) to advanced third-generation techniques such as hot carrier effects, impurity band absorption, and multiexciton generation (Green, 2003). The goal of all these advanced techniques is reducing one or both of the loss mechanisms inherent in the SQ model (i.e., transmission and carrier thermalization).

As more has been understood about the science and technology of quantum confined materials, it has been theorized that their unique properties could present a practical mechanism to implement many of the next-generation techniques. Quantum confinement refers to the fact that a dimension or dimensions associated with a given material are smaller than the Bohr exciton radius for that material. This confinement results in the electronic states associated with the material to deviate from the ordinary band theory of solids and begin to show discrete-like optoelectronic behavior (Singh, 1993).

The quantum dot (QD) is a specific type of quantum nanostructure with confinement potentials in all three Cartesian dimensions. The idealized QD can be considered to behave similarly to the famous quantum mechanical particle-in-a-box problem. This confines both the electrons and holes in three-dimensions (3D), allowing zero dimensions (0D) in their degrees of freedom, thus creating atom-like levels with discrete (Dirac delta) densities of states (see Figure 8.1) (Cress, 2008). In the simplest case, the confined energy levels, and thus the optical absorption energy, will be determined simply by the confining potential and the QD size. The minimum size of a spherical QD is determined by allowing for at least one confined energy level. This depends on the conduction band offset and effective mass and is given by (Bimberg et al., 1999)

$$ d_{min} = \frac{\pi \hbar}{\sqrt{2 m_e^* \Delta E_C}} \tag{8.1} $$

For typical InAs QD in GaAs this is in the 2 to 3 nm range. The maximum size of a QD is dictated by limiting the thermal population of higher energy levels to ~5%. This can be expressed as (Bimberg et al., 1999)

$$ kT \leq \frac{1}{3}\left(E_2 - E_1\right) \tag{8.2} $$

where E_1 and E_2 are the first and second quantum confined energy levels. For InAs QD at room temperature, this value is near 20 nm.

While there is not a single definition or type of quantum dot solar cell, the main thrust of all QD approaches seeks to use the quantized absorption or the 0D density of states of QD systems to provide means to implement "beyond the SQ limit" type solar conversion techniques. The basic physics behind the SQ limit lies in the fact that semiconducting solar cells are able to absorb light over much of the broad solar spectrum, but emit light (due to spontaneous emission) in only a narrow band near the bandgap. The detailed balance between the absorption and emission process results in the solar cell short-circuit current. A number of additional assumptions are also key to the SQ limit. The most important are that one electron-hole pair is produced for each absorbed photon, both electron and hole populations quickly lose energy and thermalize with the lattice temperature, and only photons with energy above the semiconductor bandgap (Eg) are absorbed.

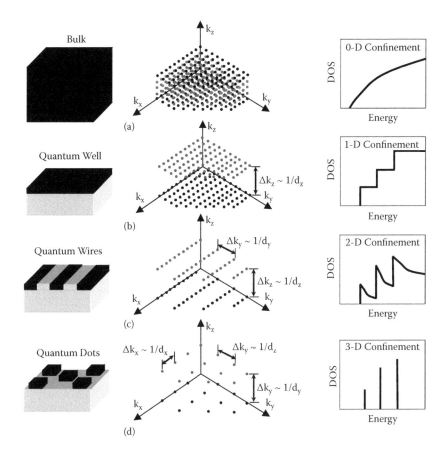

FIGURE 8.1
Schematic showing structure and the density of states (DOS) in the conduction band (CB) and valence band (VB) for (a) bulk, (b) quantum well, (c) quantum wire, and (d) quantum dot heterostructures. (Courtesy of Dr. Cory Cress; Cress, C., "Effects of Ionizing Radiation on Nanomaterials and III-V Semiconductor Devices" [PhD dissertation, Rochester Institute of Technology, Rochester, New York 2008].)

As seen in Figure 8.2, the result is that much of the sun's energy is lost to both transmission of photons with energy $E < Eg$ and thermal loss (heat) for carriers with energy $E \gg Eg$. In fact, these two processes account for loss of 23 and 33% of the sun's total power, respectively (Nelson, 2003).

The QD (and nanostructures in general) provides an enabling technology, allowing the possibility for both increasing below-bandgap (subgap) absorption and reducing thermal dissipation loss. The exact nature of how QD technology may improve upon the SQ limit depends on the nature and design of the solar converter used. In this chapter we will first review the various cell designs where QD technology could be implemented. The

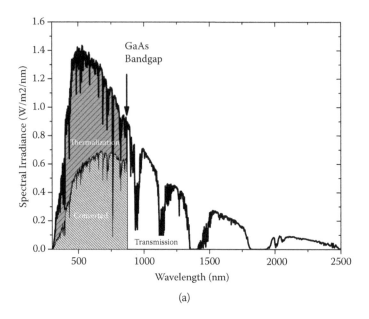

(a)

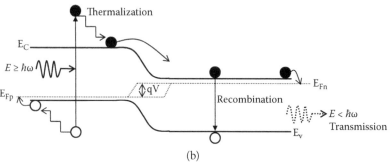

(b)

FIGURE 8.2

(a) AM1.5d solar spectrum showing the portions of light converted to electricity and lost to thermalization and transmission. (b) Band diagram of a single-junction solar cell showing thermalization and transmission loss, carrier separation, and extraction voltage (qV).

basics of the QD, growth mechanisms, and structure will then be presented. We will conclude with specific experimental behavior from current state-of-the-art cells implementing QDs.

8.2 TYPES OF QUANTUM DOTS SOLAR CELLS

As was discussed in the previous section, the QD solar cell does not refer to any single type of cell or cell design, but is a general term describing solar technology using QDs to implement advanced solar conversion techniques that go beyond the single-junction SQ limit. However, there are generally two broad classes of QD-based solar cells: (1) inorganic colloidal quantum dots in an organic host matrix and (2) epitaxial or vapor phase grown crystalline QDs. Examples of the first class are CdSe quantum dots, used for multiexciton generation (MEG) in electrolyte-based and dye-sensitized solar cells. The second class of QD solar cells use group III-V materials, such as InAs and GaSb, or group IV materials, such as Si and Si-Ge. The group III-V materials are typically grown epitaxially (usually using the Stranski–Krastanow technique or similar methods), while Si/Si-Ge QD can also be formed from chemical vapor deposition or laser ablation methods. Discussion of the latter QD growth mechanisms will be presented in Section 8.3.

8.2.1 Impurity and Intermediate Band Solar Cells

The idea of using an impurity state for improving photovoltaic energy conversion originated in the work of Martin Wolf (1960). Wolf theorized that by placing impurity states within the bandgap of the solar cell, multilevel absorption would be possible, thereby reducing transmission loss and thus increasing efficiency. The downside of this method is that impurities typically lead to increased nonradiative recombination due to the localized nature of the electron eigenfunctions within the impurity (Luque et al., 2006a). This increase in nonradiative recombination would negate any improvements to the operation of the cell. However, recent work by Brown and Green (2002) does suggest possibilities for true impurity band photovoltaics operating at the radiative limit.

A quite profound extension of the impurity effect was proposed by Luque and Marti (1997). In this approach, if the impurities were physically ordered and brought close enough that the individual wavefunctions begin to

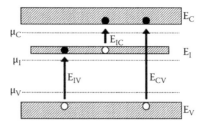

FIGURE 8.3
Energy band diagram showing the operation of the intermediate band solar cell.

overlap, a continuum "miniband" would be formed. This is a similar mechanism that results in the electronic band structure in a solid and the well-known Esaki–Tsu superlattice effect (Esaki and Tsu, 1970). Nonradiative recombination into the intermediate band is suppressed due to the continuous nature of the impurity band and a SQ type analysis can be applied.

The setup for this type of intermediate band solar cell is shown in Figure 8.3. The standard single-junction solar cells' absorption (E_{CV}) is still present. The intermediate band formed as above provides two additional absorption pathways, E_{IV} and E_{IC}. The electronic states in the intermediate band should only be accessible via direct transitions. Absorption of photon energy, E_{IV}, will pump an electron from the valence band to the intermediate band. A subsequent photon of energy, E_{CI}, then pumps an electron from the intermediate band to the conduction band. The increased performance of this design in comparison to the standard solar cell is a result of both reduced transmission loss and spectrum splitting (reduced thermalization).

In performing a detailed balance analysis of the intermediate band solar cell, there are a number of assumptions that must be applied to the SQ analysis:

1. No overlap of energy transitions for a given photon energy. In other words, referring to Figure 8.3, if E_I is placed arbitrarily closer to E_C than to E_V, there are only three energy ranges of interest: $E_{IC} \rightarrow E_{IV}$, $E_{IV} \rightarrow E_{CV}$, and $E_{CV} \rightarrow \infty$. This analysis is similar to what is done for multijunction solar cells where each subcell is analyzed only from that subcell's band edge to that of the next highest energy sub cell's band edge.

2. Current is extracted from the intermediate band only by optical pumping; i.e., electrons enter the intermediate band only by pumping from the valence band, and they leave only by subsequent pumping to the conduction band.

3. The valance, conduction, and intermediate bands each have a separate quasi-Fermi level. The difference in chemical potentials between any two bands is simply the difference between the quasi-Fermi levels of the two bands.

The spectral photon flux from the sun giving rise to the cells' short-circuit current can be found from the blackbody distribution as (Nelson, 2003)

$$\dot{N}_s = f_s \frac{2\pi}{h^3 c^2} \int_{E_g}^{\infty} \frac{E^2}{e^{E/kT_s} - 1} dE \qquad (8.3)$$

where f_s is a geometrical factor accounting for solar concentration and T_s is the temperature at the surface of the sun (6,000K). The values of f_s range from 2.1646×10^{-5} at 1 sun to 1 under full solar concentration (46,198 suns). Similarly, the spectral fluxes entering the cell from the ambient and leaving the solar cell due to spontaneous emission, thus giving rise to the dark current, are

$$\dot{N}_C = \frac{2\pi}{h^3 c^2} \int_{E_g}^{\infty} \frac{E^2}{e^{(E-\mu)/kT_C} - 1} dE \quad \text{and} \quad \dot{N}_A = \frac{2\pi}{h^3 c^2} \int_{E_g}^{\infty} \frac{E^2}{e^{(E)/kT_A} - 1} dE \quad (8.4)$$

where μ is the chemical potential, T_C is cell temperature, and T_A is ambient temperature. For the intermediate band solar cell, the current from each absorption process can be found from detailed balance as

$$\frac{J_{ij}(V)}{q} = f_s \dot{N}_s (E_i, E_j) + (1 - f_s) \dot{N}_A (E_i, E_j) - \dot{N}_C (E_i, E_j, u_{ij}) \qquad (8.5)$$

where i and j indicate the specific absorption process(J_{IV}, J_{IC}, or J_{CV}). The integration in Equations 8.3 and 8.4 should be carried out over the appropriate energy rages ($E_{IC} \rightarrow E_{IV}$, $E_{IV} \rightarrow E_{CV}$, and $E_C \rightarrow Y\infty$). The IBSC analysis has no restrictions on J_{CV}. However, as electrons or holes pumped into the intermediate band must also leave by removal to the conduction or valence bands, the currents J_{IV} and J_{IC} must be equal ($J_{IV} = J_{IC}$). The total cell current is then just $J_{CV} + J_{IV}$. Additionally, since each band must maintain a separate quasi-Fermi level,

TABLE 8.1

Maximum Efficiency and Corresponding Energy Values for
Solar Concentration from 1 to 1,000 Using a 6,000K Source

Solar Concentration	1	10	100	1,000
Efficiency limit	46.8%	50.1%	53.6%	57.3%
E_{IV} (eV)	1.49	1.43	1.36	1.31
E_{CI} (eV)	0.92	0.87	0.81	0.77
E_{CV} (eV)	2.41	2.30	2.17	2.08

$$\mu_{CV} = \mu_{IV} + \mu_{IC} = qV \tag{8.6}$$

where V is the applied voltage. Solving Equation 8.5 under these constraints and then calculating the maximum power point gives a maximum efficiency of 63.2% for $f_s = 1$ (full concentration), $E_{CI} = 0.7$ eV, and E_{IV} = 1.2 eV. Table 8.1 gives additional values for solar concentration factors from 1 to 1,000 (Aguinaldo, 2009).

Under the terrestrial AM1.5d spectrum, the water and CO_2-based absorption bands (see Figure 8.2) give rise to a number of local maximum points in IBSC bandgap vs. efficiency. Detailed balance efficiency can be found for AM1.5D by replacing the blackbody distribution in Equation 8.3 with the numerical AM1.5D spectrum given by the American Society for Testing and Materials (ASTM). The contour plots in Figure 8.4 show the 1-, 10-, 100-, and 1,000-sun AM1.5D maximum efficiency of the IBSC design for various values of E_{CI} and E_{IV} (note that the host bandgap E_{CV} = $E_{CI} + E_{IV}$).

Quantum dots have been both theoretically and experimentally evaluated as the practical implementation of this IBSC design. It was theorized that quantum dots could be coupled with each other to form a short period superlattice (Martí et al., 2002b, 2006b, 2008). The superlattice structure is realized by making an array of closely spaced quantum dots (Lazarenkova and Balandin, 2001). If sufficient wavefunction overlap occurs (coupling between adjacent well regions), then miniband formation occurs. Modeling results for miniband width of various size InAs QDs (modeled as truncated pyramids) in a GaAs matrix have been calculated by Tomic et al. (2008) and are shown in Figure 8.5. As can be seen, reducing the space between successive InAs QDs leads to stronger splitting of the energy bands.

These minibands form because the quantum wells or quantum dots form a periodic potential for charge carriers not unlike the periodic potential

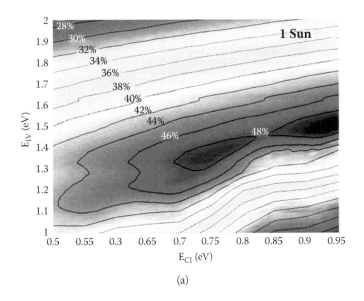

(a)

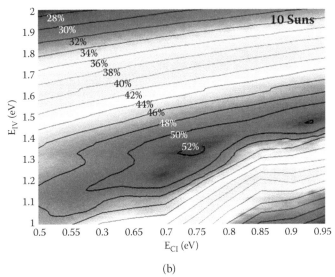

(b)

FIGURE 8.4

Efficiency contours for the IBSC design using AM1.5D spectrum and concentration factors of (a) 1, (b) 10, (c) 100, and (d) 1,000 suns. (Courtesy of Ryan Aguinaldo; Aguinaldo, R., "Modeling Solutions and Simulations for Advanced III-V Photovoltaics Based on Nanostructures," MS thesis dissertation, Rochester Institute of Technology, Rochester, New York, 2009.) *Continued*

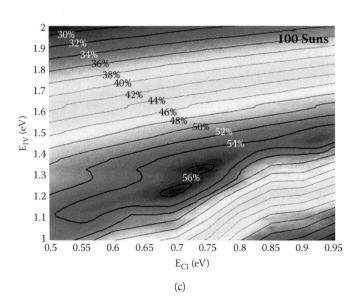

(c)

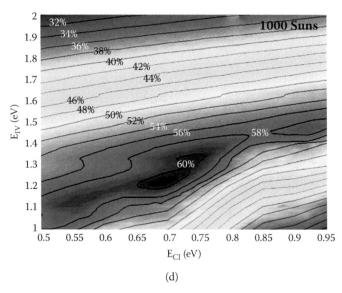

(d)

FIGURE 8.4
Continued.

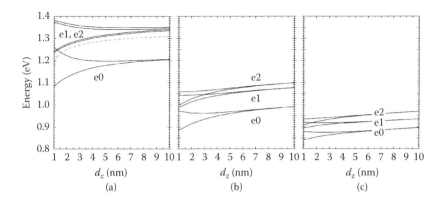

FIGURE 8.5
Miniband energy and width for truncated pyramidal InAs quantum dots. Split lines show position of the lower and upper limits of the fundamental ($e0$) and excited ($e1$, $e2$) confined states measured from the top of GaAs VB as they change with the spacer layer distance d_z. Quantum dot size is (a) 3×6 nm, (b) 6×12 nm, (c) 10×20 nm. In (a) the dashed line is the lower boundary of the WL-induced miniband. (Reprinted from Tomic, S., Jones, T. S., and Harrison, N. M., "Absorption Characteristics of a Quantum Dot Array Induced Intermediate Band: Implications for Solar Cell Design," *Applied Physics Letters* 93, p. 263105–2 [2008]. Copyright 2008, American Institute of Physics.)

formed by the crystal lattice. So just as the crystal lattice induces the formation of the usual energy bands, the superlattice induces the formation of additional minibands. The miniband formation occurs within the otherwise forbidden region of the bandgap of the host material. For an IBSC design, the QD must be doped, to allow for half filling of the IB with carriers in order to enhance the IB-to-CB absorption process.

A number of distinct advantages are apparent in a QD vs. a QW structure for IBSC operation. The strongly quantized absorption coefficient and ability to vary both materials and size would allow the use of QD superlattice arrays to form the IB at the appropriate level. The QD system provides a 0D density of states between the conduction band and the quantum confined levels. This would allow a discrete optical absorption process to occur between VB to IB and IB to CB. Additionally, this may prevent fast relaxation of carriers from the CB into the quantum confined IB levels (i.e., phonon-bottleneck effect) (Heitz et al., 2001). Finally, the symmetry selection rules for the QD system are relaxed, allowing absorption between confined states and the conduction band for normally incident light (e.g., intraband transitions in QW are forbidden for light normal to the QW stack) (Singh, 1993). Additional theoretical analysis regarding IBSCs may be found in Chapter 3.

TABLE 8.2

Detailed Balance Efficiency Limits for the InAs/GaAs
System at Several Solar Concentrations

	1×	10×	100×	1,000×
Blackbody	36.4%	41.4%	46.4%	51.0%
AM0	36.6%	41.6%	46.7%	51.3%
AM1.5d	38.6%	44.1%	49.6%	55.0%

The majority of experimental work for the intermediate band solar cell makes use of an InAs quantum dot array placed in the space charge region of a bulk GaAs solar cell (Popescu et al., 2008; Hubbard et al., 2007; Norman et al., 2005; Luque et al., 2006b). The InAs dot array induces the intermediate band by coupling of confined electronic states in the InAs conduction band; the design scheme for this system will be discussed in detail in Section 8.4. The InAs/GaAs system has the advantage that it is relatively well studied, it makes use of only binary semiconductors (as opposed to ternary or quaternary alloys), and it uses a commercially utilized solar cell material (GaAs). This InAs/GaAs (dot/host) system, however, is disadvantaged in that E_{CI} and E_{IV} are removed from the aforementioned efficiency maxima with values of approximately ~0.3 to 0.4 eV and 0.9 to 1.0 eV, respectively (Tomic et al., 2008). Additionally, the band offset is about 40% in the conduction band, leading to multiple confined hole levels. A proposed solution is to use a QD system of antimonide-based ternaries (Levy and Honsberg, 2008). These systems may have a more favorable IB energy in addition to a type II band alignment, leading to the lack of either confined hole or electron levels.

Although the InAs/GaAs system represents a nonideal material system with respect to the maximum detailed balance limits, this system may still represent an advantage over the efficiency limits of the single-junction cell. The detailed balance efficiency limits for the InAs/GaAs system, using E_{CI} = 0.4eV and E_{IV} = 1.0 eV, are determined and listed in Table 8.2 (Aguinaldo, 2009). Results are shown for a 6,000K blackbody and numerical ASTM values of both the space (AM0) and direct terrestrial (AM1.5D) spectrum.

8.2.2 Spectral Tuning Using QDS

The single-junction limiting efficiency of a solar cell is given by the detailed balance calculations of Shockley and Queisser (1961). Shown in Figure 8.6 are calculations of maximum efficiency vs. bandgap for various

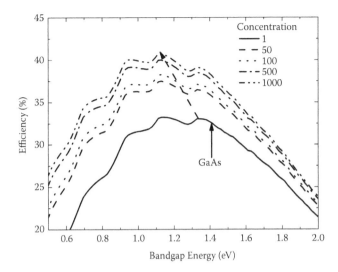

FIGURE 8.6

Calculated maximum efficiency vs. bandgap using AM1.5D spectrum. Various concentration levels are shown.

concentration levels under AM1.5D illumination. As can be seen, the maximum efficiency ranges from 33 to 41% for concentration from 1 sun up to 1,000 times (a typical operation point for concentrator photovoltaic systems). At the same time, the optimal bandgap energy varies from near 1.2 eV at 1 sun to near 1.0 eV at 1000 suns. Thus, obtaining maximum efficiency from a solar cell under concentration will require tuning the materials to match the optimal bandgap.

The spectral tuning approach to nanostructured concentrator photovoltaics can make use of the fact that the electrical and optical properties of nanomaterials can be controlled by changing the particle size. Quantum dots (QDs) and quantum wells (QWs) serve as potential wells for electrons (or holes) with dimensions on the order of a few nanometers, resulting in quantized discrete energy levels for electrons and holes. The energies of these levels are inversely related to the size of the particles, meaning the size of the particle will dictate the light energy that can be emitted or absorbed. Thus, insertion of QDs or QWs into a standard single-junction GaAs solar cell (E_g = 1.4 eV) can be used to tune the cell for operation under various concentration levels.

A second, slightly more sophisticated, approach involves application of nanomaterials to the current state of the art in III-V high-concentration photovoltaic systems, the lattice-matched triple-junction solar cell (TJSC).

The TJSC is essentially three different cells grown epitaxially on top of one another with connecting tunnel junctions in between. It is the equivalent of three individual cells connected in series. To provide efficient operation and low defect densities, each of the subsequent layers of the device are grown lattice matched to the previous layer. Figure 8.7a shows the basic device structure for a lattice-matched triple-junction solar cell. The conventional lattice-matched approach to multijunction solar cell development puts a constraint on the available bandgaps (see Figure 8.7b), and therefore how the solar spectrum is divided between the junctions (see Figure 8.8a). Optimization of the device design is done through the current-matching requirement that must be maintained through the structure.

In this particular design the middle InGaAs junction is current limiting, and therefore it is the efficiency limiting junction. Figure 8.8b shows calculations of theoretical efficiency contours based on the middle and top cell bandgap (given a fixed bottom cell bandgap of 0.66eV). This figure demonstrates how the current lattice-matched triple-junction solar cell can be improved by lowering the effective bandgap of the middle junction (Raffaelle et al., 2006).

Lowering the effective bandgap of the InGaAs middle cell in a lattice-matched triple junction can be achieved through the incorporation of an InAs QD array into the depletion region of an InGaAs cell. The InAs QDs in the InGaAs cell will provide subgap absorption and thus improve the short-circuit current of this junction. This cell could then be integrated into the three-cell stack to achieve a concentrator solar cell whose efficiency exceeds not only current state of the art (40% at 200 suns) but even the theoretical values for a quadruple-junction stack. A theoretical estimate predicts that a triple-junction cell with an improved middle cell current could have an efficiency exceeding 47% (1 sun) and even higher under concentration (52% at 200 suns) (Sinharoy et al., 2005).

8.2.3 Hot Carrier Solar Cells

The hot carrier solar cell has been theoretically studied by Würfel (1997) and Ross and Nozik (1982). The principle behind the operation of these cells involves extraction of carriers with energy $E > E_g$ before they are able to thermalize with the lattice. Thermalization of hot carriers represents a loss of 33% of the sun's energy in typical single-junction solar cells. Efficient extraction of hot carriers could lead to single-junction cell efficiency over 60% (1 sun).

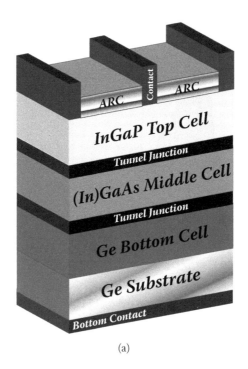

(a)

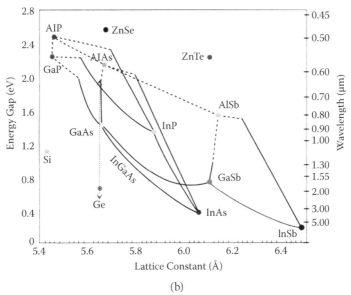

(b)

FIGURE 8.7

(a) Lattice-matched triple-junction solar cell design. (b) The crystal grower's chart showing available bandgaps vs. lattice constant for III-V materials with a dashed arrow indicating a lattice-matched triple-junction cell on Ge.

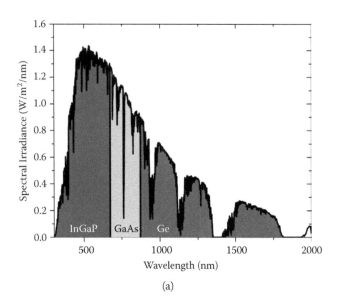

(a)

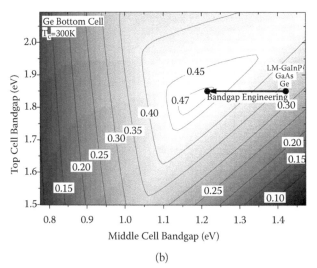

(b)

FIGURE 8.8

(a) AM1.5D spectrum with shading corresponding to the regions converted via each individual junction. (b) Theoretical efficiency contours based on the middle and top cell bandgaps of a triple-junction solar cell with a Ge bottom cell. (Figure (b) courtesy of Dr. John Andersen.)

Thermalization of carriers to the lattice temperature occurs on a timescale of picoseconds. Therefore, in order to efficiently extract hot carriers, either carrier cooling must be slowed or carriers must be extracted at a very short collection distance before thermalization can occur. In addition, as the ohmic contact of the semiconductor is in thermal equilibrium with the bulk lattice, carriers extracted using conventional contacts would be cooled and the hot carrier advantage negated. Selective energy ohmic contacts must be employed to prevent carrier cooling. If these requirements can be met, the hot carrier cell offers limiting efficiency under full concentration near 86% and under 1-sun illumination near 65% (Conibeer et al., 2008b).

The use of quantum dots in hot carrier solar cells has been proposed both for the absorber and as a selective energy contact. The so-called phonon bottleneck in QD allows for long lifetime excited states and increased thermalization time. In addition, the discrete energy levels of the QD could be used for selective energy contacts employed in a resonant tunneling scheme (see Figure 8.9). Experimental work using Si-based QDs for resonant tunneling contacts has been recently reported by Cho et al. (2008) and Conibeer et al. (2008a). Theoretical work on using QD absorbers has been reported by Conibeer et al. (2008b). Additional theoretical analysis regarding hot carrier solar cells may be found in Chapter 3.

8.3 GROWTH OF QUANTUM DOTS

Synthesis and fabrication of quantum dots can take place either through epitaxial growth or chemical synthesis (colloidal). Epitaxial III-V quantum dots have been grown using either molecular beam epitaxy (MBE) or metal organic vapor phase epitaxy (MOVPE). Crystalline group IV quantum dots are also grown using simple CVD or sputter deposition. Colloidal quantum dots are formed from chemical synthesis of either inorganic or organometallic precursors. In order to implement the QD cell designs of Section 8.2, epitaxial quantum dots are necessary and thus are the focus of this section.

8.3.1 Epitaxial Grown Quantum Dots

There have been numerous reported methods to grow epitaxial III-V quantum dots, such as e-beam lithography, nanoindentation, and epitaxial self-

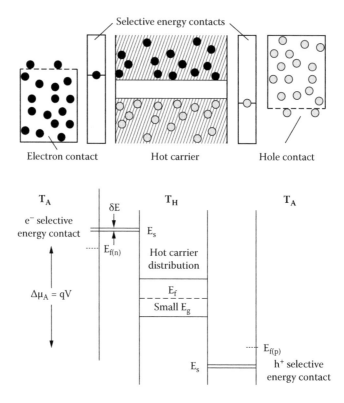

FIGURE 8.9

Schematic and band diagram of an ideal hot carrier solar cell. The absorber has a hot carrier distribution at temp T_H. Carriers cool isoentropically in the monoenergetic contacts to T_A. The difference of the Fermi levels of these two contacts is manifested as a difference in chemical potential of the carriers at each contact, and hence an external voltage, V. (Reprinted from Conibeer, G. J., Jiang, C. W., Nig, D., et al., "Selective Energy Contacts for Hot Carrier Solar Cells," *Thin Solid Films* 516 [2008a]: 6968–73. Copyright 2008, Elsevier.)

assembly. In the e-beam methods, electron beam lithography is used to form nanometer size patterns above a quantum well, and then the well is etched away to leave 3D quantum dots (Ils et al., 1993; Wang et al., 1992). This has been extended to using e-beam lithography to pattern openings in a SiO_2 mask, thus allowing later epitaxial regrowth only in the open areas (Lebens et al., 1990; Lan et al., 1999). Nanoindentation uses atomic force microscopy (AFM) or scanning probe microscopy (SPM) to locally strain an area of the substrate and thus force later growth of the quantum dots (Sleight et al., 1995; McKay et al., 2007). These techniques have the advantage of allowing growth of highly ordered and dense (>10^{11} cm^{-2}) arrays of quantum dots, potentially free of a 2D wetting layer. However, these techniques

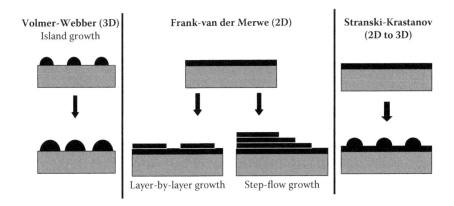

FIGURE 8.10
Epitaxial growth modes.

generally tend to produce less coherent quantum dots. Additionally, neither technique is self-organized and requires *ex situ* processing to form a single layer of quantum dots. While the above techniques may eventually be applied to the QD solar cell, current work in experimental demonstration of the designs of Section 8.2 relies on *in situ* epitaxy of self-organized QDs grown using the Stranski–Krastanov (SK) growth mode.

Thermodynamic diffusion-driven growth modes can be separated into the three categories shown in Figure 8.10: Volmer–Webber (3D), Frank–van der Merwe (2D), and Stranski–Krastanov (2D to 3D) (Ohring, 2002). Volmer–Webber growth occurs to minimize the total surface energy by increasing the surface area over the interfacial area. This is due to weak bonding between the substrate and the thin film. The result is immediate formation of coherent island-like structures with both vertical and lateral growth adding to the island size. Heteroepitaxy on a highly lattice-mismatched substrate usually results in Volmer–Webber type growth with a high density of misfit dislocations at the interface. On the other hand, when bonding between the substrate and film is high, the growing film will spread across or "wet" the surface. This leads to Frank–van der Merwe (2D) growth, where successive 2D layers are added one after the other. This method is typical of growth in lattice-matched materials resulting in 2D dislocation-free epitaxial layers.

The final growth mode (SK) occurs for strained layer epitaxy involving materials with low interface energy. Due to the low interface energy, layers start with 2D type growth. As the layer becomes thicker and strain increases, energy is minimized by formation of 3D islands in which strain is relaxed. It was initially thought that these islands were dislocated and thus noncoherent to the substrate. However, research in the mid-1980s found that using the correct growth conditions, InAs islands on GaAs could be grown coherent with the substrate (Goldstein et al., 1985). Thus began the last two decades of intensive research into QD formation by self-organized SK epitaxy (Leonard et al., 1993; Bimberg et al., 1995). The SK technique is ideal, as it can be done *in situ*, is reproducible, and may be repeated multiple times to create vertically stacked layers of QD heterostructures. This lends itself to the QD solar cell designs of Section 8.2 in that QD growth can be more easily added to an existing cell growth process, and in addition, QDs can be vertically stacked to increase absorption.

One of the most heavily studied systems of QDs grown by SK epitaxy is the InAs on GaAs system. The lattice constant of bulk InAs is 6.0583 Å (see Figure 8.7b) while that of bulk GaAs is 5.653 Å. This leads to a 7.8% compressive mismatch between the two lattice constants. The stress introduced by this large compressive mismatch will drive the SK growth process. Epitaxial growth initially results in a 2D wetting layer; however, after 1ML-2ML 2D growth, formation of coherent InAs islands (QDs) begins (Heitz et al., 1997; Leonard et al., 1994). Investigation of the InAs islands by transmission electron microscopy (TEM) or atomic force microscopy (AFM) has shown QD sizes between ~3 and 10 nm in height and ~20 and 30 nm in base diameter (Liao et al., 1998; Zou et al., 1999; Bimberg et al., 1994). The average saturation QD density ranges from 5×10^{10} to 1×10^{11} cm^{-2} (Chang et al., 2003). Surface facets of the QD vary depending on growth conditions (Ledentsov et al., 1995; Moll et al., 1998). Typical MBE growth conditions reveal the islands to be pyramidal or lens shaped with four [137] facets and two smaller [111] B facets (Eisele et al., 2008). A high-resolution scanning tunneling micrograph (STM) of a typical QD is shown in Figure 8.11 along with the surface facets derived from the micrograph.

Initial work on InAs-based QDs was done using MBE (Goldstein et al., 1985). The QD size, density, and optical properties are highly sensitive to growth conditions (Phillips, 1998). The final dot density will be highly dependent on both growth temperature and InAs coverage. Indium has a high surface diffusion length; thus, growth temperature is usually kept in the range of 470 to 530°C in order to limit this diffusion and enhance formation of InAs islands. Additionally, the InAs coverage is critical: low InAs coverage will not give the required strain energy for the 2D to 3D transition, while high InAs surface coverage leads to QD coalescence and large values of strain, resulting in defect formation. The parameter

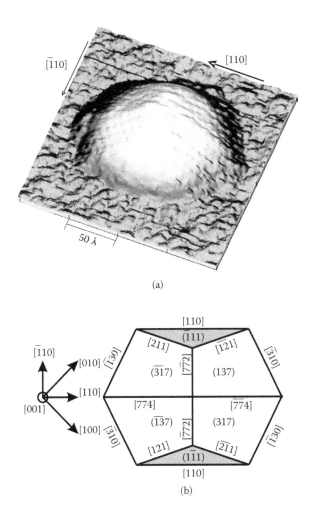

FIGURE 8.11

(a) STM image of an uncapped InAs on GaAs QD. (b) Shape model showing surface facets derived from analysis of the STM image. (Reprinted from Eisele, H., Lenz, A., Heitz, R., et al., "Change of InAs/GaAs Quantum Dot Shape and Composition during Capping," *Journal of Applied Physics* 104, p. 124301 [2008]. Copyright 2008, American Institute of Physics.)

window for high-density and good-optical-quality QDs is typically quite small (few MLs) with values highly dependent on related growth conditions (Bimberg et al., 1999).

Larger QDs can be grown using reduced strain values achieved by growth of InGaAs on GaAs. QDs can be formed with InGaAs to about 30% In content, below which the 2D to 3D transition will not occur. The lower the In content, the thicker the wetting layer, with thickness ranging from 1 ML for InAs to over 20 ML for $In_{0.30}Ga_{0.70}As$ (Berger et al., 1988).

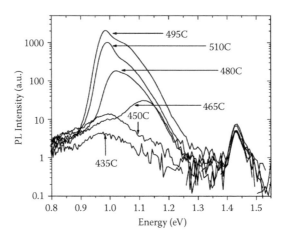

FIGURE 8.12

PL spectra for InAs QD with growth temperature variations between 435 and 510°C. (Reprinted Hubbard, S. M., Wilt, D., Bailey, S., Byrnes, D., and Raffaelle, R., "OMVPE Grown InAs Quantum Dots for Application in Nanostructured Photovoltaics," in *Proceedings of the IEEE World Conference on Photovoltaic Energy Conversion*, 118. [IEEE, Waikoloa, Hawaii, 2006]. Copyright 2006, IEEE.)

Most commercial producers of space solar cells use MOVPE for growth of cell structures. While QDs were first grown by MOVPE in the mid-1990s (Heinrichsdorff et al., 1996), progress in QD growth by MOVPE has only begun to accelerate over the last few years (El-Emawy et al., 2003; Cederberg et al., 2004; Hubbard et al., 2006); most QD optoelectronic devices are still grown by MBE. To address the need for viable production mechanisms for QD-based solar cells, growth of QDs by MOVPE is crucial.

In MOVPE, to initiate the SK growth mode, growth temperatures must be reduced to well below the values typically associated with mass transport limited growth. Thus, incomplete disassociation of the chemical precursors and kinetically limited growth tend to complicate QD formation. The net result is that QD nucleation, growth, and uniformity become much more sensitive toward MOVPE growth parameters, such as growth temperature, substrate off-cut, and V/III ratio.

The InAs growth temperature has a strong effect on overall luminescent intensity and also on peak position (see Figure 8.12). Peak intensity of QDs is increased over two orders of magnitude by increasing growth temperature from 450 to 500°C. Higher growth temperature leads to increased dot density and also reduced QD size dispersion.

At the temperatures used for InAs QD growth, the In adatom surface diffusion and growth kinetics are strongly influenced by growth temperature

(El-Emawy et al., 2003). At low growth temperatures, the reduced surface energy lowers the diffusion length of adatoms. This in turn limits redistribution of In, which would otherwise favor one or two dot sizes, thus leading to polydispersion in dot size. Higher growth temperatures give the In adatom sufficient energy to redistribute into coherent (defect-free) InAs QD. Further increasing the growth temperature will lead to In desorption and a resulting decrease in PL intensity.

In addition, the V/III ratio plays a strong role in nucleation of adatoms onto the growth surface (Sears et al., 2006). In fact, the V/III ratio plays a key role in affecting the uniformity of QD nucleation across an entire wafer. As shown in Figure 8.13, the reduction in V/III has a dramatic effect on the PL uniformity of the wafer. At higher V/III ratios, PL variations indicate that both QD size and radiative efficiency (i.e., QD coherence) are highly nonuniform. Using a reduced V/III ratio leads to improved dot nucleation across the wafer. AFM micrographs of samples grown at both

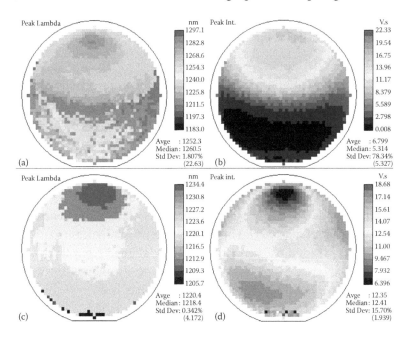

FIGURE 8.13

PL maps of QD luminescence for (a) peak wavelength at V/III = 58, (b) peak intensity at V/III = 58, (c) peak wavelength at V/III = 12, and (d) peak intensity at V/III = 12. (Reprinted Hubbard, S. M., Wilt, D., Bailey, S., Byrnes, D., and Raffaelle, R., "OMVPE Grown InAs Quantum Dots for Application in Nanostructured Photovoltaics," in *Proceedings of the IEEE World Conference on Photovoltaic Energy Conversion*, Waikoloa, Hawaii, 118. [IEEE, 2006]. Copyright 2006, IEEE.)

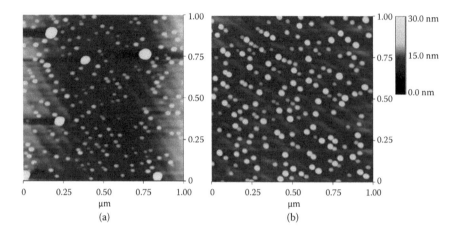

FIGURE 8.14

AFM micrographs near wafer center for samples grown using (a) V/III = 58 and (b) V/III = 12. (Reprinted Hubbard, S. M., Wilt, D., Bailey, S., Byrnes, D., and Raffaelle, R., "OMVPE Grown InAs Quantum Dots for Application in Nanostructured Photovoltaics," in *Proceedings of the IEEE World Conference on Photovoltaic Energy Conversion*, Waikoloa, Hawaii, 118. [IEEE, 2006]. Copyright 2006, IEEE.)

low and high V/III ratios are shown in Figure 8.14. The low V/III sample shows a single distribution of dot size near 7×30 nm with a dot density of 5×10^{10} cm^{-2} (Hubbard et al., 2006).

One of the other main advantages of the SK technique is the ability to vertically stack the QD structures, thus increasing the filling of QD material to host material, increasing the absorption cross-section and allowing for the vertical wavefunction overlap necessary for miniband formation. In the SK technique this is achieved inherently by way of the underlying strain field of the previous dot layer (Tersoff et al., 1996; Ledentsov et al., 1996; Darhuber et al., 1997). For thin barriers (usually less than 20 nm), QD layers can be almost perfectly vertically aligned from layer to layer. Stacks of up to twenty-five layers of InAs vertically aligned InGaAs QDs have been achieved by MBE, while MOVPE has been typically limited to stacking numbers less than 10 times (Bimberg et al., 1999). However, increasing QD stacks lead to buildup of residual net strain in the system and to defect formation unless properly strain balanced. In order to achieve the absorption necessary for the QD solar cells, multiple stacks of QD will be necessary, with low misfit dislocation density.

In addition to InAs, QD systems have been implemented in InP on GaAs (Manz et al., 2003; Georgsson et al., 1995), Si and SiGe systems (Schmidt et al., 1997; Brunner, 2002), InSb and GaSb on GaAs and InP (Bennett et al.,

TABLE 8.3

Material Systems Proposed for the QD IBSC Design

Confined	Barrier	η_{opt} (%)	ε_{CV} (eV)	$\varepsilon_{\Delta CB}$ (eV)
$InP_{0.85}Sb_{0.15}$	GaAs	53	1.42/1.52	0.49/0.51
$InA_{0.40}P_{0.60}$	GaAs	53	1.42/1.52	0.49/0.52
InAs	$GaAs_{0.88}Sb_{0.12}$	48	1.18/1.28	0.83/0.86
InP	$GaAs_{0.70}P_{0.30}$	49	1.79/1.89	0.43/0.46

Source: Reprinted from Levy, M., and Honsberg, C., "Nanostructured Absorbers for Multiple Transition Solar Cells," *IEEE Transactions on Electron Devices* 55 (2008): 706. Copyright 2008, IEEE.

1997; Tatebayashi et al., 2006), and ternary and quaternary compounds of InAlGaAs (Leon et al., 1998; Liu et al., 2002; Schlereth et al., 2008). As discussed previously, while extensively studied, the InAs/GaAs and InGaAs/GaAs QD systems do not typically give the optimal band lineup for IBSC solar cells. Other material systems have been proposed for the IBSC and are shown in Table 8.3 (Levy and Honsberg, 2008). These systems have IB levels closer to the theoretical optimum, allow for strain balancing, and are all type-II heterostructures (confining potential in only one band).

8.3.2 Strain Balancing

As discussed in Section 8.3.1, the SK technique results in accumulation of residual compressive strain as the number of QD layers in the array is increased. Above a critical thickness, the compressive strain will result in misfit dislocations and degradation of solar cell properties. The use of an additional strain-balancing layer with tensile strain (e.g., GaP) will offset compressive strain from the InAs QDs, leading to an overall strain-neutral stack. This strain-balanced technique has been effectively employed for a number of devices, including QD laser diodes and quantum well and quantum dot solar cells (Lever et al., 2004; Nuntawong et al., 2004; Ekins-Daukes et al., 2002).

In direct analogy to quantum wells, the compensation of compressive strain intrinsic to SK QD systems can be realized by introducing a layer of tensile strained material. This has been implemented using $GaAs_{1-x}P_x$ of varying P content (Popescu et al., 2008; Hubbard et al., 2008b). As seen in Figure 8.7b, this system has a lattice constant ranging from 5.653 Å (GaAs) to 5.451 Å (GaP). In order to maintain strain neutrality, the stress introduced by the $GaAs_{1-x}P_x$ layer should balance that of the compressively strained QD structures. Increasing the P content of the balancing

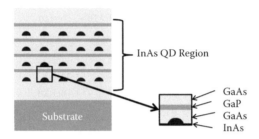

FIGURE 8.15
Strain-balancing scheme used for InAs QDs and GaP strain-balancing layer.

layer, thus reducing its lattice constant, will reduce the required thickness for proper strain balancing.

Initially, GaP was chosen for this layer, as its more aggressive tensile mismatch allows for the thinnest layers of GaP, facilitating electronic transport and possible miniband formation (Hubbard et al., 2008b). The strain balancing is implemented using a layer of InAs QDs followed by a thin GaAs capping layer, a GaP strain compensation layer, and another thin layer of GaAs, as seen in Figure 8.15.

High-resolution x-ray diffraction has been used to investigate the effects of strain-balancing layer and superlattice periodicity in QD-enhanced solar cell structures (Hubbard et al., 2008a). Scans of a nonbalanced QD solar cell show weak intensity and wide full width half maximum (FWHM) at the zero-order InAs peak, as well as low-clarity satellite peaks, indicating both lower crystalline and interface quality and disruption in superlattice periodicity. The effects of strain balancing can be seen in Figure 8.16a as improved clarity of satellite peaks, indicating improved interface quality and superlattice periodicity compared to nonbalanced samples.

The average perpendicular strain in the structure can be determined by examining the position of the zeroth-order superlattice peak (SL_0) with respect to the substrate Bragg peak (Nuntawong et al., 2004). The scan of the cell without strain balancing shows the SL_0 peak to the left of the substrate with a calculated net compressive strain of 6,000 ppm, consistent with the compressive nature of the SK growth mechanism. The scan of a 7ML thick strain-balancing layer shows the 0th-order SL peak now to the right of the GaAs Bragg peak, indicating an average tensile strain in the lattice (4,500 ppm). Finally, the scan of the 5ML strain-balancing sample shows no SL_0 peak and both SL_{-1} an SL_{+1} peaks

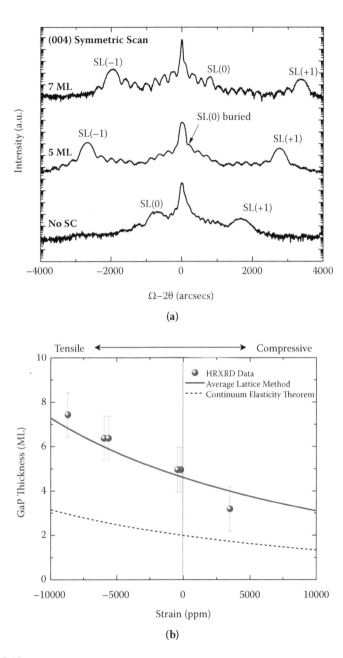

FIGURE 8.16

(a) Symmetric [004] 2-theta HRXRD scan of a 5 × Q0 solar cell showing GaAs Bragg peak and resulting superlattice peaks. Zeroth-order SL peak can be seen buried by the shoulder of the GaAs Bragg peak. (b) Raw HRXRD, simulated HRXRD data plotted with the average lattice method as a function of out-of-plane and in-plane strain.

symmetric about the substrate Bragg peak. This is a typical result for lattice-matched or strain-compensated SL structures, indicating the low residual strain.

The position of the zeroth-order superlattice peak varies with the average out-of-plane strain. The differential form of Bragg's formula is used to extract the strain (in ppm). Figure 8.16b identifies this strain in parts per million as a function of GaP thickness in ML obtained from the x-ray diffraction data. An average lattice method theory is employed to determine an optimal GaP thickness (Ekins-Daukes et al., 2002). The thickness of GaP is given as

$$t_2 = t_1 \left(\frac{a_0 - a_2}{a_1 - a_0} \right)$$

where t_1 is taken as an effective thickness of the InAs, a_0 is the lattice constant of GaAs, and a_1 and a_2 are the lattice constants of InAs and GaP, respectively. Allowing t_2 to vary depending on the strain (effective a_0 value) gives the dashed curve in Figure 8.16b.

A more rigorous method of calculating t_2 involves the continuum elasticity theorem (Ekins-Daukes et al., 2002). This method takes into account the elastic stiffness coefficients for the involved layers. The thickness of the GaP for this method is determined by

$$t_2 = t_1 \left(\frac{a_2 \left(a_1 A_1 - a_0 A_1 \right)}{a_1 \left(a_0 A_2 - a_2 A_2 \right)} \right)$$

where all constants are the same as the first method, but with A_i indicating a combination of stiffness coefficients for the ith layer/material. Further details of this method can be found in Ekins-Daukes et al. (2002) or Popescu et al. (2008). The solid curve in Figure 8.16b represents results of a modified version of this method, which incorporates the geometry and density of the QD layers (Bailey et al., 2009).

Recently, Popescu et al. (2008) demonstrated strain balancing using GaAs$_{1-x}$P$_x$ layers with lower P content. This approach allows the barrier between dots to be completely composed of the strain-balance material, eliminating the need for the GaAs spacer layers. Calculations of the strain-balanced condition using both continuum elasticity and atomistic

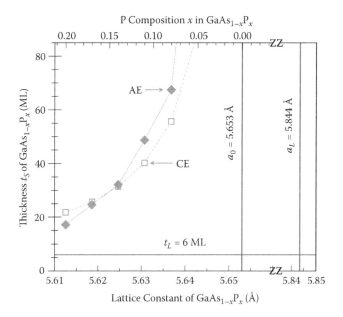

FIGURE 8.17

Thickness of GaAs$_{1-x}$P$_x$ layer necessary to attain a strain-balance condition in In$_{0.47}$Ga$_{0.53}$As/GaAs$_{1-x}$P$_x$ quantum well system on GaAs (001) substrate, as calculated from both continuum elasticity and atomistic elasticity theory. Vertical lines mark the natural lattice constants of GaAs (a_0) and In$_{0.47}$Ga$_{0.53}$As (a_L). (Reprinted from Popescu, V., Bester, G., Hanna, M. C., Norman, A. G., and Zunger, A., "Theoretical and Experimental Examination of the Intermediate-Band Concept for Strain-Balanced (In,Ga)As/Ga(As,P) Quantum Dot Solar Cells," *Phys. Rev. B* 78, p. 205321 [2008]. Copyright 2008, American Physical Society.)

elasticity theory are shown in Figure 8.17 (Popescu et al., 2008). Vertical stacks of QDs (Figure 8.18) using these conditions show no misfit dislocations and high uniformity for stacks up to fifty periods (Popescu et al., 2008).

8.4 BEHAVIOR AND CHARACTERIZATION OF QD SOLAR CELLS

As has been discussed, a practical implementation of the IBSC design involves incorporation of a multiple quantum dot superlattice (QD-SL) into standard single-junction solar cells in order to form an isolated intermediate band (IB) within the bandgap of the host (Marti et al., 2002a, 2006b). The proposed band diagram for the IBSC is shown in Figure 8.19

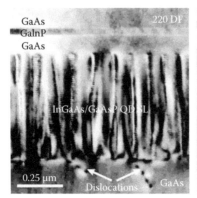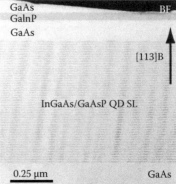

FIGURE 8.18

Left: 220 dark-field transmission electron microscopy image showing the generation of misfit dislocations in fifty-period InGaAs/GaAsP (0.8% P) QD superlattice grown on (113)B GaAs without strain balance. Right: Bright-field TEM image showing highly uniform fifty-period InGaAs/GaAsP (14% P) QD superlattice containing no misfit dislocations grown on a similar substrate. (Reprinted from Popescu, V., Bester, G., Hanna, M. C., Norman, A. G., and Zunger, A., "Theoretical and Experimental Examination of the Intermediate-Band Concept for Strain-Balanced (In,Ga)As/Ga(As,P) Quantum Dot Solar Cells," *Phys. Rev. B* 78, p. 205321 [2008]. Copyright 2008, American Physical Society.)

(Marti et al., 2006b). In this design, the QD superlattice is located between the emitter and base of the solar cell. An intermediate band would form due to wavefunction overlap and coupling between the QDs. Absorption of photons with energy below the host bandgap occurs from host valence band to IB and from IB to host conduction band. Additionally, the QD region is doped, in order to provide a separate quasi-Fermi level for the intermediate band, one of the requirements of the IBSC.

A design by Alonso-Alvarez et al. (2008) for such a cell is shown in Figure 8.20a. In this design, a 50 times superlattice of InAs QD is grown using solid source MBE. The QD-SL is added between the emitter and the base of a standard GaAs solar cell. In order to provide electrons to half fill the intermediate band (i.e., QD levels), Si δ doping is applied directly above the QD region. The δ doping directly above the QD region ensures that Si adatoms do not interfere with the sensitive QD nucleation and growth phase, while also supplying excess electrons in close proximity to the QD confining potential. A strain-balanced approach using GaP has also been employed, allowing for a low-defect-density vertically ordered 50 times QD array (see Figure 8.20b).

Figure 8.21a shows the external quantum efficiency for the cell structure grown in Alonso-Alvarez et al. (2008). The cell QE measures the

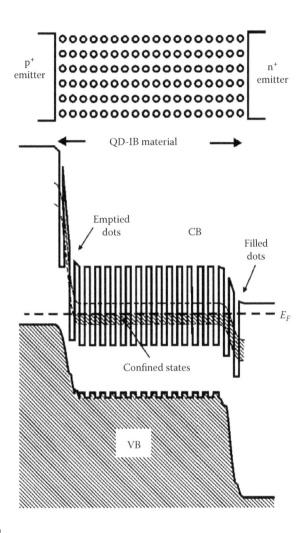

FIGURE 8.19
Band diagram of an IBSC given by Marti et al. (Reprinted from Marti, A., Lopez, N., Antolin, E., et al., "Novel Semiconductor Solar Cell Structures: The Quantum Dot Intermediate Band Solar Cell," *Thin Solid Films* 511–12 [2006b]: 638. Copyright 2006, Elsevier.)

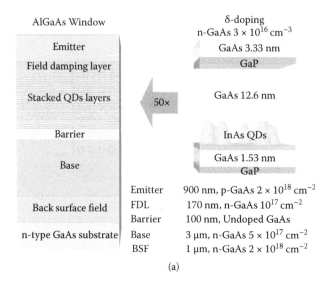

(a)

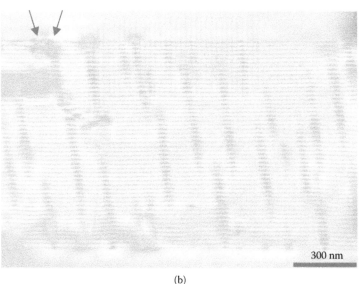

(b)

FIGURE 8.20
(a) Structure of a fifty-stacked QD solar cell with strain-balancing layers. (b) Cross-sectional TEM image of the fifty stacked QD layers. (Reprinted from Alonso-Alvarez, D., Taboada, A. G., Ripalda, J. M., et al., "Carrier Recombination Effects in Strain Compensated Quantum Dot Stacks Embedded in Solar Cells," *Applied Physics Letters* 93 [2008]: 123114. Copyright 2008, American Institute of Physics.)

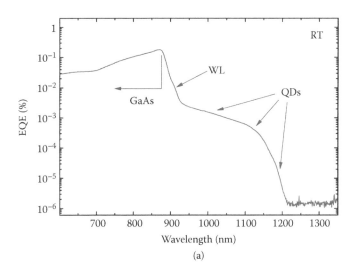

(a)

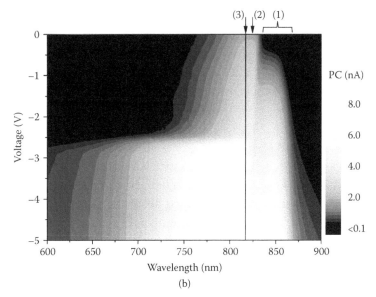

(b)

FIGURE 8.21

(a) Quantum efficiency at RT for the IBSC. (b) Photocurrent measurements at low temperature as a function of the applied bias. (Reprinted from Alonso-Alvarez, D., Taboada, A. G., Ripalda, J. M., et al., "Carrier Recombination Effects in Strain Compensated Quantum Dot Stacks Embedded in Solar Cells," *Applied Physics Letters* 93 [2008]: 123114. Copyright 2008, American Institute of Physics.)

photocurrent produced by the cell per incident photon, as a function of the photon wavelength. The quantum efficiency for a standard GaAs cell will rapidly fall off at the GaAs band edge near 870 nm. In this case, the QD cell shows an extended absorption spectrum comprised of wetting layer absorption near 900 nm and QD-related absorption up to 1,150 nm. This indicates that a portion of the short-circuit current is being generated from below bandgap absorption in the QD region, indicative of the VB to IB transition. Additionally, Marti et al. (2006a) have recently demonstrated photocurrent production solely due to the intermediate-to-conduction band transitions.

A complementary approach to the IBSC design takes advantage of the extended absorption spectrum of the lower-bandgap heterostructures (QDs) inserted into the current limiting junction of a multijunction solar cell (Raffaelle et al., 2006). In this approach, the extended absorption spectrum of the nanostructures allows for an enhancement of the overall short-circuit current and a global efficiency improvement of the triple-junction cell. As discussed in Section 8.2, modeling of an InGaP/GaAs/Ge triple-junction cell, in which QDs are used to extend the middle-junction absorption spectrum, has indicated that the efficiency could be improved to 47% under 1-sun AM0 illumination (Sinharoy et al., 2005). Recently, there have been various experimental demonstrations of QD photocurrent production in this type of solar cell using InAs (Hubbard et al., 2008b; Laghumavarapu et al., 2007; Oshima et al., 2008) and GaSb (Moscho et al., 2007) QDs in a GaAs host. These devices may also have additional benefits, such as enhanced radiation tolerance (Cress et al., 2007).

Figure 8.22 shows a simplified band diagram in the i-region of the QD-enhanced solar cell. The extended absorption spectrum (and thus enhanced short-circuit current) of the SC-QD solar cell is predicted to occur in two steps. First, incident photons with energy below the host (GaAs) bandgap result in absorption to the quantum-confined region (IB), creating a separate thermal distribution of electrons. Second, promotion (and subsequent collection) of these carriers from the IB to the conduction band can occur by either thermal (thermal-assisted extraction), tunneling, or optical (photon-assisted extraction) means. It is currently a topic of debate if purely thermal-assisted extraction can lead to an overall cell efficiency greater than the Shockley-Queisser limit for the quantum-confined material alone (Wei et al., 2007). However, for the QD tuning approach, the primary goal is enhancement of short-circuit current; some loss of voltage (and thus single-junction efficiency) may be

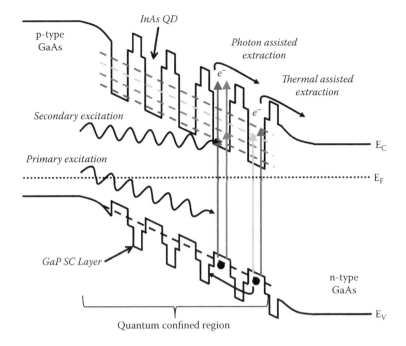

FIGURE 8.22
Schematic of band structure in the i-region of the SC-QD p-i-n solar cell. (Reprinted from Hubbard, S. M., Bailey, C., Cress, C. D., et al., "Short Circuit Current Enhancement of GaAs Solar Cells Using Strain Compensated InAs Quantum Dots," in *Proceedings of the 33rd IEEE Photovoltaic Specialists Conference* [IEEE, 2008a]. Copyright 2008, IEEE.)

acceptable while still leading to global efficiency enhancement in a triple-junction cell.

Figure 8.23 shows the design of the QD-enhanced solar cell. In this method QDs are inserted within the intrinsic region of a standard GaAs p-i-n solar cell. The high field region allows for enhancement of QD carrier collection once the electron or hole escapes from the QD either thermally or optically. This method also employs the strain-balancing technique to allow growth of a larger number of vertically ordered QDs.

Figure 8.24 shows the illuminated 1-sun AM0 current density-voltage (JV) curves for a baseline GaAs cell without QD and for a five-layer QD cell with and without strain-balance layers (Hubbard et al., 2008a). The addition of the uncompensated QD array produced substantial degradation in all device characteristics. Degradation in both short-circuit current density (J_{sc}) and open-circuit voltage (V_{oc}) are likely due to increased dark current resulting from nonradiative minority carrier recombination. Both reduction in carrier lifetime and mobility degradation in the emitter

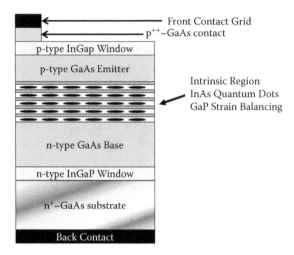

FIGURE 8.23
Design of the QD-enhanced GaAs solar cell.

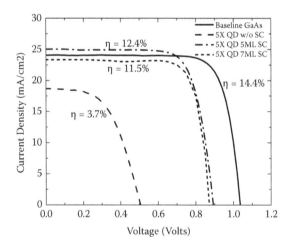

FIGURE 8.24
One sun AM0 current density vs. voltage curves for the baseline p-i-n GaAs cell and QD-enhanced p-i-n cells both with and without strain compensation (SC). (Reprinted from Hubbard, S. M., Bailey, C., Cress, C. D., et al., "Short Circuit Current Enhancement of GaAs Solar Cells Using Strain Compensated InAs Quantum Dots," in *Proceedings of the 33rd IEEE Photovoltaic Specialists Conference* [IEEE, 2008a]. Copyright 2008, IEEE.)

have been shown to be related to misfit dislocations originating at the QD layer and producing threads that create nonradiative traps and limit carrier mobility (Marti et al., 2007).

The addition of a GaP strain-balanced layer produced a dramatic improvement in all of the device characteristics. The enhancement in short-circuit current using 5ML GaP is clearly visible. Additionally, the open-circuit voltage for the 5ML SC device improved to 0.89 V. The increased J_{sc} of the device is a direct result of both photogenerated current contributed by the QDs and improved lifetime in the emitter region. The increase in V_{oc} can be related to a reduction in nonradiative recombination-driven current. Increased GaP thickness (7ML) leads to overcompensation (tensile strain) and degradation of device performance.

The QE measured for a number of QD-enhanced solar cells is shown in Figure 8.25 (Hubbard et al., 2008a). The quantum efficiency for a standard GaAs cells will rapidly fall off at the GaAs band edge near 870 nm. However, all of the QD cells show an extended QD spectrum at wavelengths below 870 nm. As in the IBSC case, this indicates that a portion of the short-circuit current is being generated from below bandgap absorption in the QD region.

Although the QD cells show an increase in QE at wavelengths above 870 nm, the effect of adding uncompensated dots to the cell structure

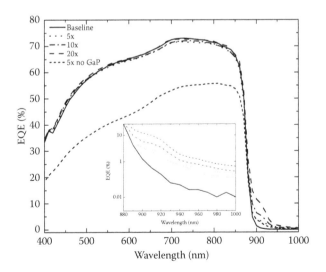

FIGURE 8.25
External quantum efficiency of the baseline solar cell, a 5× QD unbalanced QD solar cell, and 5 to 20 times QD solar cells with GaP strain-balanced layers.

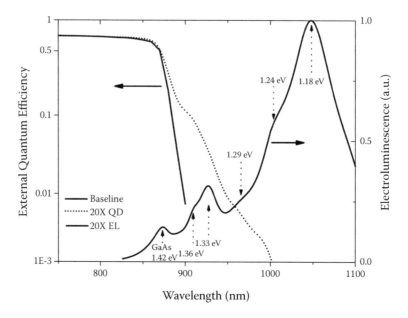

FIGURE 8.26

Electroluminescence and EQE for an InAs QD-enhanced solar cell.

degrades the QE over the λ < 870 nm portion of the spectrum. The reduction in QE below 870 nm would be consistent with increased nonradiative recombination in the emitter region of the nonbalanced QD solar cell. The QE for the strain-balanced QD cells does not show degradation in the λ < 870 nm spectral range. The QE for the strain-balanced solar cells shows that strain balancing leads to a higher-quality coherent QD superlattice. Also seen in Figure 8.25, as the number of QD layers in the superlattice is increased from 5 to 20 times, the spectral response in the QD region is increased. This is a clear indication that increasing the number of QD layers improves absorption and can thus lead to a further increase in the short-circuit current.

Room-temperature electroluminescence (EL) spectra of strain-balanced QD solar cells give further evidence of QD-related absorption and emission. As can be seen in the EL spectra of Figure 8.26, carrier injection through the i-region leads to multiple quantum dot–related peaks. Those near the GaAs emission are likely due to the wetting layer, while those at lower energy are due to radiative recombination in the QDs. Each of the EL peaks corresponds to peaks in the cell quantum efficiency. This is further evidence of light absorption and subsequent carrier collection occurring in the InAs QDs.

8.5 CONCLUSIONS AND FUTURE DIRECTIVES

While quantum dot solar cells are currently at the early stages of research and development, the authors believe this technology will play an important role in the next generation of solar cells. There is great potential in the multiple approaches available using QD technology. However, it is also clear that further work is necessary in order to improve design, material quality, and theory for QD-based solar cell systems.

Areas requiring further work involve both growth of the QD and design of the QD cell. Growth of QD systems using strain balancing to extend the number of defect-free vertically stacked QDs and thus increase the absorption of low-energy photons is crucial. Additionally, specific QD cell designs are necessary to maximize the percent of carriers removed from the QD superlattice. As was seen in the previous section, the heavily studied InAs on GaAs QD does not provide an intermediate band with proper bandgap for maximum IBSC efficiency. Thus, research into additional material systems with both proper intermediate band level and strain balancing will be necessary.

While the theory of nanostructured photovoltaics has been developed, there is still a tremendous amount of effort needed in terms of materials development and device design. This is a situation similar to that experienced by experimentalists in the early years of semiconductor device development. As was the case then, the hard work and dedication of many talented scientists was necessary to expand the field. The authors believe this could also be the case for the field of nanostructured photovoltaics. In years to come, the nanostructured solar cell may lead to a truly revolutionary shift in the power industry, leading to a potentially affordable and carbon-free bulk energy source.

ACKNOWLEDGMENTS

The authors wish to acknowledge the hard work and dedication of the students in the NanoPV group at NanoPower Research Laboratory, Rochester Institute of Technology. Specific contributions to this work were submitted from Prof. John Andersen, Chris Bailey, Stephen Polly, Ryan Aguinaldo, and Dr. Cory Cress. Authors also acknowledge support for our work from the U.S. Department of Energy (DE-FG36-08GO18012) and the U.S. government.

REFERENCES

Aguinaldo, R. 2009. Modeling solutions and simulations for advanced III-V photovoltaics based on nanostructures. MS thesis dissertation, Rochester Institute of Technology, Rochester, New York.

Alonso-Alvarez, D., Taboada, A. G., Ripalda, J. M., et al. 2008. Carrier recombination effects in strain compensated quantum dot stacks embedded in solar cells. *Applied Physics Letters* 93:123114.

Bailey, C. G., Hubbard, S. M., Forbes, D. U., and Raffaelle, R. P. 2009. Evaluation of strain balancing layer thickness for InAs/GaAs quantum dot arrays using high resolution x-ray diffraction and photoluminescence. *Appl. Phys. Lett.* 95:303110.

Bennett, B. R., Shanabrook, B. V., Thibado, P. M., Whitman, L. J., and Magno, R. 1997. Stranski-Krastanov growth of InSb, GaSb, and AlSb on GaAs: Structure of the wetting layers. *Journal of Crystal Growth* 175:888–93.

Berger, P. R., Chang, K., Bhattacharya, P., Singh, J., and Bajaj, K. K. 1988. Role of strain and growth-conditions on the growth front profile of InxGa1-xAs on gas during the pseudomorphic growth regime. *Applied Physics Letters* 53:684–86.

Bimberg, D., Grundmann, M., Ledentsov, N. N., et al. 1994. Self-organization processes in MBE-grown quantum dot structures. In *Proceedings Workshop on Molecular Beam Epitaxy-Growth Physics and Technology (MBE-GPT 94)*, 32–36. Lausanne: Elsevier Science Sa Lausanne.

Bimberg, D., Grundmann, M., Ledentsov, N. N., et al. 1995. Self-organization processes in MBE-grown quantum dot structures. *Thin Solid Films* 267:32–36.

Bimberg, D., Grundmann, M., and Ledentsov, N. N. 1999. *Quantum dot heterostructures.* New York: John Wiley.

Brown, A. S., and Green, M. A. 2002. Impurity photovoltaic effect: Fundamental energy conversion efficiency limits. *Journal of Applied Physics* 92:1329–36.

Brunner, K. 2002. Si/Ge nanostructures. *Reports on Progress in Physics* 65:27–72.

Cederberg, J. G., Kaatz, F. H., and Biefeld, R. M. 2004. The impact of growth parameters on the formation of InAs quantum dots on GaAs(1 0 0) by MOCVD. *Journal of Crystal Growth* 261:197–203.

Chang, F. Y., Wu, C. C., and Lin, H. H. 2003. Effect of InGaAs capping layer on the properties of InAs/InGaAs quantum dots and lasers. *Applied Physics Letters* 82:4477–79.

Cho, E.-C., Park, S., Hao, X., et al. 2008. Silicon quantum dot/crystalline silicon solar cells. *Nanotechnology* 19:245201.

Conibeer, G. J., Jiang, C. W., Nig, D., et al. 2008a. Selective energy contacts for hot carrier solar cells. *Thin Solid Films* 516:6968–73.

Conibeer, G. J., König, D., Green, M. A., and Guillemoles, J. F. 2008b. Slowing of carrier cooling in hot carrier solar cells. *Thin Solid Films* 516:6948–53.

Cress, C. 2008. Effects of ionizing radiation on nanomaterials and III-V semiconductor devices. PhD dissertation, Rochester Institute of Technology, Rochester, New York.

Cress, C. D., Hubbard, S. M., Landi, B. J., Raffaelle, R. P., and Wilt, D. M. 2007. Quantum dot solar cell tolerance to alpha-particle irradiation. *Applied Physics Letters* 91:183108.

Darhuber, A. A., Holy, V., Stangl, J., et al. 1997. Lateral and vertical ordering in multilayered self-organized InGaAs quantum dots studied by high resolution x-ray diffraction. *Applied Physics Letters* 70:955–57.

De Vos, A., and Pauwels, H. 1981. On the thermodynamic limit of photovoltaic energy conversion. *Applied Physics Berlin* 25:119–25.

Eisele, H., Lenz, A., Heitz, R., et al. 2008. Change of InAs/GaAs quantum dot shape and composition during capping. *Journal of Applied Physics* 104:124301.

Ekins-Daukes, N. J., Kawaguchi, K., and Zhang, J. 2002. Strain-balanced criteria for multiple quantum well structures and its signature in x-ray rocking curves. *Crystal Growth & Design* 2:287–92.

El-Emawy, A. A., Birudavolu, S., Wong, P. S., et al. 2003. Formation trends in quantum dot growth using metalorganic chemical vapor deposition. *Journal of Applied Physics* 93:3529–34.

Esaki, L., and Tsu, R. 1970. Superlattice and negative differential conductivity in semiconductors. *IBM Journal of Research and Development* 14:61.

Georgsson, K., Carlsson, N., Samuelson, L., Seifert, W., and Wallenberg, L. R. 1995. Transmission electron-microscopy investigation of the morphology of INP Stranski-Krastanow islands grown by metalorganic chemical-vapor-deposition. *Applied Physics Letters* 67:2981–82.

Goldstein, L., Glas, F., Marzin, J. Y., Charasse, M. N., and Leroux, G. 1985. Growth by molecular-beam epitaxy and characterization of InAs/GaAs strained-layer superlattices. *Applied Physics Letters* 47:1099–101.

Green, M. A. 2003. *Third generation photovoltaics: Advanced solar energy conversion.* Berlin: Springer.

Heinrichsdorff, F., Krost, A., Grundmann, M., et al. 1996. Self-organization processes of InGaAs/GaAs quantum dots grown by metalorganic chemical vapor deposition. *Applied Physics Letters* 68:3284–86.

Heitz, R., Born, H., Guffarth, F., et al. 2001. Existence of a phonon bottleneck for excitons in quantum dots. *Physical Review B* 64:4071–74.

Heitz, R., Ramachandran, T. R., Kalburge, A., et al. 1997. Observation of reentrant 2D to 3D morphology transition in highly strained epitaxy: InAs on GaAs. *Physical Review Letters* 78:4071–74.

Hubbard, S. M., Bailey, C., Cress, C. D., et al. 2008a. Short circuit current enhancement of GaAs solar cells using strain compensated InAs quantum dots. In *Proceedings of the 33rd IEEE Photovoltaic Specialists Conference.* IEEE, San Diego, CA.

Hubbard, S. M., Cress, C. D., Bailey, C. G., et al. 2008b. Effect of strain compensation on quantum dot enhanced GaAs solar cells. *Applied Physics Letters* 92:123512.

Hubbard, S. M., Raffaelle, R., Robinson, R., et al. 2007. Growth and characterization of InAs quantum dot enhanced photovoltaic devices. In *Proceedings of the Materials Research Society Symposium*, San Francisco, CA. DD13-11. MRS.

Hubbard, S. M., Wilt, D., Bailey, S., Byrnes, D., and Raffaelle, R. 2006. OMVPE grown InAs quantum dots for application in nanostructured photovoltaics. In *Proceedings of the IEEE World Conference on Photovoltaic Energy Conversion*, 118. IEEE, Waikoloa, HI.

Ils, P., Michel, M., Forchel, A., et al. 1993. Fabrication and optical-properties of InGaAs/InP quantum wires and dots with strong lateral quantization effects. *Journal of Vacuum Science & Technology B* 11:2584–87.

Laghumavarapu, R. B., El-Emawy, M., Nuntawong, N., et al. 2007. Improved device performance of InAs/GaAs quantum dot solar cells with GaP strain compensation layers. *Applied Physics Letters* 91:243115.

Lan, S., Akahane, K., Jang, K. Y., et al. 1999. Two-dimensional In0.4Ga0.6As/GaAs quantum dot superlattices realized by self-organized epitaxial growth. *Japanese Journal of Applied Physics Part 1: Regular Papers, Short Notes and Review Papers* 38:2934–43.

Lazarenkova, O. L., and Balandin, A. A. 2001. Miniband formation in a quantum dot crystal. *Journal of Applied Physics* 89:5509–15.

Lebens, J. A., Tsai, C. S., Vahala, K. J., and Kuech, T. F. 1990. Application of selective epitaxy to fabrication of nanometer scale wire and dot structures. *Applied Physics Letters* 56:2642–44.

Ledentsov, N. N., Grundmann, M., Kirstaedter, N., et al. 1995. Ordered arrays of quantum dots: Formation, electronic spectra, relaxation phenomena, lasing. In *Proceedings of the 7th International Conference on Modulated Semiconductor Structures (MSS-7)*, 785–98. Amsterdam: Pergamon-Elsevier Science Ltd.

Ledentsov, N. N., Shchukin, V. A., Grundmann, M., et al. 1996. Direct formation of vertically coupled quantum dots in Stranski-Krastanow growth. *Physical Review B* 54:8743–50.

Leon, R., Lobo, C., Clark, A., et al. 1998. Different paths to tunability in III-V quantum dots. *Journal of Applied Physics* 84:248–54.

Leonard, D., Krishnamurthy, M., Reaves, C. M., Denbaars, S. P., and Petroff, P. M. 1993. Direct formation of quantum-sized dots from uniform coherent islands of InGaAs on GaAs-surfaces. *Applied Physics Letters* 63:3203–5.

Leonard, D., Pond, K., and Petroff, P. M. 1994. Critical layer thickness for self-assembled InAs islands on GaAs. *Physical Review B* 50:11687–92.

Lever, P., Tan, H. H., and Jagadish, C. 2004. InGaAs quantum dots grown with GaP strain compensation layers. *Journal of Applied Physics* 95:5710–14.

Levy, M., and Honsberg, C. 2008. Nanostructured absorbers for multiple transition solar cells. *IEEE Transactions on Electron Devices* 55:706.

Liao, X. Z., Zou, J., Duan, X. F., et al. 1998. Transmission-electron microscopy study of the shape of buried InxGa1-xAs/GaAs quantum dots. *Physical Review B* 58:R4235–37.

Liu, H. Y., Sellers, I. R., Airey, R. J., et al. 2002. Room-temperature, ground-state lasing for red-emitting vertically aligned InAlAs/AlGaAs quantum dots grown on a GaAs(100) substrate. *Applied Physics Letters* 80:3769–71.

Luque, A., and Marti, A. 1997. Increasing the efficiency of ideal solar cells by photon induced transitions at intermediate levels. *Phys. Rev. Lett.* 78:5014.

Luque, A., Martí, A., Antolín, E., and Tablero, C. 2006a. Intermediate bands versus levels in non-radiative recombination. *Physica B: Condensed Matter* 382:320–27.

Luque, A., Marti, A., Lopez, N., et al. 2006b. Operation of the intermediate band solar cell under nonideal space charge region conditions and half filling of the intermediate band. *Journal of Applied Physics* 99:094503.

Manz, Y. M., Christ, A., Schmidt, O. G., Riedl, T., and Hangleiter, A. 2003. Optical and structural anisotropy of InP/GaInP quantum dots for laser applications. *Applied Physics Letters* 83:887–89.

Martí, A., Antolín, E., Cánovas, E., et al. 2008. Elements of the design and analysis of quantum-dot intermediate band solar cells. *Thin Solid Films* 516:6722.

Marti, A., Lopez, N., Antolin, E., Canovas, E., and Luque, A. 2007. Emitter degradation in quantum dot intermediate band solar cells. *Applied Physics Letters* 90:3.

Marti, A., Antolin, E., Stanley, C. R., et al. 2006a. Production of photocurrent due to intermediate-to-conduction-band transitions: A demonstration of a key operating principle of the intermediate-band solar cell. *Physical Review Letters* 97:247701.

Marti, A., Lopez, N., Antolin, E., et al. 2006b. Novel semiconductor solar cell structures: The quantum dot intermediate band solar cell. *Thin Solid Films* 511–12:638.

Marti, A., Cuadra, L., and Luque, A. 2002a. Quasi-drift diffusion model for the quantum dot intermediate band solar cell. *IEEE Trans. Electron Devices* 49:1632.

Martí, A., Cuadra, L., and Luque, A. 2002b. Design constraints of the quantum-dot intermediate band solar cell. *Physica EiLow-Dimensional Systems and Nanostructures* 14:157.

McKay, H., Rudzinski, P., Dehne, A., and Millunchick, J. M. 2007. Focused ion beam modification of surfaces for directed self-assembly of InAs/GaAs(001) quantum dots. *Nanotechnology* 18:6.

Moll, N., Scheffler, M., and Pehlke, E. 1998. Influence of surface stress on the equilibrium shape of strained quantum dots. *Physical Review B* 58:4566–71.

Moscho, R. B. L. A., Khoshakhlagh, A., El-Emawy, M., Lester, L. F., and Huffaker, D. L. 2007. GaSb/GaAs type II quantum dot solar cells for enhanced infrared spectral response. *Applied Physics Letters* 90:173125-1.

Nelson, J. 2003. *The physics of solar cells*. London: Imperial College Press.

Norman, A. G., Hanna, M. C., Dippo, P., et al. 2005. InGaAs/GaAs QD superlattices: MOVPE growth, structural and optical characterization, and application in intermediate-band solar cells. In *Proceedings of the 31st IEEE Photovoltaic Specialists Conference*, 43–48. IEEE, Lake Buena Vista, FL.

Nuntawong, N., Birudavolu, S., Hains, C. P., et al. 2004. Effect of strain-compensation in stacked 1.3 mu m InAs/GaAs quantum dot active regions grown by metalorganic chemical vapor deposition. *Applied Physics Letters* 85:3050–52.

Ohring, M. 2002. *The materials science of thin films*. San Diego: Academic Press.

Oshima, R., Takata, A., and Okada, Y. 2008. Strain-compensated InAs/GaNAs quantum dots for use in high-efficiency solar cells. *Applied Physics Letters* 93:083111.

Phillips, J. D. 1998. Self-assembled In(Al,Ga)As/Ga(Al)As quantum dots. PhD dissertation, University of Michigan, Ann Arbor.

Popescu, V., Bester, G., Hanna, M. C., Norman, A. G., and Zunger, A. 2008. Theoretical and experimental examination of the intermediate-band concept for strain-balanced (In,Ga)As/Ga(As,P) quantum dot solar cells. *Phys. Rev. B* 78:205321.

Raffaelle, R. P., Sinharoy, S., Andersen, J., Wilt, D. M., and Bailey, S. G. 2006. Multi-junction solar cell spectral tuning with quantum dots. In *Proceedings of the 4th IEEE World Conference on Photovoltaic Energy Conversion*, 162. IEEE, Waikoaloa, HI.

Ross, R. T., and Nozik, A. J. 1982. Efficiency of hot-carrier solar energy converters. *Journal of Applied Physics* 53:3813–18.

Schlereth, T. W., Schneider, C., Hofling, S., and Forchel, A. 2008. Tailoring of morphology and emission wavelength of AlGaInAs quantum dots. *Nanotechnology* 19:5.

Schmidt, O. G., Lange, C., Eberl, K., Kienzle, O., and Ernst, F. 1997. Formation of carbon-induced germanium dots. *Applied Physics Letters* 71:2340–42.

Sears, K., Tan, H. H., Wong-Leung, J., and Jagadish, C. 2006. The role of arsine in the self-assembled growth of InAs/GaAs quantum dots by metal organic chemical vapor deposition. *Journal of Applied Physics* 99:044908.

Shockley, W., and Queisser, H. J. 1961. Detailed balance limit of efficiency of p-n junction solar cells. *Journal of Applied Physics* 32:510.

Singh, J. 1993. *Physics of semiconductors and their heterostructures*. New York: McGraw-Hill.

Sinharoy, S., King, C. W., Bailey, S. G., and Raffaelle, R. P. 2005. InAs quantum dot for enhanced InGaAs space solar cells. In *Proceedings of the 31st IEEE Photovoltaic Specialist Conference*, 94. IEEE, Lake Buena Vista, FL.

Sleight, J. W., Welser, R. E., Guido, L. J., Amman, M., and Reed, M. A. 1995. Controlled III-V semiconductor cluster nucleation and epitaxial-growth via electron-beam lithography. *Applied Physics Letters* 66:1343–45.

Tatebayashi, J., Khoshakhlagh, A., Huang, S. H., et al. 2006. Formation and optical characteristics of strain-relieved and densely stacked GaSb/GaAs quantum dots. *Applied Physics Letters* 89:3.

Tersoff, J., Teichert, C., and Lagally, M. G. 1996. Self-organization in growth of quantum dot superlattices. *Physical Review Letters* 76:1675–78.

Tomic, S., Jones, T. S., and Harrison, N. M. 2008. Absorption characteristics of a quantum dot array induced intermediate band: Implications for solar cell design. *Applied Physics Letters* 93:263105.

Wang, P. D., Torres, C. M. S., Benisty, H., Weisbuch, C., and Beaumont, S. P. 1992. Radiative recombination in GaAs-AlxGa1-xAs quantum dots. *Applied Physics Letters* 61:946–48.

Wei, G. D., Shiu, K. T., Giebink, N. C., and Forrest, S. R. 2007. Thermodynamic limits of quantum photovoltaic cell efficiency. *Applied Physics Letters* 91:223507.

Wolf, M. 1960. Limitations and possibilities for improvement of photovoltaic solar energy converters. I. Considerations for earth's surface operation. *Proceedings of the IRE* 48:1246–63.

Wuerfel, P., and Ruppel, W. 1980. Upper limit of thermophotovoltaic solar-energy conversion. *IEEE Transactions on Electron Devices* ED-27:745–50.

Würfel, P. 1997. Solar energy conversion with hot electrons from impact ionisation. *Solar Energy Materials and Solar Cells* 46:43–52.

Zou, J., Liao, X. Z., Cockayne, D. J. H., and Leon, R. 1999. Transmission electron microscopy study of InxGa1-xAs quantum dots on a GaAs(001) substrate. *Physical Review B* 59:12279–82.

9

Luminescent Solar Concentrators

Amanda J. Chatten, Rahul Bose, Daniel J. Farrell, Ye Xiao, Ngai L. A. Chan, Liberato Manna, Andreas Büchtemann, Jana Quilitz, Michael G. Debije, and Keith W. J. Barnham

9.1 INTRODUCTION

This chapter reviews recent work by the Quantum Photovoltaic Group (QPV) at Imperial College London and their collaborators in the application of nanostructures to efficiently harvest light within the luminescent solar concentrator (LSC). In addition, the work of other groups and their means for optimizing the efficiency of the LSC are reviewed, as it is most likely that a combination of approaches will lead to the increases in system efficiency required for the commercialization of the LSC. Concentrating light with relatively inexpensive devices onto photovoltaic (PV) cells is a promising route to reducing the cost of PV energy generation and may be achieved using LSCs, which were first proposed about thirty years ago.[1,2] A LSC (see Figure 9.1) generally consists of a transparent plate or slab doped with luminescent centers. Incident light is absorbed by the luminescent centers and reemitted approximately isotropically, ideally with high luminescence quantum efficiency (LQE). While some of the emission is lost through the escape cones, a large fraction of the emission is trapped within the plate by total internal reflection (TIR) and wave guided to the concentrator edges, where it can be converted by PV cells. Concentration is achieved due to the geometrical ratio between the large surface area and the small edge areas. The LSC also converts the broad incident spectrum into a narrow emission spectrum that can be matched to the band gap of the PV cell such that the cell is operating at optimal efficiency and thermalization occurs in the LSC rather than in the more expensive cell, where the heat would reduce cell efficiency.

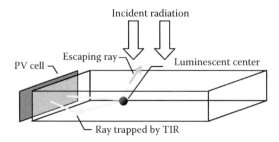

FIGURE 9.1

A typical luminescent solar concentrator with a photovoltaic cell attached, under illumination.

Major advantages of the LSC over geometric concentrators such as lenses or mirrors are that the LSC does not require expensive solar tracking and that it collects diffuse as well as direct light. The latter point is discussed in further detail in Section 9.3.3. More solar energy strikes the earth in a single hour than is consumed by all the activities of mankind in an entire year.[3] Solar energy is therefore an underused resource, and the attraction of the LSC is not that it can compete with conventional PV panels on a watt-for-watt basis, but that it offers a route to reducing the price per watt of solar-generated power. The LSC is therefore well suited to applications where generated power density can be exchanged for large area modules. This is the case for buildings, and even in the UK over seven times as much solar energy falls on buildings as is consumed inside. Because it is flat and static, the LSC is of particular interest for building-integrated PV (BIPV) applications, where it could replace conventional building materials.

The different configurations of the LSC that have been investigated in order to either optimize the efficiency or reduce costs are reviewed in the remainder of this section.

9.1.1 Stacked LSCs

A stacked arrangement of LSCs can be used to utilize the broad solar spectrum more efficiently, as proposed by Goetzberger[2] by spectrally separating the light and guiding it to appropriate PV cells. The principle of stacked LSCs is analogous to that of multijunction PV cells. LSCs with different absorption spectra are arranged on top of each other, as shown in Figure 9.2, with PV cells attached to the edges with band gaps (*Eg*) that match the respective emission spectra.

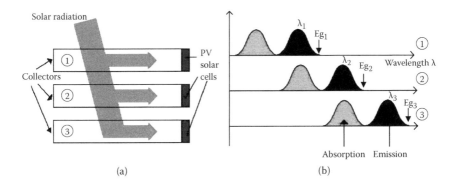

FIGURE 9.2
(a) Schematic of stacked LSCs with idealized absorption and emission spectra, as in (b).

With single LSCs there is a trade-off between absorbing a large part of the incident solar spectrum and emitting more energetic light. In general, radiation with energy lower than the excitation energy or band gap of the absorber cannot be absorbed, while radiation with higher energies can, to a certain extent. This means that a LSC with luminescent centers that absorb all the way to the low-energy regime can collect a large portion of the solar spectrum. The emission spectrum of such an absorber would, however, also be restricted to low energies. This means that although the photon current into an attached PV cell would be higher, the voltage generated by these photons would be lower, limiting the power generated. The stack solves this problem by absorbing the high-energy photons in a high-energy LSC layer. This layer would be transparent to the lower-energy photons because their energy would not be sufficient to create an excitation. These lower-energy photons can then be absorbed in another LSC with appropriate luminescent centers. Several successive layers can be prepared following this principle, so that overall, the majority of the spectrum can be absorbed without energetically downgrading the photons too much. In addition to the higher fabrication costs, the constraints on stacked LSCs are the availability of appropriate luminescent centers and PV cells with matching absorption spectra. We have investigated stacks of both dye- and quantum dot–doped LSCs[4,5] and confirmed that they utilize the broad solar spectrum more efficiently.

9.1.2 Mirrors and Number of Attached Cells

If only one solar cell is attached to the LSC, then without mirrors, emitted light at angles less than that required for TIR would be lost through all the other surfaces through the escape cones. These losses can be mitigated for all surfaces other than the top surface by the use of mirrors. An air gap is required between the LSC and any mirrors in order to conserve TIR. Specular and Lambertian reflectors have been investigated within the EC framework VI integrated project FULLSPECTRUM, and the results are presented in van Sark et al.[6] For a square LSC with only one cell attached, having both an air-gap diffuse Lambertian back reflector such as the material used within integrating spheres, which has a reflectivity of 97%, and air-gap specular edge mirrors such as 3M reflective foil, which also has a reflectivity of 97%, gives the highest efficiency.[6] As well as moderating escape cone losses, a back surface mirror provides a second pass for absorption and a diffuse reflector is advantageous because the reflected rays at wide angles have a longer path length through the LSC and are more likely to be absorbed than would be the case for a specular reflector.

The power conversion efficiency is defined as the electrical power from the attached PV cell divided by the optical power incident on the LSC. When attaching multiple PV cells to the edges, the top surface area of the LSC does not change, and thus the incident optical power remains the same. The electrical power from the PV cells, however, is increased, as more cells are delivering power and the efficiency increases. The cost of the attached cells obviously increases, but a simple geometric argument for a square LSC proves the efficiency and cost benefits of having cells attached to all four edges over having PV cells attached to two adjacent edges and mirrors on the other two. A square LSC with perfect mirrors (with a reflectivity of unity) on two adjacent sides is identical with respect to all internal processes to a square LSC of four times the area with PV cells attached to all four sides. Having the air-gap mirrors, however, leads to additional costs, and furthermore, in practice no perfect mirror material exists and the mirrors also lead to additional reflection losses, thereby degrading the LSC efficiency.

9.1.3 Thin-Film LSCs

The thin-film LSC consists of a layer of heavily doped material, such as a polymer, on top of a transparent substrate, such as glass (see Figure 9.3). The dopant concentration in the film has to be much higher than in a comparable homogeneous plate in order to achieve similar absorption.

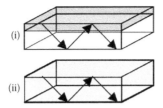

FIGURE 9.3

(i) A thin-film LSC consisting of a thin doped layer (shaded) on a transparent substrate with light rays trapped in the entire film/substrate composite. (ii) A homogeneously doped LSC of the same thickness.

Typically, the film and the substrate have matching refractive indices such that the luminescence is trapped in the entire composite.

Thin-film LSCs were originally suggested as a method to reduce reabsorption losses[7,8] with the reasoning that all reabsorption losses occur merely in the film while the trapped luminescence gains a long path length in the substrate. This effect has been debated,[9] arguing that the gain in path length in the substrate is compensated by the loss within the optically dense film. Our experimental measurements and computational simulations[10] support this view, showing no noticeable difference in performance between conventional, homogeneously doped LSCs and thin-film LSCs, and are discussed in greater detail in Section 9.5.

The main motivation now to pursue thin-film LSCs is that they are more convenient to fabricate, allowing flexibility in the choice of substrate. For example, glass that is durable and relatively inexpensive in addition to absorbing harmful UV may be utilized in the bulk of the LSC. Furthermore, the small separation between luminescent centers in the doped layer can be utilized for Förster energy transfer. This has been achieved in the work of Currie et al.[11] on thin-film organic dye-doped LSCs. This work highlights the need to consider geometric and flux gains, related to the size of the LSC. Having projected power conversion efficiencies for this system of up to 6.8%, the above factors were used to calculate commercial power generation costs and illustrate the viability of solar power generated by LSCs. This work also highlights the need to choose dyes with little overlap between the absorption and emission spectra, thereby minimizing the reabsorption that leads to both LQE and escape cone losses. Recently, efficient Förster transfer has been demonstrated between organic polymers and colloidal QDs,[12] and a critical radius of about 8 nm was deduced. The transfer was studied in the solid phase and took place from the polymer to the QDs.

9.1.4 Alternative Geometries

Whilst most fabricated LSCs today take the form of a cuboid, other geometries have been considered, including an array arrangement of cylinders. Theoretical ray-trace modeling[13] found that the optical concentration for cylindrical LSCs can reach 1.9 times that of square-planar LSCs, with identical volume and PV cell surface area. The model, however, assumes that emission occurs only at the cylindrical axis and very close to the surface, and predicts that the maximum benefit occurs only for a certain range of absorption coefficient.

The effect of varying the flat-plate geometry on LSC performance has also been analyzed using ray-trace modeling.[6,14] Square, right-angled triangular and hexagonal QD-doped LSCs of increasing top surface apertures (A_{top}) were considered and concentration ratios (CRs) predicted for increasing A_{top} for each geometry type.[14] The hexagonal geometry gave the highest CRs in the range of A_{top} considered, but for a given A_{top}, each geometry type has a different area of attached PV cells and hence different costs. Therefore, to determine the optimum geometry, relative costs per unit power output were calculated, as detailed in Kennedy et al.,[14] with the relative costs of the LSC plate and PV cells factored in. The results indicated that all geometries can attain the same minimum relative cost per unit power, and it was concluded that while the flat-plate geometry is not critical, selecting the appropriate size is.

9.2 LUMINESCENT SPECIES

The original LSC utilized organic dyes as the luminescent species, which turned out to degrade in sunlight, posing a major obstacle to the production of commercially viable concentrators. There has recently been a renewed interest in the LSC resulting from a number of factors, including the availability of photostable organic dyes[15] with high-luminescence quantum efficiency, new dyes that function in the red region of the spectrum, and polymer materials with low background absorption. In addition, novel luminescent centers such as core-shell QDs have been proposed,[16] and higher-efficiency PV cells are available with more appropriate band gaps.

We are currently evaluating the performance of both colloidal nanocrystals (QDs and nanorods) and organic dyes as the luminescent species in the LSC. Nanocrystals have advantages over organic dyes in that (1) their absorption spectra are far broader, extending into the UV; (2) their absorption properties may be tuned simply by the choice of nanocrystal size, and (3) they are inherently more stable than organic dyes. Moreover, there is a further advantage in that the red shift between absorption and luminescence is *quantitatively* related to the *spread* of nanocrystal sizes, which may be determined during the growth process, providing an additional strategy for minimizing losses due to reabsorption.[16,17] However, the LQE of nanocrystals cannot yet match the near-unity values achieved for some dyes, but encouraging results of LQE ~ 0.8 have been reported for core-shell QDs[18] and nanorods.[19] The properties of the luminescent species that have been utilized in the LSC are reviewed in the remainder of this section.

9.2.1 Light Harvesting

The advantages of nanocrystals over dyes for harvesting the solar spectrum can be illustrated by considering otherwise identical LSCs doped with typical colloidal QDs and a typical dye, respectively, with matching absorption thresholds. A 0.5-cm thick LSC illuminated by the standard AM1.5 global spectrum and doped with the QDs illustrated in Figure 9.4 would absorb 42.3% of the incident photons out to the band gap (E_g 700nm) of an InGaP cell attached to one edge, whereas one doped with the fluorescent dye illustrated in Figure 9.4 would only absorb 23.5%. Assuming

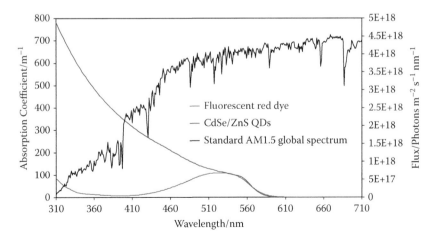

FIGURE 9.4
Spectral overlap between the standard AM1.5 global spectrum and the absorption spectra of a Bayer fluorescent red dye and core-shell QDs with matched absorption thresholds.

the record LQE of 0.84[17] for the QDs is achievable in a polymer plate and a LQE of 0.95 for the dye, thermodynamic modeling of a 10 × 10 cm LSC predicts that the short-circuit current in the attached InGaP cell of the QD-doped LSC would be almost 30% higher than that for the dye-doped LSC despite the higher LQE losses. Specular reflectors with a reflectivity of 97% on the other three edges and back surface and a PMMA host material were assumed in these calculations.

9.2.2 Nanorod-Doped LSCs

The use of nanorods (NRs) in the LSC is a novel approach for which the major motivation is that the optical characteristics of NRs are expected to further minimize reabsorption losses.

Several properties of the NRs depend on the aspect ratio. We have studied nanorod-doped LSCs[20] (see Figure 9.5a) in which the rods used had a diameter of about 5 nm and a length of about 20 nm (see Figure 9.5b), and thus an aspect ratio of 4. The LQE in solution was about 70%,[21] and the corresponding LQE in the homogeneous LSC has also been determined to be 70 ± 1%. Moreover, the nanorods are expected to exhibit anisotropic emission, with a maximal emission in the plane perpendicular to the long axis. This effect, however, is only visible when the rods collectively align along a preferred orientation. Spontaneous alignment occurs at high concentrations (see, e.g., Carbone et al.[21]), and alignment may also be achieved at lower concentrations by stretching polymer films.[22]

(a)

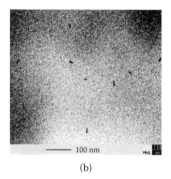
100 nm
(b)

FIGURE 9.5

(a) Photograph of a homogeneous LSC containing nanorods under UV illumination. (b) TEM image of CdSe/CdS nanorods in a polymer nanocomposite LSC. (See color insert following page 206.)

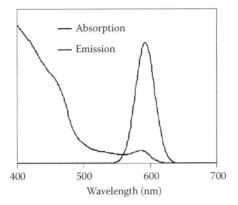

FIGURE 9.6

Absorption and emission spectra of a homogeneous nanorod-doped LSC. There is an exciton peak near the luminescence peak, but the major part of the absorption occurs at a significant shift from the luminescence. (From Bose, R. et al. 2008. *Proceedings of the 33rd IEEE Photovoltaics Specialist Conference.* With permission.)

The absorption and photoluminescence (PL) spectra of the nanorods in the concentrators were measured and found to be comparable to the spectra in solution.[20] The absorption spectrum features a peak at longer wavelengths, which arises from excitons in the core and overlaps strongly with the emission peak (see Figure 9.6). However, the overall absorption is dominated by excitons in the shell, such that the self-absorption that leads to reabsorption losses is relatively suppressed.

Our calculations show that these typical NRs double the fraction of absorbed photons emitted from the edges of the LSC, which could be converted in attached PV cells compared with comparable typical QDs.[20]

9.2.3 Multiple Dyes

As discussed in Section 9.1.1, it is important to utilize as much of the solar spectrum, at high efficiency, as possible. An alternative method for harvesting more of the solar spectrum that does not have the additional fabrication costs of a stack is to combine organic dyes in a single plate.

The idea of incorporating multiple dyes in a single LSC was presented in a letter submitted by Swartz et al.[23] The aim was to extend the conventional LSC absorption range of 300 to 600 nm to the near infrared (NIR), which would effectively double the photon flux in the collector. Swartz proposed combining dyes with different absorption spectra in a single LSC in order to achieve this, as illustrated in Figure 9.7.

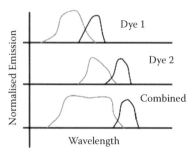

FIGURE 9.7

The basic concept of a two-mixed-dye-based collector. The absorption ranges (lighter line) of both dye 1 and dye 2 combine to cover a wider range of the solar spectrum. This results in a single emission (darker line) peak characterized by the lowest-energy dye. Cascaded emission occurs as the emission spectrum of dye 1 corresponds to the absorption spectrum of dye 2.

Cascaded emission occurs as a result of such mixing of dyes, in which photons are absorbed and emitted in series by dyes of increasing wavelength. This results in an emission spectrum dictated by the NIR dye. In effect, the system achieves a broad absorption spectrum with a single emission peak, but there are LQE losses associated with each step in addition to the thermalization losses discussed in Section 9.1.1.

Richards and McIntosh achieved a power conversion efficiency of 4.4% with a mixture of violet, yellow, orange, and red dyes.[24] Their report stated that the efficiency of the LSC depended largely on the LQE of the NIR dyes added to the collector. Due to cascaded emission, the system becomes bottlenecked by the NIR dyes, which typically have lower LQEs, at best ~0.5. As a result, a compromise must be made between extending the absorption range and maintaining a high LQE for dyes added. It was suggested that by combining visible dyes in one collector and NIR dyes in a second collector (in a stacked configuration; see Section 9.1.1), the LQE of the NIR dyes would not hinder the system efficiency as much.[24]

It is a single LSC containing two organic dyes fabricated and tested within FULLSPECTRUM that holds the current world record power conversion efficiency for a LSC.[25] A power conversion efficiency of 7.1% was achieved for a PMMA-based 5 × 5 × 0.5 cm LSC doped with Lumogen F Red 305 from BASF and Fluorescent Yellow CRS 040 from Radiant Colour. The LSC had GaAs cells attached to all four edges and connected in parallel and a diffuse back surface reflector. In practice this size is too small and GaAs cells too expensive for this design to be commercially viable, but the work provides a benchmark figure useful for illustrating the possibilities of this technology. The yellow dye used in this work has been found to be unstable, whereas the red dye is stable.[6,15] It has been shown, however, that single LSCs doped with Lumogen F Red 305 at a higher concentration than used in Slooff et al.[25] to compensate for the lack of a yellow dye to efficiently absorb the blue region of the solar spectrum can still yield reasonable efficiencies of around 4% for 1-mm thick LSCs of length about 30 cm with attached mc-Si cells,[26] which render them commercially viable.

9.2.4 Rare Earth Materials

Rare earth materials (REMs) present an alternative to the limitations of NIR organic dyes. REMs have the advantage of both being chemically stable and demonstrating large Stokes shifts.[27] However, REMs had exhibited low absorption coefficients while only making use of narrow bands of the solar spectrum.[28] To address these issues, Reisfeld and Kalisky proposed a collector doped with Nd^{3+} and Yb^{3+},[27] similar to the multiple-dye collector. The Nd^{3+} ion acted as an absorber given its broad absorption range, while the Yb^{3+} ion would emit at a NIR peak of around 970 nm, in a nature similar to that of cascaded emission. Implementing this design resulted in a collector that absorbed 20% (in tellurite glass) of the solar spectrum with a LQE of ~90%.[26] Like nanocrystals, LSC designs based on REMs are a novel idea, and initial results have been promising.

9.3 MODELING APPROACHES

It is essential to have robust models of the LSC that have been verified by comparison with experimental results in order to optimize the efficiency and cost of the LSC as well as provide insight into the processes that limit its performance. The two modeling approaches are outlined in this section, their application reviewed, and finally results using both approaches to investigate the conversion of direct and diffuse light are discussed.

The thermodynamic approach is based on the radiative transfer of angularly averaged trapped and escaping fluxes between mesh points in the concentrator plate, whereas in the ray-trace approach, which is based on Monte Carlo techniques, every incoming photon is tracked and its fate determined. The fundamental thermodynamic approach requires a minimum of input data and is quick to run, but it is limited to rectangular flat-plate LSCs homogeneously doped with a single luminescent species. The ray-trace approach is more flexible, allowing multiple dopants, thin films, and different geometries to be investigated. However, a limiting emission spectrum of the luminescent species as the concentration tends to zero is required in the ray-trace approach as the point of emission intensity distribution. In the thermodynamic approach this is taken care of by the luminescent brightness, which itself depends on the photon chemical potential distribution. If experimental results are not available, this spectrum can be reconstructed by comparison with the red-shifted emission spectra measured from the surfaces of test LSCs, but this procedure can be time-consuming. Because of the statistical nature of the ray-tracing process, large numbers of rays also need to be traced to obtain data with sufficiently small noise.

Both modeling approaches determine the spectrum of the luminescent light incident on the PV cell (and that escaping from all surfaces). If the PV cell spectral response (also often termed the external quantum efficiency [EQE]) is known, the electrical output of the LSC can be determined. The predicted electrical and spectral outputs using the two modeling approaches have been shown to be in excellent agreement,[6,20] despite the many differing processes involved in each approach. The predictions of both approaches are also in agreement with the measured electrical and spectral output of the fabricated devices. The agreement between the thermodynamic model and ray-trace models of single dye-doped test LSCs is better than 1% absolute in all cases.[6,20] This is a high level of agreement, giving confidence in using either approach as a tool for optimizing the LSC.

Owing to the greater flexibility of the ray-trace approach and its relative computational simplicity, this method for modeling the LSC has been more widely applied than the thermodynamic method. Results already presented earlier in this chapter utilize this approach,[6,10,13,14,20,26] and furthermore, it has also been used to explore the efficiency limits of the LSC.[29]

9.3.1 Ray-Trace Modeling of the LSC

We have developed a ray-trace model, which similar to the others described above,[6,10,13,14,20,26,29] uses Monte Carlo techniques and geometrical optics to trace the path of individual photons. Intersections with surfaces are computed, and experimentally measured absorption spectra are substituted into the Beer-Lambert law to determine the free path of a photon in a given direction. At each surface the reflection coefficient is calculated from the

Fresnel equations. In the case of transmission through a surface, Snell's law is applied to determine the refraction. A photon can be absorbed by the host background or by a luminescent species and reemitted, depending on the LQE. The wavelength of a reemitted photon is ideally randomly generated based on an experimental PL spectrum, as discussed above. This PL is measured by illuminating a sample with low optical density from the front and detecting the luminescent output from the back, thus minimizing the red shift due to reabsorptions such that the spectrum obtained is close to the PL emitted by a single luminescent center.

9.3.2 Thermodynamic Modeling of the LSC

We have developed self-consistent 3D flux models for planar LSCs[17,30] modules[31] and stacks[4] that show excellent agreement with experiments on test devices. Detailed balance arguments relate the absorbed light to the emission using 3D fluxes, and a Schwartzchild-Milne[32] type sampling of Chandrasekhar's radiative transfer equation[33] is performed. The resulting differential equations are integrated over the volume of the concentrator, and appropriate reflection boundary conditions are applied, giving integral equations that are applied over a mesh sampling the concentrator volume. The thermodynamic approach provides equations from which the photon chemical potential as a function of position within the concentrator may be determined by iteration. An optimal, self-consistent linearization of the depth dependence of the chemical potential for a single planar concentrator that results in only analytic expressions has also been developed. This linearized 3D flux model[34] has been validated by comparison with the results of the original 3D flux model, and the linearization is accurate to within approximately 2% of the total luminescent intensities and peak values.

9.3.3 Direct vs. Diffuse Irradiation

One of the advantages of the LSC is that it can harness diffuse radiation in addition to direct. The solar irradiance has a diffuse component that can be as high as the direct component in regions like the UK. Figure 9.8 (plotted in arbitrary units of irradiation) shows how the spectra of the direct and diffuse components of AM1.5 differ, where the diffuse is defined as the difference between the AM1.5 global spectrum and the direct. The diffuse spectrum is more blue rich, since blue light is more effectively scattered

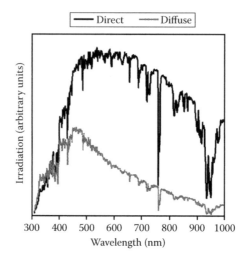

FIGURE 9.8
AM1.5 standard solar irradiation.

in the atmosphere. This could be advantageous if the LSC is used with a high-band-gap PV cell.

We modeled a 1 × 1 m thin-film LSC doped with Lumogen Red 300 dye from BASF under direct and diffuse irradiation, separately using the ray-trace approach. For the direct irradiation, a normal angle of incidence was assumed, which is not obtainable without solar tracking, but gives an upper bound for the performance. For the diffuse irradiation, a hemispherical angular distribution was assumed. Furthermore, we assumed an equal ratio between direct and diffuse components, which is typical of Northern European latitudes.[35] Table 9.1 shows some characteristics for both cases. The concentration ratio of the edge emission is only slightly higher (less than 10% relative) in the direct case. Though significantly more light (by a factor of about 5) is reflected from the top surface in the diffuse case due to the shallower angles, the total fraction of light absorbed in the concentrator is almost the same, since the shallower angles lead to longer path lengths and higher absorption probabilities. These simulations show that

TABLE 9.1

Direct vs. Diffuse Irradiation

	Direct	Diffuse
Edge concentration ratio	3.0	2.8
Incident absorbed (%)	49.2	44.9
Incident reflected (%)	3.9	20.0

the diffuse irradiation has a significant impact on the performance of even red dye-doped LSCs, which do not efficiently harvest the blue light that is abundant in the diffuse spectrum, and could be a valuable resource.

We have also examined the utilization of direct vs. diffuse light in QD-doped LSCs using the thermodynamic approach. A 40-cm long, 0.5-cm thick idealized LSC doped with CdSe/ZnS core-shell QDs with emission matched to an InGaP cell was considered. In the idealized LSC a LQE of unity was assumed as well as a perfectly transparent host material and perfect mirrors. The QD-doped LSC was again modeled under diffuse and direct AM1.5 irradiation separately, and the idealized LSC assumed for these calculations is discussed in more detail in the following section with respect to top surface escape cone losses. As in the ray-tracing work, the direct irradiation is assumed to be at normal incidence and the diffuse is assumed to be isotropic over a hemisphere. The results of this study predict that this QD-doped LSC, which owing to the high absorption coefficient in the blue region of the spectrum harvests diffuse light more efficiently than red dye-doped LSCs, would absorb 20% of the incident photons under AM1.5 direct irradiation, resulting in a power conversion efficiency of 3.9%, whereas under AM1.5 diffuse irradiation it would absorb 42% of the incident photons and achieve a power conversion efficiency of 4.7%. The spectral response (SR) or EQE of this QD-doped LSC and the incident light fluxes under direct and diffuse irradiation, respectively, are presented in Figure 9.9, where the area between the incident flux and the flux lost through the top surface illustrates the fraction of the solar spectrum harvested in the two cases. Although the efficiencies are modest, this QD-doped LSC would produce 7.3× and 12.2× the power of the cell alone under AM1.5 direct and diffuse irradiation, respectively. All nanocrystals have absorption spectra that continue to increase at high energies, and therefore it is concluded that nanocrystal-doped LSCs that can efficiently harvest the blue-rich diffuse solar spectrum will be particularly advantageous at high latitudes where significant fractions (>0.5) of the insolation are diffuse.

9.4 LOSSES

Escape cone losses, absorption losses in the host material, and reabsorption losses owing to a LQE less than unity limit the efficiency of the LSC.

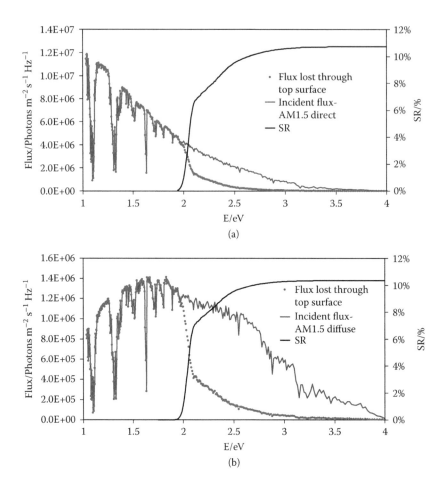

FIGURE 9.9

The spectral response of the idealized QD-doped LSC and incident light fluxes under (a) AM1.5 direct and (b) AM1.5 diffuse irradiation, respectively.

As described in the preceding section, we have developed complementary thermodynamic[4,17,30,31,34] and ray-trace[10,20] models of the LSC that allow us to investigate quantitatively the luminescent species, the doping densities, and the geometries that minimize these loss mechanisms. As discussed in Section 9.1.2, escape cone losses can be largely removed, for all the surfaces not covered by PV cells bar the top surface using conventional mirrors. However, escape cone losses through the illuminated top surface constitute the most significant loss mechanism in the LSC. In the remainder of this section the magnitude of these losses is illustrated for the idealized QD-doped LSC also used for the calculations in Section 9.3.3, and the strategies for alleviating these are reviewed. Finally, the selection of host/

substrate materials that minimize parasitic absorption by the bulk transparent component of the LSC are discussed.

9.4.1 Idealized QD-LSC and Top Surface Escape Cone Losses

We have utilized the thermodynamic approach to study an idealized, mirrored, 40 × 0.5 cm LSC doped with CdSe/ZnS QDs with emission matched to a GaInP cell.[36] These calculations illustrate the importance of reducing the top surface escape losses that occur through the escape cone through both primary emission and subsequent reabsorption and reemission. As described earlier, the idealized LSC has perfect mirrors on three edges and the bottom surface, a perfectly transparent host material, and a LQE of unity. The LSC absorption, incident AM1.5g spectrum, and the average concentrated flux escaping the bare right-hand edge of the idealized LSC are illustrated in Figure 9.10. Note that a logarithmic scale is necessary to compare the narrow concentrated escaping flux and the incident flux. This idealized LSC absorbs 24% of the incident photons in the AM1.5 global solar spectrum.

The CR is 4.18, but since the concentrated flux escaping the right-hand surface is a narrow band matched to the spectral response of the cell, it can all be converted and the idealized LSC would produce eight times the current of the cell alone exposed to AM1.5g. However, 78% of the lumi-

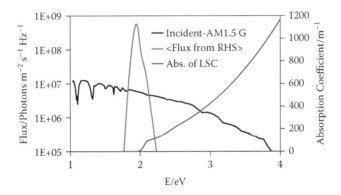

FIGURE 9.10
Absorption of the LSC material used in the calculations for the idealized system together with the flux incident on the top surface and the predicted concentrated average luminescent flux escaping the right-hand surface of the idealized LSC that would be coupled into the PV cell.

nescence is lost through the large top surface area, and only 22% may be collected in an attached PV cell.

Therefore, there is a need to reduce these large top surface losses in order to design more efficient devices. An idealized notch filter on the top surface, which has a reflectivity of 99.9% in the notch (covering the luminescence between 1.8 and 2.2 eV) and a reflectivity of 0.01% everywhere else, would increase the output by a factor of 3.1 times. Such a top coating reflects a portion of the incident light before it can be absorbed in the LSC, so the coated idealized LSC absorbs only 21% of the incident light but has a predicted CR of 12.98. Real coatings do not have the idealized properties used in these predictions, but these calculations serve to illustrate the maximum gains that could be achieved.

The idealized notch filter described above has the properties of a photonic band stop (PBS) layer and, in their paper exploring the efficiency limits of the LSC, Rau et al.[29] state that it is essential to use a PBS layer to maximize the theoretical power conversion efficiency. Using their ray-trace model, they calculate a maximum power conversion efficiency of 23.9% when a PBS layer is included.

9.4.2 Distributed Bragg Reflectors

One candidate for a PBS layer or wavelength-selective coating is a distributed Bragg reflector (DBR). It comprises layers of dielectric material alternating with a high–low refractive index profile. Fresnel reflection occurs at the layer interfaces, causing forward and backward propagating electromagnetic waves. When the layer thickness is one-quarter of the propagating wavelength destructive interference occurs. This results in light, centered around a designated wavelength referred to as the stop band, being reflected. At oblique angles, different wavelengths satisfy the interference condition because of the changing optical path length through the layers. For this reason DBRs have an angular-dependent reflectivity.

Richards et al.[37] first proposed reducing top surface escape losses by using a DBR with a NIR stop band. In this design the DBR, or hot mirror, was placed on the top surface of a plate containing multiple dyes and a rare earth metal. The rare earth, $Nb^{3+}:Yb^{3+}$, is the primary emitter and has a luminescent quantum efficiency of 75% and emission at 970 nm, slightly below the absorption edge of a silicon cell. The dye ensemble absorbs the incident spectrum and fluorescently pumps the rare earth, causing the majority of the propagating luminescence to be in the NIR such that it is reflected by the coating. A 25% performance gain and system efficiency of 6.5% were predicted due to elimination of top surface escape cone losses.

9.4.3 Rugate Filters

A rugate filter is essentially a DBR with a sinusoidally varying periodic refractive index profile through the structure. This structure and subsequent index matching to the external media suppress side lobes that occur on either side of the stop band.[38] Rugate filters are good candidates for top surface coatings because they strongly reflect luminescence while not reflecting the incident solar irradiation in the part of the spectrum where the luminescent species absorbs.

Goldschmidt et al.[39,40] have proposed stacked and multiple-species LSCs with a bottom surface silicon cell and a top surface rugate filter. Experimental results on a small 2 × 2 cm test device, containing a single red emitting dye, showed an 11% increase in optical efficiency. However, the rugate filter was not optimized because a significant proportion of the incident light was rejected. As a result, no overall system efficiency gain was achieved.

9.4.4 Cholesteric Coatings

Recently, use has been made of wavelength-selective cholesteric liquid crystal coatings applied to the top surface in order to reduce escape cone losses.[41] These coatings, which also act as PBS layers, have been considered by a number of workers,[6,26,30] and our detailed evaluation of the ability of real cholesteric coatings to produce performance gains[30] is described in greater detail in Section 9.6.

9.4.5 Aligned Dyes

Escape cone losses may also be reduced by a larger Stokes shift (as illustrated for NRs in Section 9.2.2) or directional emission. Through the alignment of dye layers using a liquid crystalline host material, directional emission has been achieved in thin-film LSCs.[42] It was found that the preferred axes of emission and absorption were the same. Therefore, a layer with dyes at a tilted angle was suggested as a trade-off between reduced absorption and increased directional emission. In a planar alignment 25% more light was emitted from the edge perpendicular to the dye alignment than from the edge parallel, and 15% more light was emitted from an aligned sample than from an isotropic one.[42]

9.4.6 Host Losses

As well as optimizing the luminescent species, the optimization of the host material is also of great importance. Ideally the host material should be transparent throughout the spectral range of both the absorption and emission of the luminescent species, but transparency over the spectral range of the emission is critical owing to the long path lengths in the LSC for collection of this light in the attached solar cells. PMMA has emerged as a leading contender for use with organic dyes, being cheap, optically inactive, and chemically stable.[43] However, it can be difficult to disperse nanocrystals at sufficient concentrations to efficiently harvest light in PMMA, and we have found that PLMA is advantageous in this respect. A recent study by Gallagher et al.[44] has also shown epoxy resin to be a suitable host for QDs, with the QDs retaining 77.5% of their LQE. In addition, the thin-film architecture allows the use of glass as a substrate, and high-quality solar glass has absorption properties comparable to those of commercial PMMA (absorption coefficient ~ 0.3 m^{-1}) in the visible region of the spectrum. Polymer optical fibers can achieve absorption coefficients as low as 0.001 m^{-1}, and if such low host absorption could be achieved in the LSC, significant performance gains on the order of 20% could be achieved.

9.5 RAY-TRACE MODELING OF THIN-FILM LSCS

In Section 9.1.3 we stated that our simulations and experiments comparing the thin-film and homogeneously doped architectures[10] indicated that no performance gains were achieved through reduced reabsorption losses using thin films. This work is described in greater detail in this section.

9.5.1 Samples

Two sets of homogeneous and thin-film LSCs, fabricated at the Fraunhofer IAP, were examined. The dimensions were $50 \times 50 \times 3$ mm and the luminescent species was Lumogen Red 300 dye manufactured by BASF. The

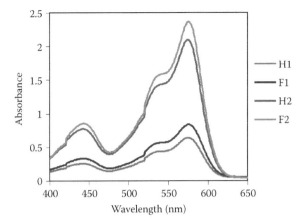

FIGURE 9.11
Absorbencies of homogeneous LSCs (H1, H2) and thin-film (F1, F2) composites of approximately equal dopant amounts, respectively. The absorbencies for each set are reasonably similar.

homogeneous concentrators consisted of PMMA made from Plexit, and the thin films were prepared from a PMMA/CHCl$_3$ solution that was drop cast with a thickness of approximately 50 μm onto Plexiglas GS233 substrates. For comparison, the two kinds of concentrators in each set were required to be doped with approximately identical amounts of dye. The close match of the respective absorbance (see Figure 9.11) shows that this was achieved to a suitable degree and implies precise control over the fabrication process.

9.5.2 Experimental Measurements

A short-circuit current measurement technique was applied,[17] in which the top surface of the sample was illuminated uniformly using a lamp with a broad spectrum (400 to 700 nm). For calibration, a solar cell of known spectral response was used to map out the light intensity reaching the top surface. The same cell was used to scan one of the short edges of the concentrator and detect the luminescent output. The arising short-circuit current was registered and the associated current density (J_{SC}) deduced.

Based on the incident intensity on the front surface, the experimental PL leaving the edge, and the spectral and angular responses of the PV cell, the photon concentration ratio (CR) was calculated and used for comparison. The CR is defined as the photon flux leaving the detection edge divided by the incident flux. The experiment yielded similar performances, to within

TABLE 9.2

Experimental Measurement

		CR	Error
Homogeneous LSC	(H1)	0.81	0.08
Thin-film composite	(F1)	0.90	0.09
Homogeneous LSC	(H2)	1.61	0.13
Thin-film composite	(F2)	1.60	0.13

TABLE 9.3

Ray-Trace Model

		CR	Uncertainty
Homogeneous LSC	(H1)	0.81	0.05
Thin-film composite	(F1)	0.80	0.05

experimental error, between the homogeneous and thin-film LSCs (see Table 9.2).

9.5.3 Ray-Trace Simulation

The ray-trace model was applied to the first set of samples. In order to provide for an accurate comparison, the ray-trace simulation was based on identical absorbencies within each set of samples (using the absorbance spectrum of the homogeneous sample). The model was found to be in agreement with the experimental results (see Table 9.3) and indicated no significant difference between the two different LSC configurations.

9.6 THERMODYNAMIC MODELING OF CHOLESTERIC COATINGS

As outlined earlier in Section 9.4.4, wavelength-selective cholesteric liquid crystal coatings[41] have been designed to reduce escape cone losses through the top surface of the LSC, and our studies to evaluate the performance of such coatings are discussed in greater detail in this section. The thermodynamic approach may be used to quantify the effects of such coatings[36] by using the measured reflectivities of the liquid crystal coatings as the boundary conditions for the incident and luminescent light at the coated surfaces.

9.6.1 Test Coatings

Two dye-doped LSCs were characterized both with and without appropriate focal-conic cholesteric coatings[36] and with and without back surface air-gap 3M multilayer dielectric foil mirrors. The two dye-doped test LSCs were comprised of Plexit doped with red and yellow Coumarin fluorescent dyes purchased from Bayer. In order to predict the luminescence, the energy dependence of the absorption of each dye at threshold is fitted to a Gaussian.[17] These fits represent the experimental data very well for both dyes, as seen in Figure 9.12, which also shows good agreement between the shape and position of the predicted and observed luminescence for both dyes.

The cholesteric coatings have a reflectivity band that is highly angular dependent. The band positions for each applied coating were tuned to give the maximum $\cos(\theta)$ weighted overlap integral between the reflectivity ($R \sim 1 - T$) of the coating and the luminescence of each test LSC. Figure 9.13 illustrates the angularly dependent transmission of one such right-handed cholesteric coating exposed to right-circularly polarized light.

Short-circuit currents, J_{SC}, resulting from the radiation escaping the right-hand surfaces of the dye and QD-doped test LSCs, both with and without the coatings (tested on the top and bottom surfaces) and with and without a 3M multilayer dielectric foil back surface air-gap mirror, were measured and compared with the values predicted by the thermodynamic model.[36] The model and predictions showed excellent agreement for both concentrators, and the results for the yellow dye concentrator are illustrated in Figure 9.14.

At this time the cholesteric coatings tested led to a predicted reduction in output owing to both coating transmission losses (~10%) and reflection of the incident light by the coatings. This was not compensated for by the increase in trapping of the luminescent light inside the test LSCs by the coatings. The increase in trapping was seen in the modeling through a higher average photon chemical potential.[36] Cholesteric coatings with much reduced transmission losses and improved spectral coverage are now available and will be evaluated shortly.

9.7 CONCLUSIONS

We have presented an overview of the recent development of LSCs at our laboratories and by other workers in the field. The renewed interest in

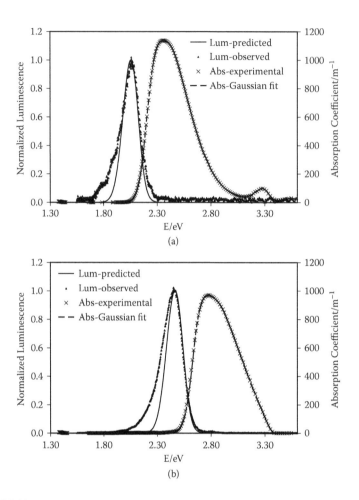

FIGURE 9.12

Measured absorption coefficient and Gaussian fit together with the normalized predicted and observed luminescence escaping the bottom surfaces of the (a) 2.5-mm-thick test slab of Fluorescent Red in Plexit and (b) 2.7-mm-thick test slab of Fluorescent Yellow in Plexit. (From Chatten, A. J. et al. 2005. *Proceedings of the 31st IEEE Photovoltaic Specialists Conference*. With permission.)

LSCs, since the pioneering work in the late 1970s, has led to the development of thermodynamic and ray-trace models, which equally well describe the performance of LSCs. The models provide tools for both the analysis of experimental work on test devices and the optimization of practical concentrators. Furthermore, detailed cost analyses indicate that Fluorescent Red 305 doped LSCs are already competitive with other conventional PV panels, and this system is ripe for commercial development if the stability of the dye over the timescales (>15 years) required for practical application

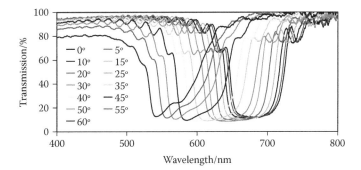

FIGURE 9.13
Variation of a cholesteric coating transmission, *T*, with angle of incidence.

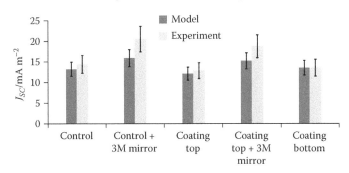

FIGURE 9.14
Measured and predicted short-circuit currents, J_{SC}, for the yellow dye-doped test LSC.

can be confirmed. Moreover, new dyes and semiconductor nanocrystals have been developed and included in homogeneous and thin-film concentrators. In addition, many different approaches are currently under investigation in order to reduce the losses that currently limit the performance of LSCs, and it is likely that a combination of these will lead to further increases in performance and reductions in costs. The ability of the LSC to utilize the often large diffuse component of solar irradiation and its suitability for building integration should lead to the exploitation of this promising technology in the near future.

ACKNOWLEDGMENTS

The authors thank the European Commission (through the sixth framework program Integrated Project FULLSPECTRUM, contract SES6-CT-2003-502620) and the EPSRC for financial support.

REFERENCES

1. W H Weber and J Lambe. 1976. *Appl. Opt.* 15:2299.
2. A Goetzberger and W Greubel. 1977. *Appl. Phys.* 14:123.
3. N Lewis. 2007. *Science* 315:798.
4. A J Chatten, D J Farrell, C Jermyn, P Thomas, B F Buxton, A Büchtemann, R Danz, and K W J Barnham. 2005. In *Proceedings of the 31st IEEE Photovoltaic Specialists Conference,* Orlando, FL, p. 82.
5. D J Farrell, A J Chatten, A Büchtemann, and K W J Barnham. 2006. In *Proceedings of the IEEE 4th World Conference on Photovoltaic Energy Conversion*, Waikoloa, Hawaii, p. 217.
6. W G J H M van Sark, K W J Barnham, L H Slooff, A J Chatten, A Büchtemann, A Meyer, S J McCormack, R Koole, D J Farrell, R Bose, E Bende, A R Burgers, T Budel, J Quilitz, M Kennedy, T Meyer, S H Wadman, G van Klink, G van Koten, A Meijerink, and D Vanmaekelbergh. 2008. *Opt. Express* 16:21773.
7. W Viehmann and R L Frost. 1979. *Nucl. Instr. Meth.* 167:405.
8. C F Rapp and N L Boling. 1978 *Proceedings of the 13th IEEE Photovoltaic Specialists Conference*, Washington, DC, p. 690.
9. A Zastrow. 1981. PhD thesis, University of Freiburg, Germany.
10. R Bose, D J Farrell, A J Chatten, M Pravettoni, A Büchtemann, and K W J Barnham. 2007. In *Proceedings of the 22nd European Photovoltaic Solar Energy Conference*, Milan, Italy, p. 210.
11. M J Currie J K Mapel, T D Heidel, S Goffri, and M A Baldo. 2008. *Science* 32:226.
12. M Anni, L Manna, et al. 2004. *Appl. Phys. Lett.* 85:4169.
13. K R McIntosh, N Yamada, and B S Richards. 2007. *Appl. Phys. B* 88:285.
14. M Kennedy, et al. 2007. In *Proceedings of the ISES World Solar Congress*, Beijing, China.
15. R Kinderman, et al. 2007. *Journal of Solar Energy Engineering* 129:277.
16. K W J Barnham, J L Marques, J Hassard, and P O'Brien. 2000. *Appl. Phys. Lett.* 76:1197.
17. A J Chatten, K W J Barnham, B F Buxton, N J Ekins-Daukes, and M A Malik. 2004. *Semiconductors* 38:949.
18. A P Alivisatos. 1998. *MRS Bull.* 23:2.
19. D V Talapin, J H Nelson, E V Shevchenko, S Aloni, B Sadtler, and A P Alivisatos. 2007. *Nano Lett.* 7:2951.
20. R Bose, D J Farrell, A J Chatten, M Pravettoni, A Büchtemann, J Quilitz, A Fiore, L Manna, and K W J Barnham. 2008. In *Proceedings of the 33rd IEEE Photovoltaics Specialist Conference*, San Diego.
21. L Carbone, C Nobile, M De Giorg, F D Sala, G Morello, P Pompa, M Hytch, E Snoeck, A Fiore, I R Franchini, M Nadasan, A F Silvestre, L Chiodo, S Kudera, R Cingolani, R Krahne, and L Manna. 2007. *Nano Lett.* 7:2942.
22. D V Talapin, R Koeppe, S Gotzinger, A Kornowski, J M Lupton, A L Rogach, O Benson, J Feldmann, and H Weller. 2003. *Nano Lett.* 3:1677.
23. B A Swartz, T Cole, and A H Zewail. 1977. *Opt. Lett.* 1:73.
24. B S Richards and K R McIntosh. 2006. In *Proceedings of the 21st European Photovoltaic Solar Energy Conference and Exhibition,* Dresden, Germany, p. 185.
25. L H Slooff, E E Bende, A R Burgers, T Budel, M Pravettoni, R P Kenny, E D Dunlop, and A Büchtemann. 2008. *Phys. Stat. Solidi* 2:257.

26. E E Bende, L H Slooff, A R Burgers, W G J H M van Sark, and M Kennedy. 2008. In *Proceedings of the 23rd European Photovoltaic Solar Energy Conference and Exhibition*, Valencia, Spain.

27. R Reisfeld and Y Kalisky. 1981. *Chem. Phys. Lett.* 80:178.

28. B C Rowan, L R Wilson, and B S Richards. 2008 *IEEE J. Selected Topics Quantum Electronics* 14:1312.

29. U Rau, F Einsele and G C Glaeser. 2006. *Appl. Phys. Lett.* 88:176102.

30. A J Chatten, K W J Barnham, B F Buxton, N J Ekins-Daukes, and M A Malik. 2003. In *Proceedings of the 3rd World Conference on Photovoltaic Energy Conversion*, Osaka, Japan, p. 2657.

31. A J Chatten, K W J Barnham, B F Buxton, N J Ekins-Daukes, and M A Malik. 2004. In *Proceedings of the 19th European Photovoltaic Solar Energy Conference and Exhibition*, Paris, p. 109.

32. E A Milne. 1921. *Monthly Notices of the Royal Astronomy Society of London* 81:361.

33. S Chandrasekhar. 1950. *Radiative transfer*. Oxford: Clarendon.

34. A J Chatten, D J Farrell, B F Buxton, A Büchtemann, and K W J Barnham. 2006. In *Proceedings of the 21st European Photovoltaic Solar Energy Conference*, Dresden, Germany, p. 315.

35. A Goetzberger. 1978. *Appl. Phys.* 16:399.

36. A J Chatten, D J Farrell, R Bose, A Büchtemann, and K W J Barnham. 2007. In *Proceedings of the 22nd European Photovoltaic Solar Energy Conference*, Milan, Italy, p. 349.

37. B S Richards, A Shalav, and R P Corkish. 2004. In *Proceedings of the 19th European Photovoltaic Solar Energy Conference and Exhibition*, Paris, p. 4.

38. W H Southwell. 1989. *Appl. Optics* 28:5091.

39. J C Goldschmidt, S W Glunz, A Gombert, and G Willeke. 2006. In *Proceedings of the 21st European Photovoltaic Solar Energy Conference*, Dresden, Germany, p. 107.

40. J C Goldschmidt, M Peters, P Loper, O Schultz, F Dimroth, S W Glunz, A Gombert, and G Willeke. 2007. In *Proceedings of the 22nd European Photovoltaic Solar Energy Conference*, Milan, Italy, p. 608.

41. M G Debije, R H L van der Blom, D J Broer, and C W M Bastiaansen. 2006. In *Proceedings of the World Renewable Energy Congress IX*, Florence, Italy.

42. M G Debije, D J Bruer, and C W Bastiaansen. 2007. In *Proceedings of the 22nd European Photovoltaic Solar Energy Conference*, Milan, Italy, p. 87.

43. N N Barashkov, et al. 1991. Translated from *Zhurnal Prikladnoi Spektroskopii* 55:906.

44. S J Gallagher, et al. 2007. *Solar Energy* 81:540.

10

Nanoparticles for Solar Spectrum Conversion

W. G. J. H. M. van Sark, A. Meijerink, and R. E. I. Schropp

10.1 Introduction

In this chapter a review is presented on the use of nanometer-sized particles (including quantum dots) in the conversion of parts of the solar spectrum incident on solar cells to more usable regions. The modification of the solar spectrum ideally would lead to a narrow-banded incident spectrum at a center wavelength corresponding to an energy that is slightly larger than the band gap of the semiconductor material employed in the solar cell, which would lead to an enhancement of the overall solar energy conversion efficiency. Modification of the spectrum requires down- or upconversion or -shifting of the spectrum, meaning that the energy of photons is modified either to lower (down) or higher (up) energy. Nanostructures such as quantum dots, luminescent dye molecules, and lanthanide-doped glasses are capable of absorbing photons at a certain wavelength and emitting photons at a different (shorter or longer) wavelength. We will discuss down- and upconversion and shifting by quantum dots, luminescent dyes, and lanthanide compounds, and assess their potential in contributing to ultimately lowering the cost per kWh of solar-generated power.

10.2 BACKGROUND

10.2.1 General

Conventional single-junction semiconductor solar cells only effectively convert photons of energy close to the semiconductor band gap, E_g, as a result of the mismatch between the incident solar spectrum and the spectral absorption properties of the material (Green 1982; Luque and Hegedus 2003). Photons with an energy, E_{ph}, smaller than the band gap are not absorbed and their energy is not used for carrier generation. Photons with energy, E_{ph}, larger than the band gap are absorbed, but the excess energy, $E_{ph} - E_g$, is lost due to thermalization of the generated electrons. These fundamental spectral losses in a single-junction silicon solar cell can be as large as 50% (Wolf 1971), while the detailed balance limit of conversion efficiency for such a cell was determined at 31% (Shockley and Queisser 1961). Several routes have been proposed to address spectral losses, and all of these methods or concepts obviously concentrate on a better exploitation of the solar spectrum, e.g., multiple stacked cells (Law et al., 2009), intermediate band gaps (Luque and Marti 1997), multiple exciton generation (Klimov 2006; Klimov et al. 2007), quantum dot concentrators (Barnham et al., 2009; Chatten et al. 2003a), down- and upconverters (Trupke et al. 2002a, 2002b), and downshifters (Richards 2006a; Van Sark 2005). In general they are referred to as third- or next-generation photovoltaics (PV) (Green 2003; Luque et al. 2005; Martí and Luque 2004). Nanotechnology is essential in realizing most of these concepts (Soga 2006; Tsakalakos 2008), and semiconductor nanocrystals have been recognized as building blocks of nanotechnology for use in next-generation solar cells (Kamat 2008). Being the most mature approach, it is not surprising that the current world record conversion efficiency is above 40% for a stack of GaInP/GaAs/Ge solar cells (Green et al. 2009), although this is reached at a concentration of about two hundred times.

As single-junction solar cells optimally perform under monochromatic light at wavelength $\lambda_{opt} \sim 1240 / E_g$ (with λ_{opt} in nm and E_g in eV), an approach "squeezing" the wide solar spectrum (300 to 2,500 nm) to a single small-band spectrum without too many losses would greatly enhance solar cell conversion efficiency. Such a quasi-monochromatic solar cell could in principle reach efficiencies over 80%, which is slightly dependent on band gap (Luque and Martí 2003). For (multi)crystalline silicon ((m)c-Si) solar cells, $\lambda_{opt} = 1100$ nm (with $E_g = 1.12$ eV); for hydrogenated amorphous silicon (a-Si:H), the optimum wavelength is $\lambda_{opt} = 700$ nm (with $E_g = 1.77$ eV). However, as amorphous silicon solar cells only contain a thin absorber layer, the optimum spectrum response occurs at about 550 nm (Schropp and Zeman 1998; Van Sark 2002).

Modification of the spectrum by means of so-called down- or upconversion or -shifting is presently being pursued for single-junction cells (Richards 2006a), as illustrated in Figure 10.1, as a relatively easy and

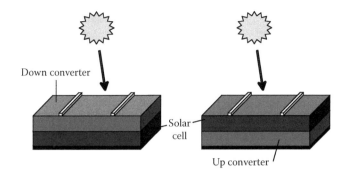

FIGURE 10.1
Schematic drawings of (left) a solar cell with downconverter layer on top and (right) a solar cell on top of an upconverter layer.

cost-effective means to enhance conversion efficiency. In addition, so-called luminescent solar converters (LSCs; see Chapter 9 for more details) employ spectrum modification as well (Goetzberger 2008; Goetzberger and Greubel 1977). Downconverters or -shifters are located on top of solar cells, as they are designed to modify the spectrum such that UV and visible photons are converted, leading to a more red-rich spectrum that is converted at higher efficiency by the solar cell. Upconverters modify the spectrum of photons that are not absorbed by the solar cell to effectively shift the infrared (IR) part of the transmitted spectrum to the near-IR (NIR) or visible part; a back reflector usually is applied as well.

In the case of downconversion (DC), an incident high-energy photon is converted into two or more lower-energy photons, which can lead to quantum efficiency of more than 100%; therefore, it is also termed quantum cutting (Timmerman et al. 2008; Wegh et al. 1999); for upconversion (UC), two or more low-energy photons (sub-band gap) are converted into one high-energy photon (Strümpel et al. 2007); see also Figure 10.2. Downshifting (DS) is similar to downconversion, where an important difference is that only one photon is emitted and that the quantum efficiency of the conversion process is lower than unity (Richards 2006a), although close to unity is preferred to minimize losses. Downshifting is also termed photoluminescence (Strümpel et al. 2007). DC, UC, and DS layers only influence solar cell performance optically. As DC and DS both involve one incident photon per conversion, the intensity of converted or shifted emitted photons linearly scales with incident light intensity. UC involves two photons; therefore, the intensity of converted light scales quadratically with incident light intensity.

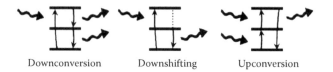

| Downconversion | Downshifting | Upconversion |

FIGURE 10.2

Energy diagrams showing photon absorption and subsequent downconversion, down-shifting, and upconversion.

10.2.2 Downconversion

Downconversion was theoretically suggested first by Dexter in the 1950s (Dexter 1953, 1957), and shown experimentally twenty years later using the lanthanide ion praseodymium, Pr^{3+}, in an yttrium fluoride, YF_3, host (Piper et al. 1974; Sommerdijk et al. 1974). A VUV photon (185 nm) is absorbed in the host, and its energy is transferred into the $^1S_0$ state of the Pr^{3+} ion, from where two photons (408 and 620 nm) are emitted at higher than unity quantum efficiency. Another frequently used ion is gadolinium, Gd^{3+} (Wegh et al. 1997), either single or co-doped (Wegh et al. 1999). These lanthanide ions are characterized by a rich and well-separated energy-level structure in the so-called Dieke energy diagrams (Dieke 1968), and have been identified as perfect "photon managers" (Meijerink et al. 2006). The energy levels arise from the interactions between electrons in the inner 4f shell. Trivalent lanthanides have an electronic configuration of $[Xe]4f^n5s^25p^6$. Inside the filled 5s and 5p shells, there is a partially filled 4f shell where the number of electrons (n) can vary between 0 and 14. The number of possible arrangements for n electrons in 14 available f-orbitals is large (14 over n), which gives rise to a large number of different energy levels that are labeled by so-called term symbols. The energy-level diagrams for the different lanthanides are shown in Figure 10.3. Transitions between the energy levels give rise to sharp absorption and emission lines. Energy transfer between neighboring lanthanide ions is also possible and helps converting photons that are absorbed to photons of different energy. Based on their unique and rich energy-level structure, lanthanide ions are promising candidates to realize efficient downconversion, and recent research in this direction will be discussed below.

Downconversion in solar cells was theoretically shown to lead to a maximum conversion efficiency of 36.5% (Trupke et al. 2002a) for non-concentrated sunlight when applied in a single-junction solar cell configuration, such as shown in Figure 10.1. These detailed balance calculations (Shockley and Queisser 1961) were performed as a function of the band

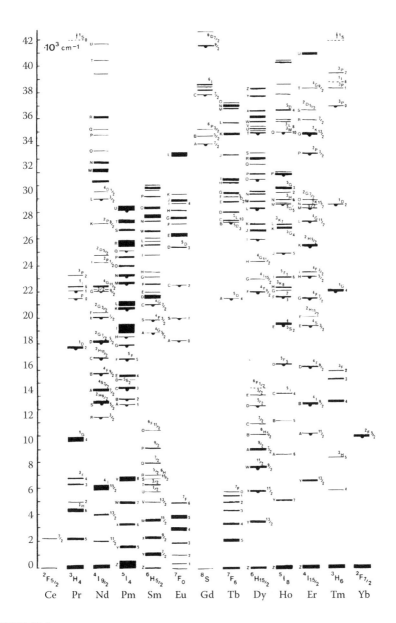

FIGURE 10.3

Energy-level schemes of all trivalent lanthanide ions from $4f^1$ (Ce^{3+}) to $4f^{13}$ (Yb^{3+}); also known as the Dieke diagram. (From Dieke, G. H., *Spectra and Energy Levels of Rare Earth Ions in Crystals.* 1968. New York: Wiley Interscience. With permission.)

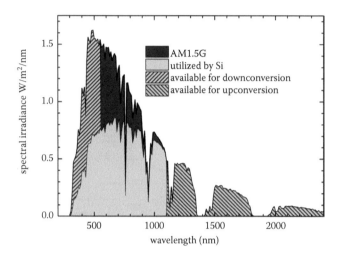

FIGURE 10.4

Potential gains for down- and upconversion for a silicon solar cell, depicted as fractions of the AM1.5G spectrum. A silicon cell effectively utilizes only 468 W/m² (58% of the energy between 300 and 1,150 nm); the available extra fractions for DC and UC are 149 and 164 W/m², respectively. (Based on Richards, B. S., 2006a. "Enhancing the Performance of Silicon Solar Cells via the Application of Passive Luminescence Conversion Layers," *Solar Energy Materials and Solar Cells* 90:2329–37. With permission.)

gap and refractive index of the solar cell material, for a 6,000K blackbody spectrum. The efficiency limit is reached for a band gap of 1.05 eV, and asymptotically approaches 39.63% for very high refractive indices larger than 10. For c-Si, with refractive index of 3.6, the limit efficiency is 36.5%. Analysis of the energy content of the incident standard air mass 1.5 global (AM1.5G) spectrum (ASTM 2003) and the potential gain DC can have shows that with a DC layer, an extra amount of 32% is incident on a silicon solar cell (Richards 2006a), which can be converted at high internal quantum efficiency. Figure 10.4 illustrates the potential gains for DC and UC.

10.2.3 Downshifting

Downshifting or photoluminescence is a property of many materials, and is similar to downconversion; however, only one photon is emitted and energy is lost due to nonradiative relaxation (see Figure 10.2). Therefore, the quantum efficiency is lower than unity. DS can be employed to overcome poor blue response of solar cells (Hovel et al. 1979) due to, e.g., noneffective front surface passivation for silicon solar cells. Shifting the incident spectrum to wavelengths where the internal quantum efficiency

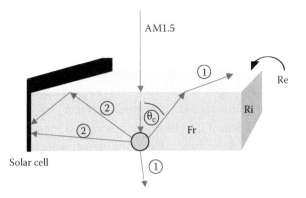

FIGURE 10.5
Schematic 3D view of a luminescent concentrator. AM1.5 light is incident from the top. The light is absorbed by a luminescent particle. The luminescence from the particle is randomly emitted. Part of the emission falls within the escape cone (determined by the angle, θ_c) and is lost from the luminescent concentrator at the surfaces (1). The other part (2) of the luminescence is guided to the solar cell by total internal reflection. (From Van Sark, W. G. J. H. M., K. W. J. Barnham, L. H. Slooff, et al., 2008. "Luminescent Solar Concentrators—A Review of Recent Results," *Optics Express* 16:21773–92. With permission.)

of the solar cell is higher than in the blue can effectively enhance the overall conversion efficiency by ~10% (Van Sark et al. 2005). Improvement of front passivation may make downshifters obsolete, or at least less beneficial. Downshifting layers can also be used to circumvent absorption of high-energy photons in heterojunction window layers, e.g., CdS on CdTe cells (Hong and Kawano 2003).

Downshifting was suggested in the 1970s to be used in so-called luminescent solar concentrators (LSCs) that were attached on to a solar cell (Garwin 1960; Goetzberger and Greubel 1977; Rapp and Boling 1978; Weber and Lambe 1976). In these concentrators (Figure 10.5), organic dye molecules absorb incident light and reemit this at a red-shifted wavelength. Internal reflection ensures collection of all the reemitted light in the underlying solar cells. As the spectral sensitivity of silicon is higher in the red than in the blue, an increase in solar cell efficiency was expected. Also, it was suggested to use a number of different organic dye molecules of which the reemitted light was matched for optimal conversion by different solar cells. This is similar to using a stack of multiple solar cells, each sensitive to a different part of the solar spectrum. The expected high efficiency of ~30% (Smestad et al. 1990; Yablonovitch 1980) in practice was not reached as a result of not being able to meet the stringent requirements to the organic dye molecules, such as high quantum efficiency and stability, and the transparency of collector materials in which the dye molecules were dispersed (Garwin 1960; Goetzberger and Greubel 1977; Rapp and Boling 1978; Weber and Lambe 1976).

Nowadays, new organic dyes can have extremely high luminescence quantum efficiency (LQE) (near unity) and are available in a wide range of colors at better reabsorption properties that may provide necessary UV stability. Quantum dots (QDs) have been proposed for use in luminescent concentrators to replace organic dye molecules: the quantum dot concentrator (Barnham et al. 2000, 2009; Chatten et al. 2003a, 2003b; Gallagher et al. 2007). Quantum dots are nanometer-sized semiconductor crystals of which the emission wavelength can be tuned by their size, as a result of quantum confinement (Alivisatos 1996; Gaponenko 1998). Recently, both QDs and new organic dyes have been evaluated for use in LSCs (Van Sark et al. 2008a). QDs have advantages over dyes in that (1) their absorption spectra are far broader, extending into the UV; (2) their absorption properties may be tuned simply by the choice of nanocrystal size; and (3) they are inherently more stable than organic dyes (Bruchez et al. 1998). Moreover, there is a further advantage in that the red shift between absorption and luminescence is *quantitatively* related to the *spread* of QD sizes, which may be determined during the growth process, providing an additional strategy for minimizing losses due to reabsorption (Barnham et al. 2000). However, as yet QDs can only provide reasonable LQE: a LQE > 0.8 has been reported for core-shell QDs (Peng et al. 1997). Performance in LSCs has been modeled using thermodynamic as well as ray-trace models (Burgers et al. 2005; Chatten et al. 2003b, 2004a, 2004b, 2005; Gallagher et al. 2004; Kennedy et al. 2008), and results are similar and also compare well with experimental values (Kennedy et al. 2008; Van Sark et al. 2008a). Calculated efficiencies vary between 2.4% for an LSC with mc-Si cell at certain mirror specifications and 9.1% for an LSC with InGaP cell for improved specifications.

Alternatives for dye molecules used in LSCs are luminescent ions. Traditionally, efficient luminescent materials rely on the efficient luminescence of transition metal ions and lanthanide ions. In the case of transition metal ions intraconfigurational $3d^n$ transitions are responsible for the luminescence, while in the case of lanthanide ions both intraconfigurational $4f^n$-$4f^n$ transitions and interconfigurational $4f^n$-$4f^{n-1}5d$ transitions are capable of efficient emission. In most applications efficient emission in the visible is required and emission from lanthanide ions and transition metal ions is responsible for almost all the light from artificial light sources (e.g., fluorescent tubes, displays [flat and cathode ray tube] and white light-emitting diodes [LEDs]) (Blasse and Grabmaier 1995). For LSCs to be used in combination with c-Si solar cells, efficient emission in the NIR is needed. The optimum wavelength is between 700 and 1,000 nm, which is close to the band gap of c-Si and in the spectral region where c-Si solar cells have their optimum conversion efficiency. Two types of schemes can be utilized to achieve efficient conversion of visible light into narrow band NIR emission. A single ion can be used if the ion shows a strong broad band absorption in the visible spectral range followed by relaxation to the lowest excited state from which efficient narrow band or line emission in the NIR occurs (see Figure 10.6, top left). Alternatively, a combination of two ions can be used where one ion (the sensitizer) absorbs the light and subsequently transfers the energy to a second ion (the activator), which

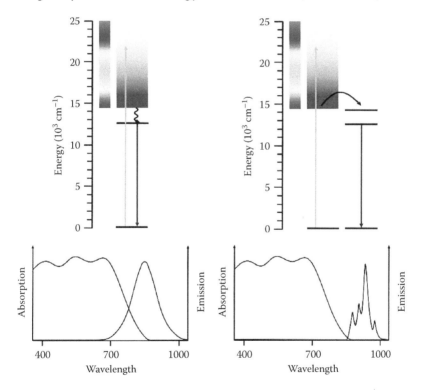

FIGURE 10.6
Schematic representation of energy-level diagrams for spectral conversion of the full solar spectrum into red/infrared radiation in an LSC involving a single activator (left-hand side) or energy transfer from a sensitizer to an activator. In the figures below idealized absorption and emission spectra are shown corresponding to the two processes. (Artwork courtesy of L. Aarts, Utrecht University.)

emits efficiently in the NIR as shown in Figure 10.6 (top right). Both concepts have been investigated for LSCs by incorporating luminescent lanthanides and transition metal ions in glass matrices. The stability of these systems is not a problem, in contrast to LSCs based on dye molecules; however, the quantum efficiency of luminescent ions in glasses appeared to be much lower than in crystalline compounds, especially in the infrared, thus hampering use for LSCs.

10.2.4 Upconversion

Like DC, upconversion was suggested in the 1950s, by Bloembergen (1959), and was related to the development of IR detectors: IR photons would be detected through sequential absorption, as would be possible by the arrangement of energy levels of a solid. However, as Auzel (2004) has pointed out, the essential role of energy transfer was only recognized nearly twenty years later. Several types of upconversion mechanisms exist (Auzel 2004), of which the APTE (addition de photon par transferts d'energie) or, in English, ETU (energy transfer upconversion) mechanism is the most efficient; it involves energy transfer from an excited ion, named sensitizer, to a neighboring ion, named activator (see Figure 10.7). Others are two-step absorption, being a ground-state absorption (GSA) followed by an excited-state absorption (ESA), and second harmonics generation (SHG). The latter mechanism would require extremely high intensities, of about 10^{10} times the sun's intensity on a sunny day, to take place (Strümpel et al. 2007). This may explain why research in this field with a focus on enhancing solar cell efficiency was started only recently (Shalav et al. 2007; Strümpel et al. 2007).

Upconverters usually combine an active ion, of which the energy-level scheme is employed for absorption, and a host material, in which the active ion is embedded. The most efficient upconversion has been reported for the lanthanide ion couples (Yb, Er) and (Yb, Tm); the corresponding upconversion schemes are shown in Figure 10.8. The first demonstration of such an UC layer on the back of solar cells comprised an ultra-thin (3 μm) GaAs cell (band gap 1.43 eV) that was placed on a 100-μm thick vitroceramic containing Yb^{3+} and Er^{3+} (Gibart et al. 1996): it showed 2.5% efficiency upon excitation of 256 kW/m^2 monochromatic sub-band-gap (1.391 eV) laser light (1 W on 0.039 cm^2 cell area), as well as a clear quadratic dependence on incident light intensity. Successful application for silicon solar cells has been reported as well: 2.5% efficiency was determined at

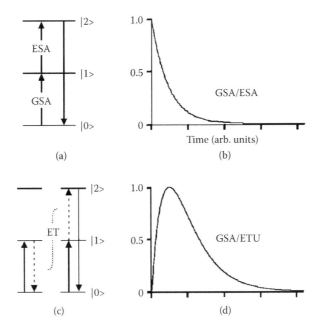

FIGURE 10.7

Schematic representation of two upconversion processes and the characteristic time response of the upconverted emission after a short excitation pulse. (a, b) Ground-state absorption (GSA) followed by excited-state absorption (ESA). This is a single-ion process and takes place during the excitation pulse. (c, d) Upconversion is achieved by GSA followed by energy transfer between ions, and the delayed response is characteristic of the energy transfer upconversion (ETU). (Reproduced from Suyver, J. F., A. Aebischer, D. Biner, et al., 2005a. "Novel Materials Doped with Trivalent Lanthanides and Transition Metal Ions Showing Near-Infrared to Visible Photon Upconversion," *Optical Materials* 27:1111–30. With permission.)

1,523-nm monochromatic laser light for a cell attached to a 1.5-mm-thick layer of sodium yttrium fluoride ($NaYF_4$) containing 20% Er^{3+} mixed with a transparent acrylic medium (Shalav et al. 2005).

Upconversion in solar cells was calculated to potentially lead to a maximum conversion efficiency of 47.6% (Trupke et al. 2002b) for nonconcentrated sunlight using a 6,000K blackbody spectrum in detailed balance calculations. This optimum is reached for a solar cell material of ~2 eV band gap. Applied on the back of silicon solar cells, as in Figure 10.1, the efficiency limit would be about 37% (Trupke et al. 2002b). The analysis of the energy content of the incident AM1.5G spectrum presented in Figure 10.3 revealed that cells with an UC layer would benefit from an extra amount of 35% light incident in the silicon solar cell (Richards 2006a).

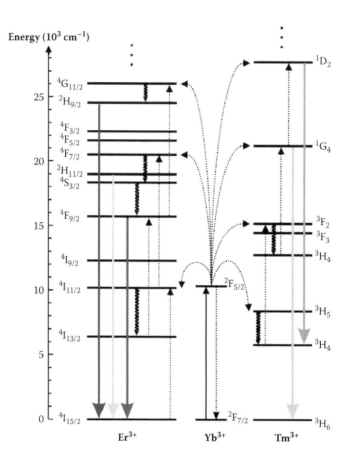

FIGURE 10.8

Upconversion by energy transfer between Yb^{3+} and Er^{3+} (left) and Yb^{3+} and Tm^{3+} (right). Excitation is around 980 nm in the Yb^{3+} ion, and by a two-step or three-step energy transfer process higher excited states of Er^{3+} or Tm^{3+} are populated giving rise to visible (red, green, or blue) emission. Full, dotted, and curly arrows indicate radiative energy transfer, nonradiative energy transfer, and multiphonon relaxation processes, respectively. (Reproduced from Suyver, J. F., A. Aebischer, D. Biner, et al., 2005. "Novel Materials Doped with Trivalent Lanthanides and Transition Metal Ions Showing Near-Infrared to Visible Photon Upconversion," *Optical Materials* 27:1111–30. With permission.)

10.3 STATE OF THE ART

In this section we will describe a selection of recent developments regarding theoretical and experimental work on down- and upconversion and downshifting.

10.3.1 Modeling

An extension to the models described above was presented in a study by Trupke et al. (2006), in which realistic spectra were used to calculate limiting efficiency values for upconversion systems. Using an AM1.5G spectrum leads to a somewhat higher efficiency of 50.69% for a cell with a band gap of 2.0 eV. For silicon, the limiting efficiency would be 40.2%, or nearly 10% larger than the value of 37% obtained for the 6,000K blackbody spectrum (Trupke et al. 2002b). This increase was explained by the fact that absorption in the earth's atmosphere at energies lower than 1.5 eV (as evident in the AM1.5G spectrum) leads to a decrease in light intensity. Badescu and Badescu (2009) have presented an improved model, which according to them now appropriately takes into account the refractive index of solar cell and converter materials. Two configurations are studied: cell and rear converter (C-RC), the usual upconverter application, and front converter and cell (FC-C). They confirm the earlier results of Trupke et al. (2002b) in that the limiting efficiency is larger than that of a cell alone, with higher efficiencies at high concentration. Also, the FC-C combination, i.e., upconverter layer on top of the cell, does not improve the efficiency, which is obvious. Further, by studying the variation of refractive indexes of cell and converter separately, as opposed to Trupke et al. (2002b), it was found that the limiting efficiency increases with refractive index of both cell and upconverter. In practice, a converter layer may have a lower refractive index (1.5, for a transparent polymer, viz. polymethylmethacrylate [PMMA] [Richards and Shalav 2005]) than that of a cell (3.4). Using a material with similar refractive index as the cell would improve the efficiency by about 10%.

In a series of papers the groups of Badescu, De Vos, and co-workers (Badescu and De Vos 2007; Badescu et al. 2007; De Vos et al. 2009) have reexamined the model for downconversion as proposed by Trupke et al. (2002a) and have added the effects of nonradiative recombination and radiation transfer through interfaces. Analogous to the model for upconverters, they studied FC-C and C-RC configurations, with down-conversion or -shifting properties. First, neglecting nonradiative recombination, Badescu et al. (2007) qualitatively confirm the results presented by Trupke et al. (2002a). For both configurations, the efficiency of the combined system is larger than that of a single monofacial cell, albeit that the efficiency is smaller (~26%) due to inclusion of front reflections. Second, including radiative recombination for both converter and cell only increases the efficiency for high (near-unity) radiative recombination efficiency values. Interestingly, they report in this case that the C-RC combination cell–rear converter yields a higher efficiency than the FC-C combination for high-quality solar cells, while for low-quality solar cells, this is reversed. More realistic device values and allowing for different refractive indices in cell and converter was studied in Badescu and De Vos (2007), leading to the conclusion that in reality downconverters may not always be beneficial. However, extending the model once more, with the inclusion of antireflection coating and light trapping texture, showed a limiting efficiency of 39.9%, as reported by De Vos et al. (2009).

Del Cañizo et al. (2008) presented a Monte Carlo ray-tracing model, in which photon transport phenomena in the converter/solar cell system are coupled to nonlinear rate equations that describe luminescence. The model was used to select candidate materials for up- and downconversion, but was set up for use with rare earth ions. Results show that for both converters, the potential gain in short-circuit current is small, and may reach 6 to 7 mA/cm^2 at intensities as high as 1,000 suns, in correspondence with earlier work by Shalav et al. (2007).

Modeling downshifting layers on solar cells was also extended for non-AM1.5G spectra, including varying air mass between 1 and 10, and diffuse and direct spectra (Van Sark 2005). Here, the PC1D model (Basore and Clugston 1996) was used to model quantum dots dispersed in a PMMA layer on top of a multicrystalline silicon cell (mc-Si) as a function of the concentration of quantum dots. Figure 10.9 shows the enhancement of short-circuit current as a function of air mass for the global, direct, and diffuse AM1-10 spectra, at an optimum quantum dot concentration of 100 µM. Spectra were modeled using SPCTRAL2 (Bird and Riordan 1986). Annual performance has been modeled by using modeled spectra from the model SEDES2 (Houshyani Hassanzadeh et al. 2007; Nann and Riordan 1991); these spectra can be considered realistic, as actual irradiation data are used as input. It was found that the simulated short-current

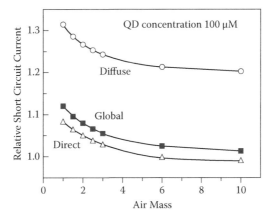

FIGURE 10.9

Enhancement of short-circuit current for a downshifting layer on top of a mc-Si solar cell as in Figure 10.1, as a function of air mass for global, direct, and diffuse spectra. The downshifting layer is comprised of a 1-mm thick PMMA layer in which CdSe quantum dots are dispersed at a concentration of 100 μM.

enhancement, which varies between about 7 and 23%, is linearly related with the average photon energy (APE) of the spectra from considering hourly spectra of four typical and other randomly selected days throughout the year, and of monthly spectra. The annual short-circuit increase was determined at 12.8% using the annual distribution of APE values and their linear relation (Van Sark 2007), which is to be compared with the 10% increase in case of the AM1.5G spectrum. For mc-Si cells with improved surface passivation and a concomitant improved blue response, the relative short-current increase has been calculated to be lower (Van Sark 2006).

Modeling large area LSCs has indicated the importance of top-surface losses that occur through the escape cone (Chatten et al. 2007) both through primary emission and through emission of luminescence that has been reabsorbed and might otherwise have been trapped via total internal reflection or by mirrors. As an example, for an idealized (perfectly transparent host, LQE = 1), mirrored (perfect mirrors on one short and two long edges, and the bottom surface), 40 × 5 × 0.5 cm LSC doped with CdSe/ZnS core-shell QDs with emission matched to a GaInP cell (Figure 10.10a), 24% of the AM1.5G spectrum is absorbed. The LSC absorption, incident AM1.5G spectrum, and the average concentrated flux escaping the bare right-hand edge of the idealized LSC are depicted in Figure 10.10b. This idealized LSC absorbs 24% of the incident photons in the AM1.5G spectrum. The

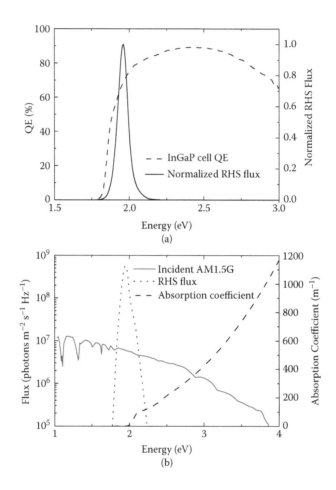

FIGURE 10.10

(a) Quantum efficiency (spectral response) of the GaInP cell used in modeling the ideal-ized LSC together with the modeled luminescence escaping the right-hand surface (RHS) of the LSC that would be coupled into the cell. (b) Absorption of the LSC material used in the calculations for the idealized system together with the flux incident on the top surface and the predicted concentrated average luminescent flux escaping the right-hand surface of the idealized LSC. (From Van Sark, W. G. J. H. M., K. W. J. Barnham, L. H. Slooff, et al., 2008a. "Luminescent Solar Concentrators—A Review of Recent Results," *Optics Express* 16:21773–92. With permission.)

photon concentration ratio, C, which is the ratio of the concentrated flux escaping the right-hand surface of the LSC to the flux incident on the top surface, is 4.18. However, since the concentrated flux escaping the right-hand surface is a narrow band matched to the spectral response of the cell, it can all be converted and the idealized LSC would produce eight times the current compared to the cell alone exposed to AM1.5G. However, 78% of the luminescence is lost through the large top-surface area, and only 22% may be collected at the right-hand surface. This leads to the use of wavelength-selective cholesteric liquid crystal coatings applied to the top surface in order to reduce the losses (Debije et al. 2006), as these coatings are transparent to incoming light but reflect the emitted light.

Ray tracing for LSCs uses basic ray-tracing principles, which means that a ray, which represents light of a certain wavelength traveling in a certain direction, is traced until it leaves the system, e.g., by absorption or reflection at the interface (Burgers et al. 2005; Gallagher et al. 2004). The main extension to the standard ray-tracing model is the handling of the absorption and emission by the luminescent species in the LSC. Ray tracing has been used to perform parameter studies for a 5×5 cm^2 planar LSC to find attainable LSC efficiencies. The concentrator consists of a PMMA plate (refractive index $n = 1.49$, absorption 1.5 m^{-1}) doped with two luminescent dyes, CRS040 from Bayer and Lumogen F Red 305 from BASF, with a FQE of 95% (Slooff et al. 2006). With the ray-tracing model the efficiency of this plate together with a mc-Si solar cell was determined to be 2.45%. Results for attaining efficiencies for other configurations are shown in Table 10.1 (Van Sark et al. 2008a). Replacing the mc-Si cell by a GaAs cell or an InGaP cell will increase the efficiency from 3.8 to 6.5 and

TABLE 10.1

Calculated Efficiencies (in Percent) for an LSC for Various Optimized Configurations and Parameters

mc-Si	GaAs	InGaP	Parameters
2.4	4.2	5.9	Fixed mirrors, 85% reflectivity, dyes with 95% LQE
2.9	5.1	7.1	97% reflectivity "air-gap mirrors" on sides, and 97% reflectivity Lambertian mirror at bottom
3.4	5.9	8.3	Reduce background absorption of polymer matrix from 1.5 to 10^{-3} m^{-1}
3.8	6.5	9.1	Increase of refractive index from 1.49 to 1.7

Source: Van Sark, W. G. J. H. M., K. W. J. Barnham, L. H. Slooff, et al., "Luminescent Solar Concentrators—A Review of Recent Results," 2008a. *Optics Express* 16:21773–92. With permission.

9.1%, respectively (based on V_{oc} [FF] values of 0.58 [0.83], 1.00 [0.83], and 1.38 [0.84] V, for mc-Si, GaAs, and InGaP, respectively). Thus, the use of GaAs or InGaP cells will result in higher efficiencies, but these cells are more expensive. A cost calculation must be performed to determine if the combination of the luminescent concentrator with this type of cells is an interesting alternative to mc-Si-based solar technology.

Currie et al. (2008) projected conversion efficiencies as high as 6.8%, for a tandem LSC based on two single LSCs that consist of a thin layer of organic dye molecules deposited onto a glass plate to which a GaAs cell was attached. Using CdTe or Cu(In,Ga)Se$_2$, solar cell conversion efficiencies of 11.9 and 14.5%, respectively, were calculated.

Annual performance has been modeled using an LSC of which the properties and geometry resulted from a cost-per-unit-of-power optimization study by Bende et al. (2008). A square plate of 23.7 × 23.7 × 0.1 cm^3 was used and in the ray-trace model attached to four c-Si solar cells (18.59% efficiency) on all sides. Using a Lumogen F Red 305 (BASF) dye, it was calculated that 46.5% of all photons in the wavelength range of 370 to 630 nm was collected, leading to an LSC efficiency of 4.24% (Van Sark et al. 2008b). The annual yield of this LSC was determined using realistic spectra representative for the Netherlands (Houshyani Hassanzadeh et al. 2007) and amounted to 41.3 kWh/m^2, not taking into account temperature effects; this is equivalent to an effective annual efficiency of 3.81%.

10.3.2 Experimental

10.3.2.1 Downconversion

The most promising systems for downconversion rely on lanthanide ions. The unique and rich energy-level structures of these ions allow for efficient spectral conversion, including up- and downconversion processes mediated by resonant energy transfer between neighboring lanthanide ions (Auzel 2004; Wegh et al. 1999). Considering the energy levels of all lanthanides, as shown in the Dieke energy-level diagram (Dieke 1968; Peijzel et al. 2005; Wegh et al. 2000), it is immediately evident that the energy-level structure of Yb^{3+} is ideally suited to be used in downconversion for use in c-Si solar cells. The Yb^{3+} ion has a single excited state (denoted by the term symbol $^2F_{5/2}$) some 10,000 cm^{-1} above the $^2F_{7/2}$ ground state, corresponding to emission around 1,000 nm. The absence of other energy

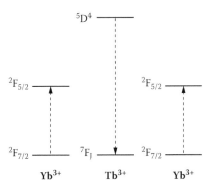

FIGURE 10.11

Cooperative energy transfer from Tb to two Yb ions. From the $^5D_4$ state of one Tb ion, two neighboring Yb ions are excited to the $^2F_{5/2}$ state from where emission of 980 nm photons can occur. (From Meijerink, A., R. Wegh, P. Vergeer, and T. Vlugt, "Photon Management with Lanthanides," *Optical Materials* 28 [2006]: 575–81, and Vergeer, P., T. J. H. Vlugt, M. H. F. Kox, et al., 2005. "Quantum Cutting by Cooperative Energy Transfer in YbxY1–xPO4:Tb3+," *Physical Review B* 71:014119-1–11. With permission.)

levels allows Yb^{3+} to exclusively "pick up" energy packages of 10,000 cm^{-1} from other lanthanide ions and emitting ~1,000 nm photons that can be absorbed by c-Si. Efficient downconversion using Yb^{3+} as acceptor requires donor ions with an energy level around 20,000 cm^{-1} and an intermediate level around 10,000 cm^{-1}. From inspection of the Dieke diagram one finds that potential couples are (Er^{3+}, Yb^{3+}), (Nd^{3+}, Yb^{3+}), and (Pr^{3+}, Yb^{3+}) for a resonant two-step energy transfer process. Also, cooperative sensitization is possible where energy transfer occurs from a high excited state of the donor to two neighboring acceptor ions without an intermediate level.

The first report on efficient downconversion for solar cells was based on cooperative energy transfer from Tb^{3+} to two Yb^{3+} ions in $Yb_xY_{1-x}PO_4$:Tb^{3+} (Vergeer et al. 2005), as shown schematically in Figure 10.11. The $^5D_4$ state of Tb^{3+} is around 480 nm (21,000 cm^{-1}), and from this state cooperative energy transfer to two Yb^{3+} neighbors occurs, both capable of emitting a 980-nm photon.

The same efficient downconversion process was observed in other host lattices: Zhang et al. (2007) observed cooperative quantum cutting in (Yb,Gd)Al$_3$(BO$_3$)$_4$:Tb^{3+}, and quantum efficiencies up to 196% were reported. Cooperative downconversion for other couples of lanthanides was also claimed. For the couple (Pr, Yb) and (Tm, Yb) co-doped into borogermanate glasses, a decrease of the emission from higher energy levels of Pr and Tm was observed (Liu et al. 2008). In case of Pr^{3+} the starting level for the cooperative downconversion is the $^3P_0$ level, while for Tm^{3+} the $^1G_4$ is at twice the energy of the $^2F_{5/2}$ level of Yb^{3+} (see also the Dieke diagram, Figure 10.3). The efficiency of the downconversion process was estimated from the decrease of the lifetime of the donor state. For the Tm^{3+} emission a decrease from 73 μs (for the sample doped with Tm only) to 45 μs (for a sample co-doped with 20% of Yb) was observed, giving a transfer efficiency of 38%. For the same glass co-doped with Pr and Yb a 65% decrease of the lifetime was observed, implying a 165% quantum yield. Similar results were found for the (Pr, Yb) couple in aluminosilicate glasses (Lakshminarayana et al. 2008). More recent work on the (Pr, Yb) couple confirmed the presence of downconversion, but the mechanism involved was pointed out to be a resonant two-step energy transfer rather than a cooperative energy transfer (Van der Ende et al. 2009). For the (Pr, Yb) couple an intermediate level ($^1G_4$) is available around 10,000 cm^{-1}, which makes a two-step resonant energy transfer process possible. This first-order process will have a much higher probability than the second-order (cooperative) transfer process. The results demonstrated efficient quantum cutting of one visible photon into two NIR photons in SrF_2:Pr^{3+}, Yb^{3+}. Comparison of absorption and excitation spectra provided direct evidence that the downconversion efficiency is close to 200%, in agreement with a two-step energy transfer process that can be expected based on the energy-level diagrams of Pr^{3+} and Yb^{3+}. This first-order energy transfer process is effective at relatively low Yb^{3+} concentrations (5%), where concentration quenching of the Yb^{3+} emission is limited. Comparison of emission spectra, corrected for the instrumental response, for SrF_2:Pr^{3+} (0.1%) and SrF_2:Pr^{3+} (0.1%), Yb^{3+} (5%) revealed an actual conversion efficiency of 140%.

Downconversion of two near-infrared photons per absorbed blue photon was reported by Chen et al. (2008a) in transparent glass ceramics with embedded Pr^{3+}/Yb^{3+}:β-YF^3 nanocrystals. The Pr^{3+} ions are excited with a visible photon (482 nm), and subsequently two near-infrared photons (976 nm) are emitted by the Yb^{3+} ions through an efficient cooperative energy transfer from Pr^{3+} to Yb^{3+}, with quantum efficiency close to 200%.

Quantum cutting downconversion was shown in borate glasses using co-doping of Ce^{3+} and Yb^{3+} (Chen et al. 2008b). A UV photon (330 nm) excites a Ce^{3+} ion, and cooperative energy transfer between the Ce^{3+} and Yb^{3+} ion leads to the emission of two NIR photons (976 nm), at 174% quantum efficiency, owing to the 74% efficiency energy transfer.

Timmerman et al. (2008) demonstrated so-called space-separated quantum cutting within SiO_2 matrices containing both silicon nanocrystals and

Er^{3+}. Energy transfer from photoexited silicon nanocrystals to Er^{3+} had been observed earlier (Fujii et al. 1997), but Timmerman et al. (2008) show that upon absorption of a photon by a silicon nanocrystal, a fraction of the photon energy is transferred, generating an excited state within either an erbium ion or another silicon nanocrystal. As also the original silicon NC relaxes from a highly excited state toward the lowest-energy excited state, the net result is two electron-hole pairs for each photon absorbed.

10.3.2.2 Downshifting

Chung et al. (2007) reported downshifting phosphor coatings consisting of $Y_2O_3:Eu^{3+}$ or $Y_2O_2S:Eu^{3+}$ dispersed in either polyvinyl alcohol or polymethylmethacrylate on top of mc-Si solar cells: an increase in conversion efficiency was found of a factor of 14 under UV illumination by converting the UV radiation (for which the response of c-Si is low) to 600-nm emission from the Eu^{3+} ion. The solar cells used were encapsulated in an epoxy that absorbs photons with energy higher than ~3 eV. This protective coating can remain in place and downconverters or -shifters can easily be added.

Svrcek et al. (2004) demonstrated that silicon nanocrystals incorporated into spin-on-glass (SOG) on top of c-Si solar cells are successful as downshifters, leading to a potential efficiency enhancement of 1.2%, while they experimentally showed an enhancement of 0.4% (using nanocrystals of 7 nm diameter with a broad emission centered around 700 nm). McIntosh et al. (2009) recently presented results on encapsulated c-Si solar cells, of which the PMMA encapsulant contained downshifting molecules, i.e., Lumogen dyes. These results indicate a ~1% relative increase in the module efficiency, based on a 40% increase in external quantum efficiency for wavelengths of <400 nm. Mutlugun et al. (2008) claimed a twofold increase in efficiency applying a so-called nanocrystal scintillator on top of a c-Si solar cell; it comprises a PMMA layer in which CdSe/ZnS core-shell quantum dots (emission wavelength 548 nm) are embedded. However, the quality of their bare c-Si cells was very poor.

Stupca et al. (2007) demonstrated the integration of ultra-thin films (2 to 10 nm) of monodisperse luminescent Si nanoparticles on polycrystalline Si solar cells. One-nanometer-sized blue emitting and 2.85 nm red emitting particles enhanced the conversion efficiency by 60% in the UV, and 3 to 10% in the visible for the red and blue emitting particles, respectively. These numbers are similar to what has been predicted (Van Sark et al. 2005). Van Sark et al. (2004) failed to observe their predicted 10% increase of conversion efficiency in their experiments using drop-casted PMMA in which CdSe/ZnS core-shell quantum dots (emission wavelength 603 nm) are embedded on top of mc-Si cells. This was explained by the fact that most of the shifted light did not enter the solar cell, but escaped from the sides and top, and proper use of mirrors could have prevented that.

An 8% relative increase in conversion efficiency was reported (Maruyama and Kitamura 2001) for a CdS/CdTe solar cell, where the coating in which the fluorescent coloring agent was introduced increased the sensitivity in the blue; a maximum increase in efficiency was calculated to be 30 to 40%. Others showed results that indicate a 6% relative increase in conversion efficiency (Maruyama and Bandai 1999) upon coating a mc-Si solar cell. The employed luminescent species has an absorption band around 400 nm and a broad emission between 450 and 550 nm. As QDs have a much broader absorption, it is expected that potentially the deployment of QDs in planar converters could lead to relative efficiency increases of 20 to 30%. Downshifting employing QDs in a polymer composite has been demonstrated in a light-emitting diode (LED), where a GaN LED was used as an excitation source (λ_m) for QDs emitting at 590 nm (Lee et al. 2000). Besides QDs, other materials have been suggested, such as rare earth ions (Wegh et al. 1999) and dendrimers (Serin et al. 2002). A maximum increase of 22.8% was calculated for a thin-film coating of $KMgF_3$ doped with Sm on top of a CdS/CdTe solar cell, while experimental results show an increase of 5% (Hong and Kawano 2003).

Recent efforts to surpass the historical 4% efficiency limit of LSCs (Goetzberger 2008; Wittwer et al. 1984; Zastrow 1994), albeit for smaller area size, have been successful. For example, Goldschmidt et al. (2009) showed that for a stack of two plates with different dyes, to which four GaInP solar cells were placed at the sides, the conversion efficiency is 6.7%; the plate was small (4 cm²), and the concentration ratio was only 0.8. It was argued that the conversion efficiency was limited by the spectral range of the organic dyes used, and that if the same quantum efficiency as was reached for the 450- to 600-nm range could be realized for the range 650 to 1,050 nm, an efficiency of 13.5% could be within reach. They also discuss the benefits of a photonic structure on top of the plate, to reduce the escape cone loss (Goldschmidt et al. 2009). The proposed structure is a so-called rugate filter; this is characterized by a varying refractive index in contrast to standard Bragg reflectors, which suppresses the side loops that could lead to unwanted reflections. The use of these filters would increase the efficiency by ~20%, as was determined for an LSC consisting of one plate and dye. Slooff et al. (2008) presented results on $50 \times 50 \times 5$ mm³ PMMA plates in which both CRS040 and Red305 dyes were dispersed at 0.003 and 0.01 wt%, respectively. The plates were attached to either mc-Si, GaAs, or InGaP cells, and a diffuse reflector (97% refection) was used at the rear side of the plate. The highest efficiency measured was 7.1% for four GaAs cells connected in parallel (7% if connected in series).

As stated above, quantum dots are potential candidates to replace organic dye molecules in an LSC, for their higher brightness, better stability, and wider absorption spectrum (Barnham et al. 2000; Chatten et al. 2003a, 2003b; Gallagher et al. 2007). In fact, the properties and availability of QDs started renewed interest in LSCs around 2000 (Barnham et al. 2000), with the main focus on modeling, while more recently some experimental results have been presented. Schüler et al. (2007) proposed to make LSCs by coating transparent glass substrates with QD-containing composite films, using a potentially cheap sol-gel method. They reported on the successful fabrication of thin silicon oxide films that contain CdS QDs using a sol-gel dip-coating process, whereby the 1- to 2-nm sized CdS QDs are formed during thermal treatment after dip coating. Depending on the anneal temperature, the colors of the LSC ranged from green for 250°C to yellow for 350°C and orange for 450°C.

Reda (2008) also prepared CdS QD concentrators, using sol-gel spin coating, followed by annealing. The annealing temperature was found to affect absorption and emission spectrum: luminescent intensity and Stokes' shift both decreased for four weeks' outdoor exposure to sunlight, which probably was caused by aggregation and oxidation. It is known that oxidation leads to blue shifts in emission (Van Sark et al. 2002). Blue shifts have also been observed by Gallagher et al. (2007), who dispersed CdSe QDs in several types of resins (urethane, PMMA, epoxy), for fabrication of LSC plates.

Quilitz et al. (2009b) have addressed several problems regarding incorporation of QDs in an organic polymer matrix, viz. phase separation, agglomeration of particles leading to turbid plates, and luminescence quenching due to exciton energy transfer (Koole et al. 2006). They have synthesized QDSCs using CdSe core-multishell QDs (Koole et al. 2008) (QE = 60%) that were dispersed in laurylmethacrylate (LMA) (see also Lee et al. 2000; Walker et al. 2003). UV polymerization was employed to yield transparent PLMA plates with QDs without any sign of agglomeration. To one side of this plate, a mc-Si solar cell was placed, and aluminum mirrors to all other sides. Compared to the bare cell (5×0.5 cm²) that generated a current density of 40.28 mA/cm² at 1,000 W AM1.5G spectrum, the best QDC made generated a current of 77.14 mA/cm², nearly twice as much. The QDC efficiency is 3.5%. In addition, exposure to a 1,000 W sulfur lamp for 280 hours continuously showed very good stability: the current density decreased by 4% only, on average. However, reabsorption may still be a problem, as is demonstrated by a small red shift in the emission spectrum for long photon pathways; also, absorption by the matrix is occurring. Besides QDs, nanorods (NRs) have also been dispersed in PLMA, showing excellent transmittance of 93%; for long rods (aspect ratio of 6) a QE of 70% is observed, which is only slightly smaller than the QE in solution, implying that these rods are stable throughout the polymerization process (Quilitz et al. 2009a). In addition, these NRs have also been dispersed in cellulosetriacetate (CTA), and a ~10-μm thin film on a glass substrate was made showing bright orange luminescence (Quilitz et al. 2009a).

Hyldahl et al. (2009) used commercially available CdSe/ZnS core-shell QDs with QE = 57% in LSCs, both liquid (QDs dissolved in toluene, between two 6.2 × 6.2 × 0.3 cm glass plates) and solid (QDs dispersed in epoxy), and they obtained efficiencies of 3.98 and 1.97%, respectively. They also used the organic dye Lumogen F Red300, and obtained an efficiency of 2.6% in toluene. They conclude that QDSCs outperform LSCs with organic dyes.

10.3.2.3 Upconversion

Lanthanides have also been employed in upconverters attached to the back of bifacial silicon solar cells. Trivalent erbium is ideally suited for upconversion of NIR light due to its ladder of nearly equally spaced energy levels that are multiples of the $^4I_{15/2}$ to $^4I_{13/2}$ transition (1,540 nm) (see also Figure 10.3). Shalav et al. (2005) have demonstrated a 2.5% increase of external quantum efficiency due to upconversion using $NaYF_4$:20% Er^{3+}. By depicting luminescent emission intensity as a function of incident monochromatic (1,523 nm) excitation power in a double-log plot, they showed that at low light intensities a two-step upconversion process ($^4I_{15/2} \rightarrow ^4I_{13/2} \rightarrow ^4I_{11/2}$) dominates, while at higher intensities a three-step upconversion process ($^4I_{15/2} \rightarrow ^4I_{13/2} \rightarrow ^4I_{11/2} \rightarrow ^4S_{3/2}$ level) is involved.

Strümpel et al. (2007) have identified the materials of possible use in up- (and down-) conversion for solar cells. In addition to the $NaYF_4$:(Er, Yb) phosphor, they suggest the use of $BaCl_2$:(Er^{3+}, Dy^{3+}) (Strümpel et al. 2005), as chlorides were thought to be a better compromise between having a low phonon energy and a high excitation spectrum than the NaYF$_4$ (Gamelin and Güdel 2001; Ohwaki and Wang 1994; Shalav et al. 2007). These lower phonon energies lead to lower nonradiative losses. In addition, the emission spectrum of dysprosium is similar to that of erbium, but the content of Dy^{3+} should be <0.1%, to avoid quenching (Auzel 2004; Strümpel et al. 2007).

NaYF$_4$ co-doped with (Er^{3+}, Yb^{3+}) is to date the most efficient upconverter (Suyver et al. 2005a, 2005b), with ~50% of all absorbed NIR photons upconverted and emitted in the visible wavelength range. However, the (Yb, Er) couple is not considered beneficial for upconversion in c-Si cells, as silicon also absorbs in the 920 to 980 nm wavelength range. These phosphors can be useful for solar cells based on higher-band-gap materials such as the Grätzel cell (O'Regan and Grätzel 1991), a-Si(Ge):H, or CdTe. In that case, the $^2F_{5/2}$ level of Yb^{3+} would serve as an intermediate step for upconversion (Shalav et al. 2007), and IR radiation between ~700 and 1,000 nm that is not absorbed in these wider-band-gap solar cells can be converted into green (550 nm) light that can be absorbed.

A typical external collection efficiency (ECE) graph of a standard single-junction p-i-n a-Si:H solar cell is shown in Figure 10.12. These cells are manufactured on textured SnO$_2$:F-coated glass substrates and routinely

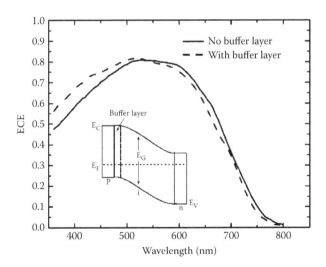

FIGURE 10.12

The spectral response for cells with and without a buffer layer at the p/i interface in the as deposited state obtained at –1 V bias voltage. Inset: The p-i-n structure showing the position of the buffer layer. (From Munyeme, G., M. Zeman, R. E. I. Schropp, and W. F. Van der Weg, 2004. "Performance Analysis of a-Si:H p-i-n Solar Cells with and without a Buffer Layer at the p/i Interface," *Physica Status Solidi C* 9:2298–303. With permission.)

have >10% initial efficiency. Typically, the active Si layer in the cell has a thickness of 300 nm and the generated current is 14.0 to 14.5 mA/cm², depending on the light-trapping properties of the textured metal oxide and the back reflector. After light-induced introduction of the stabilized defect density (Staebler-Wronski effect [Schropp and Zeman 1998]), the stabilized efficiency is 8.2 to 8.5%. From Figure 10.12 it can be seen that the maximum ECE is 0.8 at ~550 nm, and the cutoff occurs at 700 nm, with a response tailing toward 800 nm. The response is shown with and without the use of a buffer layer at the front of the cell. The buffer layer already causes an improvement at the short wavelengths, and a downconverter or -shifter may not be beneficial. The purpose of an upconverter should be to tune the energy of the emitted photons to the energy where the spectral response shows a maximum. If the energy of the emitted photons is too close to the absorption limit (the band gap edge), then the absorption coefficient is too low and the upconverted light would not be fully used.

The photogenerated current could be increased by 40% if the spectral response was sustained at a high level up to the band gap cutoff at 700 nm, and by even more if light with wavelengths $\lambda > 700$ nm could be more fully absorbed. These two effects can be achieved with the upconversion layer, combined with a highly reflecting back contact. While the upconversion layer converts sub-band-gap photons to super-band-gap photons that can thus be absorbed, a nonconductive reflector is a much better alternative than any metallic mirror, thus sending back both the unabsorbed super-band-gap photons and the "fresh" super-band-gap photons into the cell. It is estimated that the stabilized efficiency of the 8.2 to 8.5% cell can be enhanced to ~12%.

When applying an upconverter, it is thus actually advantageous to use it with a cell that has a rather high band gap, such as protocrystalline Si (1.8 eV) (instead of amorphous Si [1.7 eV]) (Trupke et al. 2002b). Another important property of protocrystalline Si is that it is more stable under prolonged light exposure (performance stability is within 10% of the initial performance) and that the band gap is slightly higher (1.8 eV). Other frequently used materials in thin-film solar cells have a much lower band gap, such as CIGS (1.5 eV) or CdTe (1.4 eV).

For $NaYF_4$ co-doped with (Er^{3+}, Yb^{3+}), absorption of 980 nm (by the Yb^{3+} ion) leads to efficient emission of 653 nm (red) and 520 to 540 nm (green) light (by the Er^{3+}) after a two-step energy transfer process. Figure 10.13 schematically shows at which wavelengths the processes take place, relative to the absorption regime (in blue) of the a-Si cell. The narrow absorption band around 980 nm for Yb^{3+} limits the spectral range

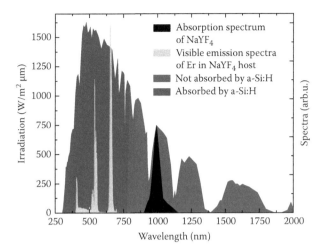

FIGURE 10.13

Part of the solar spectrum that is transmitted and can be converted by NaYF$_4$ upconverter doped with Er and Yb as sensitizer around 980 nm. The blue part of the solar spectrum is the absorption region of a protocrystalline silicon single-junction solar cell. (Figure courtesy of J. de Wild, Utrecht University.)

of the IR that can be used for upconversion. By using a third ion (for example, Ti^{3+}) as a sensitizer, the full spectral range between 700 and 980 nm can be efficiently absorbed and converted to red and green light by the Yb-Er couple. The resulting light emission in the green and red region is very well absorbed by the cell with very good quantum efficiency for electron-hole generation.

Upconversion systems consisting of lanthanide nanocrystals of YbPO$_4$ and LuPO$_4$ have been demonstrated to be visible by the naked eye in transparent solutions, but at an efficiency lower than than for solid-state upconversion phosphors (Suyver et al. 2005a; Heer et al. 2003). Other host lattices (NaXF$_4$, X = Y, Gd, La) have been used and co-doping with Yb^{3+} and Er^{3+}, or Yb^{3+} and Tm^{3+} appeared successful, where Yb^{3+} acts as sensitizer. Nanocrystals of <30 nm in size, to prevent scattering in solution, have been prepared, and they can be easily dissolved in organic solvents forming colloidal solutions, without agglomeration. Figure 10.14 demonstrates visible upconversion in transparent 1 wt% colloidal solutions in dimethylsulfoxide (DMSO) excited by moderate-intensity NIR laser light, thus illustrating the high luminescence efficiency. Further efficiency increase is possible by growing a shell of undoped NaYF$_4$ around the nanocrystal; in addition, surface modification is needed to allow dissolution in water, for use in biological labeling.

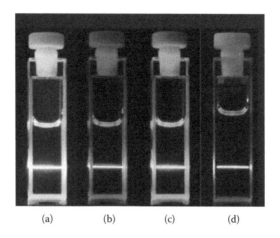

(a)   (b)   (c)   (d)

FIGURE 10.14
Photographs of the upconversion luminescence in 1 wt% colloidal solutions of nanocrystals in dimethylsulfoxide excited at 10,270 cm^{-1} (invisible) with a laser power density of 5.9 kW/cm^2. (a) Total upconversion luminescence of the NaYF4:20% Yb^{3+}, 2% Er^{3+} sample. (b, c) The same luminescence through green and red color filters, respectively. (d) Total upconversion luminescence of the NaYF4: 20% Yb^{3+}, 2% Tm^{3+} sample. (Courtesy of Prof. M. Haase.) (See color insert following page 206.)

Porous silicon layers are investigated for use as upconverter layers as host for rare earth ions, because these ions can easily penetrate the host due to the large surface area and porosity. A simple and low-cost dipping method has been reported (Díaz-Herrera et al. 2009), in which a porous silicon layer is dipped into a nitrate solution of erbium and ytterbium in ethanol, $(Er(NO_3)_3:Yb(NO_3)_3:C_2H_5OH)$, which is followed by a spin-on procedure and a thermal activation process at 900°C. Excitation of the sample at 980 nm revealed upconversion processes as visible and NIR photoluminescence is observed; co-doping of Yb with Er is essential, and doping only with Er shows substantial quenching effects (González-Díaz et al. 2008).

Sensitized triplet-triplet annihilation (TTA) using highly photostable metal-organic chromophores in conjunction with energetically appropriate aromatic hydrocarbons has been shown to be another alternative upconversion system (Singh-Rachford et al. 2008). This mechanism was shown to take place under ambient laboratory conditions, i.e., low light intensity conditions, clearly of importance for outdoor operation. These chromophores (porphyrins in this case) can be easily incorporated in a solid polymer so that the materials can be treated as thin-film materials (Islangulov et al. 2007).

10.4 DISCUSSION

Two issues remain to be solved before downconverters can be applied in solar cells: the absorption strength needs to be increased as the transitions involved for the trivalent lanthanides are sharp and weak (parity forbidden). A second issue is concentration quenching. High Yb concentrations are needed for efficient energy transfer, as every donor (Tb, Pr, or Tm) needs to have two Yb neighbors for energy transfer. For these high concentrations, energy migration over the Yb sublattice occurs and trapping of the migrating excitation energy by quenching sites strongly reduces the emission output.

At present, research is conducted to resolve these issues. The limited absorption can be solved by the inclusion of a sensitizer for the $^3P_0$ level of Pr^{3+}, which is able to absorb efficiently over a broad wavelength range (300 to 500 nm) and subsequent energy transfer to the $^3P_0$ level of Pr^{3+} (see Figure 10.15). In principle, such sensitization can be realized in an efficient and cost-effective manner by the inclusion of a sensitizer ion. The 4f-5d

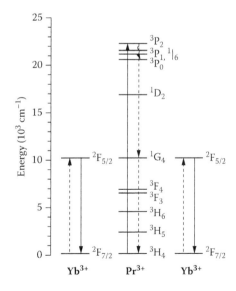

FIGURE 10.15
Energy levels and quantum cutting mechanism for the Pr^{3+}-Yb^{3+} couple. Two-step energy transfer occurs upon excitation into the ^3P_J ($J = 0, 1, 2$) and $^1I_6$ levels of Pr^{3+}. A single visible photon absorbed by these levels is thereby converted into two ~1,000 nm photons. Solid arrows are optical transitions, dotted arrows represent nonradiative energy transfer processes, while curved arrows symbolize nonradiative relaxation.

luminescence of Ce^{3+}, for example, is often used to sensitize the Tb^{3+} luminescence in phosphors for fluorescent tubes. Concentration quenching may be limited by optimization of the synthesis conditions (less quenching sites), while the synthesis of nanocrystals may also be beneficial. In nanocrystals the volume probed by energy migration is limited (due to the small size of the nanocrystal), and for defect-free nanocrystals high quantum yields can be expected, similar to the increase in quantum yield observed for quantum dots vs. bulk semiconductors.

For upconverters based on lanthanides also, absorption strengths need to be increased and quenching decreased. In addition, upconversion could be useful for solar cells with a band gap higher than that of crystalline silicon, and presently, research is directed toward optimum matching of NIR absorption and visible emission with the band gap of the solar cell to which the upconverter is attached.

Modeling studies on incorporation of conversion layers on top (downconverter or downshifter) or at the bottom (upconverter) of single-junction solar cells have shown that the conversion efficiency may increase by about 10% (Glaeser and Rau 2007; Richards 2006b; Strümpel et al. 2007; Trupke et al. 2006; Van Sark 2005; Van Sark et al. 2005). This is still an experimental challenge. Also, stability of converter materials is a critical issue, as their lifetime should be >20 years to comply with present solar module practice.

Notwithstanding the impressive progress that is evident, the usefulness of down- and upconversion and downshifting depends on the incident spectrum and intensity. While solar cells are designed and tested according to the ASTM standard (ASTM 2003), specifying intensity of 1,000 W/m^2 and the AM1.5G global spectrum, and temperature of 25°C, these conditions are rarely met outdoors, where clear skies are interchanged with (partly) overcast skies, in addition to diurnal and seasonal variations. Spectral conditions for solar cells vary from AM0 (extraterrestrial) via AM1 (equator, summer and winter solstice) to AM10 (sunrise, sunset). The weighted average photon energy (APE) (Minemoto et al. 2007) can be used to parameterize this; the APE (using the range 300 to 1,400 nm) of AM1.5G is 1.674 eV, while the APEs of AM0 and AM10 are 1.697 and 1.307 eV, respectively (see also Figure 10.16.) Further, overcast skies cause higher scattering leading to diffuse spectra, which are blue rich; e.g., the APE of the AM1.5 diffuse spectrum is calculated to be 2.005 eV—indeed much larger than the APE of the AM1.5 direct spectrum of 1.610 eV.

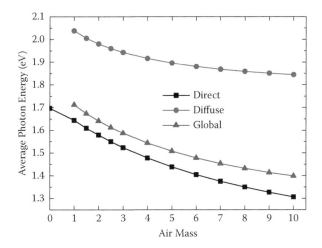

FIGURE 10.16

Variation of average photon energy as a function of air mass, for direct, diffuse, and global spectra.

As DC and DS effectively red-shift spectra, the more blue an incident spectrum contains (high APE), the more gain that can be expected (Van Sark 2005, 2008). Application of DC layers will therefore be more beneficial for regions with high diffuse irradiation fraction, such as Northwestern Europe, where this fraction can easily be 50%. Here also LSCs are expected to be deployed successfully (Van Sark et al. 2008b). In contrast, solar cells with UC layers will be performing well in countries with high direct irradiation fractions, such as around the equator, or in early morning and evening due to the high air mass, but to a lesser extent due to the nonlinear response to intensity.

The variation of the incident spectra is of particular concern for series-connected multiple-junction cells, such as triple a-Si:H (Krishnan et al., 2009) and GaAs-based cells. Current-mismatch due to spectral differences with respect to the AM1.5G standard leads to lowering of the conversion efficiency, as the cell with the lowest current determines the total current. It has been shown that the calculated limiting efficiency of a GaAs-based triple cell is reduced to 32.6% at AM5, while its AM1.5G efficiency was 52.5% (Trupke et al. 2006). A single cell with an UC layer, having an AM1.5G efficiency of 50.7%, does not suffer that much from an increase in air mass: at AM5 the efficiency is lowered to 44.0% (Trupke et al. 2006). This so-called spectral robustness is due to the current-matching constraints, which are much more relaxed in the cell/UC layer case.

10.5 SUMMARY

Spectrum modification by means of down- and upconversion and down-shifting for application in solar cells has been reviewed. Nanoparticles, based on semiconductors or lanthanides, embedded in predominantly polymer layers on top or at the back of solar cells are an essential ingredient in attaining higher conversion efficiencies.

Progress in material synthesis has been fast in the past years, as implications for solar cell efficiency improvement became clearer, not only to the R&D community that traditionally focused on lighting applications. It is imperative for our future on earth that renewable energy is used, in particular solar energy due to its vast resource. Increasing the conversion efficiency, while keeping manufacture cost as low as possible, thus is the main R&D target in solar photovoltaic energy for the coming decades. Nanotechnology will play a key role in this.

10.6 FUTURE PERSPECTIVE

Successfully optimizing absorption strength and quenching in lanthanide-based downconverters will bring the theoretical limits within reach. This also holds for upconversion, where, in addition, TTA and mixed transition metal-lanthanide systems (Suyver et al. 2005a) constitute new material systems.

Future use of DS layers on top of solar cells may be limited as blue response of present cells will advance to higher levels (Van Sark 2006). On the other hand, these improvements might require additional expensive processing, while application of a DS layer is expected to be low cost, as it involves coating of a plastic with dispersed luminescent species only.

LSC development will focus on material systems that should (Rowan et al. 2008) (1) absorb all photons with a wavelength of >950 nm, and emit them red shifted at ~1,000 nm, for use with c-Si solar cells; (2) have as low as possible spectral overlap between absorption and emission spectra to minimize reabsorption losses; (3) have a near-unity luminescence quantum yield; (4) have low escape cone losses; (5) be stable outdoors for longer than ten years; and (6) be easy to manufacture at low cost. Much progress has occurred, which is illustrated by the recent efficiency record of 7.1% (Slooff et al. 2008) for organic dyes in PMMA. The present lack of NIR dyes will prohibit further increase of conversion efficiency toward the 30% limit. Here, quantum dots or nanorods may have to be used, as their broad absorption spectrum is very favorable. However, they should be emitting in the NIR at high quantum efficiency, larger than the present ~70 to 80%, and their Stokes' shift should be larger. The latter would be possible as the size distribution of a batch of QDs influences the Stokes' shift (Barnham et al. 2000). Another strategy could be the use of so-called type-II QDs, as their Stokes' shift could be very large (~300 nm); however, their stability and QE are not good enough yet. Stability could be improved using multishell QDs (Koole et al. 2008), while interfacial alloying can be optimized to obtain type II QDs with desired properties (Chin et al. 2007), i.e., a Stokes' shift of ~50 to 100 nm, without spectral overlap. Alternatively, the originally proposed three-plate stack (Goetzberger and Greubel 1977) could be further developed using perhaps a combination of organic dyes and nanocrystals, or even rare earth ions (Rowan et al. 2008), with optimized dedicated solar cells for each spectral region. Additional details regarding quantum dot– and rod-based LSCs are provided in Chapter 9.

ACKNOWLEDGMENTS

The authors gratefully acknowledge numerous colleagues at Utrecht University and elsewhere who contributed to the presented work. This work was financially supported by the European Commission as part of the Framework 6 integrated project FULLSPECTRUM (contract SES6-CT-2003-502620), SenterNovem as part of their Netherlands Nieuw Energie Onderzoek (New Energy Research) programme, the Netherlands Foundation for Fundamental Research on Matter (FOM), and the Netherlands Organisation for Scientific Research (NWO).

REFERENCES

Alivisatos, A. P. 1996. Perspectives on the physical chemistry of semiconductor nanocrystals. *Journal of Physical Chemistry* 100:13226–39.
ASTM. 2003. *Standard tables for reference solar spectral irradiances: Direct normal and hemispherical on 37° tilted surface*. Standard G173-03e1. West Conshohocken, PA: American Society for Testing and Materials.

Auzel, F. 2004. Upconversion and anti-Stokes processes with f and d ions in solids. *Chemical Reviews* 104:139–73.

Badescu, V., and A. M. Badescu. 2009. Improved model for solar cells with up-conversion of low-energy photons. *Renewable Energy* 34:1538–44.

Badescu, V., and A. De Vos. 2007. Influence of some design parameters on the efficiency of solar cells with down-conversion and down shifting of high-energy photons. *Journal of Applied Physics* 102:073102-1–7.

Badescu, V., A. De Vos, A. M. Badescu, and A. Szymanska. 2007. Improved model for solar cells with down-conversion and down-shifting of high-energy photons. *Journal of Physics D: Applied Physics* 40:341–52.

Barnham, K., J. L. Marques, J. Hassard, and P. O'Brien. 2000. Quantum-dot concentrator and thermodynamic model for the global redshift. *Applied Physics Letters* 76:1197–99.

Barnham, K. W. J., I. M. Ballard, B. C. Browne, et al. 2009. Recent progress in quantum well solar cells. *In Nanotechnology for photovoltaics*, ed. L. Tsakalakos. Boca Raton, FL: Taylor and Francis. 187–210.

Basore, P. A., and D. A. Clugston. 1996. PC1D version 4 for Windows: From analysis to design. In *Proceedings of 25th IEEE Photovoltaic Specialists Conference*, 377–81. IEEE. Washington, DC

Bende, E. E., A. R. Burgers, L. H. Slooff, W. G. J. H. M. Van Sark, and M. Kennedy. 2008. Cost and efficiency optimisation of the fluorescent solar concentrator. In *Proceedings of Twenty-third European Photovoltaic Solar Energy Conference*, 461–69. Munich: WIP.

Bird, R. E., and C. Riordan. 1986. Simple solar spectral model for direct and diffuse irradiance on horizontal and tilted planes at the earth's surface for cloudless atmospheres. *Journal of Climate and Applied Meteorology* 25:87–97.

Blasse, G., and B. C. Grabmaier. 1995. *Luminescent materials*. Berlin: Springer.

Bloembergen, N. 1959. Solid state infrared quantum counters. *Physical Review Letters* 2:84–85.

Bruchez Jr., M., M. Moronne, P. Gin, S. Weiss, and A. P. Alivisatos. 1998. Semiconductor nanocrystals as fluorescent biological labels. *Science* 281:2013–16.

Burgers, A. R., L. H. Slooff, R. Kinderman, and J. A. M. van Roosmalen. 2005. Modeling of luminescent concentrators by ray-tracing. In *Proceedings of Twentieth European Photovoltaic Solar Energy Conference*, 394–97. Munich: WIP.

Chatten, A. J., K. W. J. Barnham, B. F. Buxton, N. J. Ekins-Daukes, and M. A. Malik. 2003a. A new approach to modelling quantum dot concentrators. *Solar Energy Materials and Solar Cells* 75:363–71.

Chatten, A. J., K. W. J. Barnham, B. F. Buxton, N. J. Ekins-Daukes, and M. A. Malik. 2003b. The quantum dot concentrator: Theory and results. In *Proceedings of Third World Conference on Photovoltaic Energy Conversion (WPEC-3)*, 2657–60. Osaka.

Chatten, A. J., K. W. J. Barnham, B. F. Buxton, N. J. Ekins-Daukes, and M. A. Malik. 2004a. Quantum dot solar concentrators. *Semiconductors* 38:909–17.

Chatten, A. J., K. W. J. Barnham, B. F. Buxton, N. J. Ekins-Daukes, and M. A. Malik. 2004b. Quantum dot solar concentrators and modules. In *Proceedings of 19th European Photovoltaic Solar Energy Conference*, 109–12. Munich: WIP.

Chatten, A. J., D. J. Farrell, C. M. Jermyn, et al. 2005. Thermodynamic modelling of the luminescent solar concentrator. In *Proceedings of 31st IEEE Photovoltaic Specialists Conference*, 82–85. IEEE. Orlando, FL.

Chatten, A. J., D. J. Farrell, R. Bose, et al. 2007. Thermodynamic modelling of luminescent solar concentrators with reduced top surface losses. In *Proceedings of Twenty-second European Photovoltaic Solar Energy Conference,* 349–53. Munich: WIP.

Chen, D., Y. Wang, N. Yu, P. Huang, and F. Weng. 2008a. Near-infrared quantum cutting in transparent nanostructured glass ceramics. *Optics Letters* 33:1884–86.

Chen, D., Y. Wang, Y. Yu, P. Huang, and F. Weng. 2008b. Quantum cutting downconversion by cooperative energy transfer from Ce3+ to Yb3+ in borate glasses. *Journal of Applied Physics* 104:116105-1–3.

Chin, P. T. K., C. De Mello Donegá, S. S. Van Bavel, et al. 2007. Highly luminescent CdTe/CdSe colloidal heteronanocrystals with temperature-dependent emission color. *Journal of the American Chemical Society* 129:14880–86.

Chung, P., H.-H. Chung, and P. H. Holloway. 2007. Phosphor coatings to enhance Si photovoltaic cell performance. *Journal of Vacuum Science and Technology A* 25:61–66.

Currie, M. J., J. K. Mapel, T. D. Heidel, S. Goffri, and M. A. Baldo. 2008. High-efficiency organic solar concentrators for photovoltaics. *Science* 321:226–28.

De Vos, A., A. Szymanska, and V. Badescu. 2009. Modelling of solar cells with down-conversion of high energy photons, anti-reflection coatings and light trapping *Energy Conversion and Management* 50:328–36.

Debije, M. G., R. H. L. Van der Blom, D. J. Broer, and C. W. M. Bastiaansen. 2006. Using selectively-reflecting organic mirrors to improve light output from a luminescent solar concentrator. In *Proceedings of World Renewable Energy Congress IX.*

Del Cañizo, C., I. Tobias, J. Pérez-Bedmar, A. C. Pan, and A. Luque. 2008. Implementation of a Monte Carlo method to model photon conversion for solar cells. *Thin Solid Films* 516:6757–62. Florence, Italy: WREC.

Dexter, D. L. 1953. A theory of sensitized luminescence in solids. *Journal of Chemical Physics* 21:836–50.

Dexter, D. L. 1957. Possibility of luminescent quantum yields greater than unity. *Physical Review* 108:630–33.

Díaz-Herrera, B., B. González-Díaz, R. Guerrero-Lemus, et al. 2009. Photoluminescence of porous silicon stain etched and doped with erbium and ytterbium. *Physica E* 41:525–28.

Dieke, G. H. 1968. *Spectra and energy levels of rare earth ions in crystals.* New York: Wiley Interscience.

Fujii, M., M. Yoshida, Y. Kanzawa, S. Hayashi, and K. Yamamoto. 1997. 1.54 μm photoluminescence of Er3+ doped into SiO2 films containing Si nanocrystals: Evidence for energy transfer from Si nanocrystals to Er3+. *Applied Physics Letters* 71:1198–200.

Gallagher, S. J., P. C. Eames, and B. Norton. 2004. Quantum dot solar concentrator behaviour predicted using a ray trace approach. *Journal of Ambient Energy* 25:47–56.

Gallagher, S. J., B. C. Rowan, J. Doran, and B. Norton. 2007. Quantum dot solar concentrator: Device optimisation using spectroscopic techniques. *Solar Energy* 81:540.

Gamelin, D. R., and H. U. Güdel. 2001. Upconversion processes in transition metal and rare earth metal systems. *Topics in Current Chemistry* 214:1–56.

Gaponenko, S. V. 1998. *Optical properties of semiconductor nanocrystals.* Cambridge: Cambridge University Press.

Garwin, R. L. 1960. The collection of light from scintillation counters. *Review of Scientific Instruments* 31:1010–11.

Gibart, P., F. Auzel, J.-C. Guillaume, and K. Zahraman. 1996. Below band-gap IR response of substrate-free GaAs solar cells using two-photon up-conversion. *Japanese Journal of Applied Physics* 351:4401–2.

Glaeser, G. C., and U. Rau. 2007. Improvement of photon collection in Cu(In,Ga)Se2 solar cells and modules by fluorescent frequency conversion. *Thin Solid Films* 515:5964.

Goetzberger, A. 2008. Fluorescent solar energy concentrators: Principle and present state of development. In *High-efficient low-cost photovoltaics—Recent developments*, ed. V. Petrova-Koch, R. Hezel, and A. Goetzberger, 159–76. Heidelberg: Springer.

Goetzberger, A., and W. Greubel. 1977. Solar energy conversion with fluorescent collectors. *Applied Physics* 14:123–39.

Goldschmidt, J. C., M. Peters, A. Bösch, et al. 2009. Increasing the efficiency of fluorescent concentrator systems. *Solar Energy Materials and Solar Cells* 93:176–82.

González-Díaz, B., B. Díaz-Herrera, R. Guerrero-Lemus, et al. 2008. Erbium doped stain etched porous silicon. *Materials Science and Engineering B* 146:171–74.

Green, M., K. Emery, Y. Hishikawa, and W. Warta. 2009. Solar cell efficiency tables (version 33). *Progress in Photovoltaics: Research and Applications* 17:85–94.

Green, M. A. 1982. *Solar cells; operating principles, technology and systems application.* Englewood Cliffs, NJ: Prentice-Hall.

Green, M. A. 2003. *Third generation photovoltaics, advanced solar energy conversion.* Berlin: Springer Verlag.

Heer, S., O. Lehmann, M. Haase, and H.-U. Güdel. 2003. Blue, green, and red upconversion emission from lanthanwide-doped $LuPO_4$ and $YbPO_4$ noncrystals in a transparent colloidal solution. *Angewom Dete Chemie*, International edition. 42:3179–82.

Hong, B.-C., and K. Kawano. 2003. PL and PLE studies of KMgF3:Sm crystal and the effect of its wavelength conversion on CdS/CdTe solar cell. *Solar Energy Materials and Solar Cells* 80:417–32.

Houshyani Hassanzadeh, B., A. C. De Keizer, N. H. Reich, and W. G. J. H. M. Van Sark. 2007. The effect of a varying solar spectrum on the energy performance of solar cells. In *Proceedings of 21st European Photovoltaic Solar Energy Conference*, 2652–58. Munich: WIP.

Hovel, H. J., R. T. Hodgson, and J. M. Woodall. 1979. The effect of fluorescent wavelength shifting on solar cell spectral response. *Solar Energy Materials* 2:19–29.

Hyldahl, M. G., S. T. Bailey, and B. P. Wittmershaus. 2009. Photo-stability and performance of CdSe/ZnS quantum dots in luminescent solar concentrators. *Solar Energy* 83:566–73.

Islangulov, R. R., J. Lott, C. Weder, and F. N. Castellano. 2007. Noncoherent low-power upconversion in solid polymer films. *Journal of the American Chemical Society* 129:12652–53.

Kamat, P. V. 2008. Quantum dot solar cells. Semiconductor nanocrystals as light harvesters. *Journal of Physical Chemistry C* 112:18737–53.

Kennedy, M., A. J. Chatten, D. J. Farrell, et al. 2008. Luminescent solar concentrators: A comparison of thermodynamic modelling and ray-trace modelling predictions. In *Proceedings of Twenty-third European Photovoltaic Solar Energy Conference*, 334–37. Munich: WIP.

Klimov, V. I. 2006. Mechanisms for photogeneration and recombination of multiexcitons in semiconductor nanocrystals: Implications for lasing and solar energy conversion. *Journal of Physical Chemistry B* 110: 16827–45.

Klimov, V. I., S. A. Ivanov, J. Nanda, et al. 2007. Single-exciton optical gain in semiconductor nanocrystals. *Nature* 447:441–46.

Koole, R., P. Liljeroth, C. De Mello Donegá, D. Vanmaekelbergh, and A. Meijerink. 2006. Electronic coupling and exciton energy transfer in CdTe quantum-dot molecules. *Journal of the American Chemical Society* 128:10436–41.

Koole, R., M. Van Schooneveld, J. Hilhorst, et al. 2008. On the incorporation mechanism of hydrophobic quantum dots in silica spheres by a reverse microemulsion method. *Chemistry of Materials* 20:2503–12.

Krishnan, P., J. W. A. Schüttauf, C. H. M. Van der Werf, et al. 2009. Response to simulated typical daily outdoor irradiation conditions of thin-film silicon-based triple-band-gap, triple-junction solar cells. *Solar Energy Materials and Solar Cells*, 93:691–7.

Lakshminarayana, G., H. Yang, S. Ye, Y. Liu, and J. Qiu. 2008. Cooperative downconversion luminescence in Pr3+/Yb3+:SiO2–Al2O3–BaF2–GdF3 glasses. *Journal of Materials Research* 23:3090–95.

Law, D. C., R. R. King, H. Yoon, et al. 2009. Future technology pathways of terrestrial III-V multijunction solar cells for concentrator photovoltaic systems. *Solar Energy Materials and Solar Cells*, in press.

Lee, J., V. C. Sundar, J. R. Heine, M. G. Bawendi, and K. F. Jensen. 2000. Full color emission from II-VI semiconductor quantum dot-polymer composites. *Advanced Materials* 12:1102–5.

Liu, X., Y. Qiao, G. Dong, et al. 2008. Cooperative downconversion in Yb3+–RE3+ (RE = Tm or Pr) codoped lanthanum borogermanate glasses. *Optics Letters* 33:2858–60.

Luque, A., and S. Hegedus, eds. 2003. *Handbook of photovoltaic science and engineering*. Chichester: Wiley.

Luque, A., and A. Marti. 1997. Increasing the efficiency of ideal solar cells by photon induced transitions at intermediate levels. *Physical Review Letters* 78:5014–17.

Luque, A., and A. Martí. 2003. Theoretical limits of photovoltaic conversion. In *Handbook of photovoltaic science and engineering*, 113–49. Chichester: Wiley.

Luque, A., A. Martí, A. Bett, et al. 2005. FULLSPECTRUM: A new PV wave making more efficient use of the solar spectrum. *Solar Energy Materials and Solar Cells* 87:467–79.

Martí, A., and A. Luque, eds. 2004. *Next generation photovoltaics, high efficiency through full spectrum utilization*. Series in Optics and Optoelectronics. Bristol, UK: Institute of Physics Publishing.

Maruyama, T., and J. Bandai. 1999. Solar cell module coated with fluorescent coloring agent. *Journal of the Electrochemical Society* 146:4406–9.

Maruyama, T., and R. Kitamura. 2001. Transformations of the wavelength of the light incident upon CdS/CdTe solar cells. *Solar Energy Materials and Solar Cells* 69:61–68.

McIntosh, K. R., G. Lau, J. N. Cotsell, K. Hanton, and D. L. Bätzner. 2009. Increase in external quantum efficiency of encapsulated silicon solar cells from a luminescent down-shifting layer. *Progress in Photovoltaics: Research and Applications* 17:191–97.

Meijerink, A., R. Wegh, P. Vergeer, and T. Vlugt. 2006. Photon management with lanthanides. *Optical Materials* 28:575–81.

Minemoto, T., M. Toda, S. Nagae, et al. 2007. Effect of spectral irradiance distribution on the outdoor performance of amorphous Si/thin-film crystalline Si stacked photovoltaic modules. *Solar Energy Materials and Solar Cells* 91:120–22.

Munyeme, G., M. Zeman, R. E. I. Schropp, and W. F. Van der Weg. 2004. Performance analysis of a-Si:H p-i-n solar cells with and without a buffer layer at the p/i interface. *Physica Status Solidi C* 9:2298–303.

Mutlugun, E., I. M. Soganci, and H. V. Demir. 2008. Photovoltaic nanocrystal scintillators hybridized on Si solar cells for enhanced conversion efficiency in UV. *Optics Express* 16:3537–45.

Nann, S., and C. Riordan. 1991. Solar spectral irradiance under clear and cloudy skies: Measurements and a semiempirical model. *Journal of Applied Meteorology* 30:447–62.

Ohwaki, J., and Y. Wang. 1994. Efficient 1.5 μm to visible upconversion in Er3+ doped halide phosphors. *Japanese Journal of Applied Physics* 33:L334–37.

O'Regan, B., and M. Grätzel. 1991. A low-cost, high-efficiency solar cell based on dye-sensitized colloidal TiO2 films. *Nature* 353:737–40.

Peijzel, P. S., A. Meijerink, R. T. Wegh, M. F. Reid, and G. W. Burdick. 2005. A complete 4fn energy level diagram for all trivalent lanthanide ions. *Journal of Solid State Chemistry* 178:448–53.

Peng, X., M. C. Schlamp, A. V. Kadavanich, and A. P. Alivisatos. 1997. Epitaxial growth of highly luminescent CdSe/CdS core/shell nanocrystals with photostability and electronic accessibility. *Journal of the American Chemical Society* 119:7019–29.

Piper, W. W., J. A. DeLuca, and F. S. Ham. 1974. Cascade fluorescent decay in Pr3+-doped fluorides: Achievement of a quantum yield greater than unity for emission of visible light. *Journal of Luminescence* 8:344–48.

Quilitz, J., A. Fiore, L. Manna, et al. 2009a. Fabrication and spectroscopic studies on highly luminescent CdSe nanorod polymer composites. *Nano Letters*, submitted.

Quilitz, J., R. Koole, L. H. Slooff, et al. 2009b. Fabrication and full characterization of state-of-the-art CdSe quantum dot luminescent solar concentrators. *Nanotechnology*, submitted.

Rapp, C. F., and N. L. Boling. 1978. Luminescent solar concentrators. In *Proceedings of 13th IEEE Photovoltaic Specialists Conference*, 690–93. IEEE. Washington, DC.

Reda, S. M. 2008. Synthesis and optical properties of CdS quantum dots embedded in silica matrix thin films and their applications as luminescent solar concentrators. *Acta Materialia* 56:259–64.

Richards, B. S. 2006a. Enhancing the performance of silicon solar cells via the application of passive luminescence conversion layers. *Solar Energy Materials and Solar Cells* 90:2329–37.

Richards, B. S. 2006b. Luminescent layers for enhanced silicon solar cell performance: Down-conversion. *Solar Energy Materials and Solar Cells* 90:1189.

Richards, B. S., and A. Shalav. 2005. The role of polymers in the luminescence conversion of sunlight for enhanced solar cell performance. *Synthetic Metals* 154:61–64.

Rowan, B. C., L. R. Wilson, and B. S. Richards. 2008. Advanced material concepts for luminescent solar concentrators. *IEEE Journal of Selected Topics in Quantum Electronics* 14: 1312–22.

Schropp, R. E. I., and M. Zeman. 1998. *Amorphous and microcrystalline silicon solar cells: Modeling, materials, and device technology*. Boston: Kluwer Academic Publishers.

Schüler, A., M. Python, M. Valle del Olmo, and E. de Chambrier. 2007. Quantum dot containing nanocomposite thin films for photoluminescent solar concentrators. *Solar Energy* 81:1159–65.

Serin, J. M., D. W. Brousmiche, and J. M. J. Frechet. 2002. A FRET-based ultraviolet to near-infrared frequency convertor. *Journal of the American Chemical Society* 124:11848–49.

Shalav, A., B. S. Richards, and M. A. Green. 2007. Luminescent layers for enhanced silicon solar cell performance: Up-conversion. *Solar Energy Materials and Solar Cells* 91:829–42.

Shalav, A., B. S. Richards, T. Trupke, K. W. Krämer, and H. U. Güdel. 2005. Application of NaYF4:Er3+ up-converting phosphors for enhanced near-infrared silicon solar cell response. *Applied Physics Letters* 86:013505-1–3.

Shockley, W., and H. J. Queisser. 1961. Detailed balance limit of efficiency of p-n junction solar cells. *Journal of Applied Physics* 32:510.

Singh-Rachford, T. N., A. Haefele, R. Ziessel, and F. N. Castellano. 2008. Boron dipyrromethene chromophores: Next generation triplet acceptors/annihilators for low power upconversion schemes. *Journal of the American Chemical Society* 130:16164–65.

Slooff, L. H., E. E. Bende, A. R. Burgers, et al. 2008. A luminescent solar concentrator with 7.1% power conversion efficiency. *Physica Status Solidi—Rapid Research Letters* 2:257–59.

Slooff, L. H., R. Kinderman, A. R. Burgers, et al. 2006. The luminescent concentrator illuminated. *Proceedings of SPIE* 6197:61970k1–8.

Smestad, G., H. Ries, R. Winston, and E. Yablonovitch. 1990. The thermodynamic limits of light concentrators. *Solar Energy Materials* 21:99–111.

Soga, T., ed. 2006. *Nanostructured materials for solar energy conversion*. Amsterdam: Elsevier.

Sommerdijk, J. L., A. Bril, and A. W. De Jager. 1974. Two photon luminescence with ultraviolet excitation of trivalent praseodymium. *Journal of Luminescence* 8:341–43.

Strümpel, C., M. McCann, G. Beaucarne, et al. 2007. Modifying the solar spectrum to enhance silicon solar cell efficiency—An overview of available materials. *Solar Energy Materials and Solar Cells* 91:238.

Strümpel, C., M. McCann, C. Del Cañizo, I. Tobias, and P. Fath. 2005. Erbium-doped up-converters on silicon solar cells: Assessment of the potential. In *Proceedings of Twentieth European Photovoltaic Solar Energy Conference,* 43–46. Munich: WIP.

Stupca, M., M. Alsalhi, T. Al Saud, A. Almuhanna, and M. H. Nayfeh. 2007. Enhancement of polycrystalline silicon solar cells using ultrathin films of silicon nanoparticle. *Applied Physics Letters* 91:063107-1–3.

Suyver, J. F., A. Aebischer, D. Biner, et al. 2005a. Novel materials doped with trivalent lanthanides and transition metal ions showing near-infrared to visible photon upconversion. *Optical Materials* 27:1111–30.

Suyver, J. F., J. Grimm, K. W. Krämer, and H. U. Güdel. 2005b. Highly efficient near-infrared to visible up-conversion process in NaYF4:Er3+, Yb3+. *Journal of Luminescence* 114:53–59.

Svrcek, V., A. Slaoui, and J.-C. Muller. 2004. Silicon nanocrystals as light converter for solar cells. *Thin Solid Films* 451–52:384–88.

Timmerman, D., I. Izeddin, P. Stallinga, I. N. Yassievich, and T. Gregorkiewicz. 2008. Space-separated quantum cutting with silicon nanocrystals for photovoltaic applications. *Nature Photonics* 2:105–9.

Trupke, T., M. A. Green, and P. Würfel. 2002a. Improving solar cell efficiencies by down-conversion of high-energy photons. *Journal of Applied Physics* 92:1668–74.

Trupke, T., M. A. Green, and P. Würfel. 2002b. Improving solar cell efficiencies by up-conversion of sub-band-gap light. *Journal of Applied Physics* 92:4117–22.

Trupke, T., A. Shalav, B. S. Richards, P. Wurfel, and M. A. Green. 2006. Efficiency enhancement of solar cells by luminescent up-conversion of sunlight. *Solar Energy Materials and Solar Cells* 90:3327.

Tsakalakos, L. 2008. Nanostructures for photovoltaics. *Materials Science and Engineering R: Reports* 62:175–89.

Van der Ende, B. M., L. Aarts, and A. Meijerink. 2009. Near infrared quantum cutting for photovoltaics. *Advanced Materials,* 21:3073–7.

Van Sark, W. G. J. H. M. 2002. Methods of deposition of hydrogenated amorphous silicon for device applications. In *Thin films and nanostructures,* ed. M. H. Francombe, 1–215. San Diego: Academic Press.

Van Sark, W. G. J. H. M. 2005. Enhancement of solar cell performance by employing planar spectral converters. *Applied Physics Letters* 87:151117.

Van Sark, W. G. J. H. M. 2006. Optimization of the performance of solar cells with spectral down converters. In *Proceedings of Twenty-first European Photovoltaic Solar Energy Conference*, 155–59. Munich: WIP.

Van Sark, W. G. J. H. M. 2007. Calculation of the performance of solar cells with spectral down shifters using realistic outdoor solar spectra. In *Proceedings of Twenty-second European Photovoltaic Solar Energy Conference*, 566–70. Munich: WIP.

Van Sark, W. G. J. H. M. 2008. Simulating performance of solar cells with spectral down-shifting layers. *Thin Solid Films* 516:6808–12.

Van Sark, W. G. J. H. M., P. L. T. M. Frederix, A. A. Bol, H. C. Gerritsen, and A. Meijerink. 2002. Blinking, blueing, and bleaching of single CdSe/ZnS quantum dots. *ChemPhysChem* 3:871–79.

Van Sark, W. G. J. H. M., C. De Mello Donegá, C. Harkisoen, et al. 2004. Improvement of spectral response of solar cells by deployment of spectral converters containing semiconductor nanocrystals. In *Proceedings of 19th European Photovoltaic Solar Energy Conference*, 38–41. Munich: WIP.

Van Sark, W. G. J. H. M., A. Meijerink, R. E. I. Schropp, J. A. M. Van Roosmalen, and E. H. Lysen. 2005. Enhancing solar cell efficiency by using spectral converters. *Solar Energy Materials and Solar Cells* 87:395–409.

Van Sark, W. G. J. H. M., K. W. J. Barnham, L. H. Slooff, et al. 2008a. Luminescent solar concentrators—A review of recent results. *Optics Express* 16:21773–92.

Van Sark, W. G. J. H. M., G. F. M. G. Hellenbrand, E. E. Bende, A. R. Burgers, and L. H. Slooff. 2008b. Annual energy yield of the fluorescent solar concentrator. In *Proceedings of Twenty-third European Photovoltaic Solar Energy Conference*, 198–202. Munich: WIP.

Vergeer, P., T. J. H. Vlugt, M. H. F. Kox, et al. 2005. Quantum cutting by cooperative energy transfer in YbxY1–xPO4:Tb3+. *Physical Review B* 71:014119-1–11.

Walker, G. W., V. C. Sundar, C. M. Rudzinski, et al. 2003. Quantum-dot optical temperature probes. *Applied Physics Letters* 83:3555–57.

Weber, W. H., and J. Lambe. 1976. Luminescent greenhouse collector for solar radiation. *Applied Optics* 15:2299–300.

Wegh, R., H. Donker, A. Meijerink, R. J. Lamminmäki, and J. Hölsä. 1997. Vacuum-ultraviolet spectroscopy and quantum cutting for Gd3+ in LiYF4. *Physical Review B* 56:13841–48.

Wegh, R. T., H. Donker, K. D. Oskam, and A. Meijerink. 1999. Visible quantum cutting in LiGdF$_4$:Eu^{3+} through downconversion. *Science* 283:663–66.

Wegh, R. T., A. Meijerink, R. J. Lamminmäki, and J. Hölsä. 2000. Extending Dieke's diagram. *Journal of Luminescence*. 87–89:1002–4.

Wittwer, V., W. Stahl, and A. Goetzberger. 1984. Fluorescent planar concentrators. *Solar Energy Materials* 11:187–97.

Wolf, M. 1971. New look at silicon solar cell performance. *Energy Conversion* 11:63–73.

Yablonovitch, E. 1980. Thermodynamics of the fluorescent planar concentrator. *Journal of the Optical Society of America* 70:1362–63.

Zastrow, A. 1994. The physics and applications of fluorescent concentrators: A review. *Proceedings of SPIE* 2255:534–47.

Zhang, Q. Y., G. F. Yang, and Y. X. Pan. 2007. Cooperative quantum cutting in one-dimensional (YbxGd1–x)Al3(BO3)4):Tb3+ nanorods. *Applied Physics Letters* 90:021107-1–3.

11

Nanoplasmonics for Photovoltaic Applications

E. T. Yu

11.1 INTRODUCTION

The last several years have witnessed an explosion of interest in understanding and exploiting the optical properties of metals and metallic nanostructures, most specifically those associated with plasmonic resonances that can give rise to pronounced optical absorption, field localization, and scattering effects. For example, surface plasmon resonances and related phenomena in metallic nanostructures are currently being exploited for a variety of applications, including molecular sensing[1-3] and tagging,[4,5] focusing of light,[6] near-field optical microscopy,[7,8] subwavelength photonics,[9] and optical metamaterials.[10] The appeal of plasmon excitations for such applications typically arises from the large electromagnetic field enhancement that occurs in the vicinity of the metal surface, and the dependence of the resonance wavelength on the nanoparticle's size, shape, and local dielectric environment, or from very strong scattering effects that occur at wavelengths in the vicinity of plasmonic resonances.

A variety of studies have also been directed toward the incorporation of such effects into the performance of semiconductor-based photovoltaic devices, both for organic semiconductors in which absorption and field localization effects are most prominent, and more recently for solid-state inorganic semiconductors, in which scattering effects are more likely to be of greatest utility. In this chapter we review various approaches that have emerged, starting in the mid-1990s, for the exploitation of plasmonic and related effects in solid-state nanostructures in high-performance photovoltaic devices.

11.2 PLASMONIC FIELD LOCALIZATION AND FORWARD SCATTERING EFFECTS

11.2.1 Plasmonic Field Localization and Amplitude Enhancement

Localization, and the accompanying amplitude enhancement, of electromagnetic fields associated with the excitation of surface plasmon polariton resonances in metal nanoparticles has been employed extensively in surface-enhanced Raman spectroscopy (SERS),[1–3] in which the very large field amplitudes resulting from excitation of surface plasmon polaritons in Au nanoparticles are exploited to increase signal strength in Raman scattering experiments to a level sufficient to enable detection of signals from an individual molecule. The wavelength of the surface plasmon polariton resonance in an isolated metal nanoparticle depends on the inherent dielectric response of the metal, and for a metal sphere of radius a the free-space resonance wavelength λ for the lowest-order normal mode, also referred to as the Fröhlich mode, is determined by the condition[11]

$$\varepsilon = -\left(2 + \frac{12}{5}\left(\frac{2\pi n_m a}{\lambda}\right)^2\right)\varepsilon_m \tag{11.1}$$

where ε is the dielectric function of the metal, ε_m is the dielectric function of the surrounding medium, and n_m is the index of refraction of the surrounding medium. For nanoparticles that are small compared to the free-space wavelength, this condition reduces to $\varepsilon = -2\varepsilon_m$, and the frequency at which this condition is satisfied is referred to as the Fröhlich frequency. The electric field amplitude at the surface of such a metal particle can be increased, compared to the amplitude of the incident field exciting the surface plasmon polariton resonance, by nearly two orders of magnitude for a low-loss metal such as Ag.[12] The corresponding increase in optical transition rate, and consequently signal intensity, can be much larger: the signal intensity in a SERS measurement is proportional to

$$\left\langle \left|\mathbf{E}(\omega)\right|^2 \left|\mathbf{E}(\omega')\right|^2 \right\rangle$$

where **E**(ω) is the local electric field at the excitation frequency ω, ω' is the Stokes-shifted frequency, and the angle brackets indicate an average over the surface of the nanoparticle.

In analogous fashion, attempts have been made starting over a decade ago to exploit field amplification in the vicinity of metal nanoparticles to increase optical transition rates, and consequently photon absorption and photocurrent generation, in semiconductor-based photovoltaic devices, with initial efforts focusing on organic semiconductors. In a typical semiconductor, adopting the terminology associated with solid-state inorganic semiconductors, efficiency in photon absorption is determined by the electronic band structure, and consequently the absorption coefficient, of the semiconductor, and the electromagnetic field distribution created within the semiconductor by the incident electromagnetic radiation. Specifically, the optical transition rate, W, in a semiconductor is given, using Fermi's golden rule, by[13]

$$W \propto \left|E_0\right|^2 \times \left|\mathbf{e} \cdot \mathbf{p}_{cv}\right|^2 \times \int d^3 k \delta \left(E_c\left(\mathbf{k}\right) - E_v\left(\mathbf{k}\right) - \hbar\omega \right) \qquad (11.2)$$

where E_0 is the electric field amplitude, **e** is the field polarization unit vector, $\mathbf{p}_{cv}$ is the momentum operator matrix element between the initial (valence band) and final (conduction band) states, **k** is a wave vector, $E_c(\mathbf{k})$ and $E_v(\mathbf{k})$ are the conduction band and valence band dispersion relations, respectively, $\hbar$ is the reduced Planck's constant, and ω is the frequency of the incident photon. Of the factors present in Equation 11.2, the band structure and matrix element can be engineered via techniques such as semiconductor heterostructure and nanostructure design, while the electric field amplitude and spatial distribution can be tailored through the use of structures such as waveguides, resonant cavities, thin-film coatings, etc. Most approaches for exploiting plasmonic effects in photovoltaic devices have focused on altering local field amplitudes and distributions to increase optical absorption in appropriate device regions.

11.2.1.1 Application to Organic Semiconductor Photovoltaics

Initial efforts in this regard focused on the integration of metal nanoparticles into phthalocyanine-based organic semiconductor photovoltaic devices, specifically at the interface between the organic semiconductor

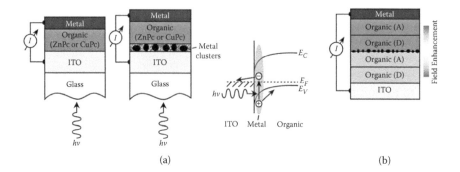

FIGURE 11.1

(a) Schematic diagram of metal phthalocyanine (CuPc or ZnPc) organic photovoltaic device formed by a Schottky ITO-organic contact, without (left) and with (middle) metal nanoclusters formed at the ITO-organic interface by thin-film deposition and thermal processing. The schematic band diagram (right) illustrates the mechanism proposed in Westphalen et al.[16] for photocurrent enhancement at incident photon energies below the bandgap of the metal phthalocyanine. (b) Schematic diagram of tandem organic photovoltaic device of Rand et al.[17] in which Ag nanocluster formation at the tunnel junction between two organic photovoltaic cells leads to photocurrent enhancement.

and a transparent conducting indium tin oxide (ITO) contact, as illustrated schematically in Figure 11.1a. Incorporation of Ag, Au, and Cu metal nanoclusters in a copper phthalocyanine-based (CuPc) organic photovoltaic device in this manner resulted in an observed increase in photocurrent generation by as much as a factor of 2.7, with increased photocurrent response measured over a wide range of incident photon energies, ranging from 1.75 eV to over 4 eV; these observations were attributed to a combination of interband transitions in the metal clusters and surface plasmon polariton resonance excitation.[14,15] A later study of zinc phthalocyanine-based (ZnPc) solar cells in which Ag metal clusters were incorporated at the interface between the organic semiconductor and a transparent ITO contact also revealed significant improvements in photocurrent generation, over a broad range of wavelengths, upon incorporation of the Ag clusters; in this case, the increased photocurrent response was attributed to electron-hole pair excitation at the semiconductor-ITO interface mediated by surface plasmon polariton excitations in the metal cluster, as illustrated schematically in Figure 11.1a.[16] A key observation supporting this interpretation was that a large enhancement in photocurrent response was observed at wavelengths of ~400 to 550 nm, at which the ZnPc is not absorbing, suggesting that a direct increase in absorption in the ZnPc layer due to field enhancement in the vicinity of the metal nanoparticle was not the dominant mechanism for the observed increase in overall photocurrent in the device.

The presence of Ag nanoclusters has also been exploited, more recently, to increase photocurrent response in tandem organic solar cells. Incorporation of an array of Ag nanoclusters ~5 nm in diameter at the interface between the individual junctions constituting the tandem cell, as illustrated schematically in Figure 11.1b, was observed to substantially increase experimentally measured photovoltaic power conversion efficiency, with increased photocurrent response being observed over a broad range of wavelengths. On the basis of detailed numerical simulations for chains of Ag nanoparticles, these effects, including the broad observed spectral response and a predicted spatial extent of optical absorption enhancement of ~10 nm—considerably greater than that expected for isolated Ag nanoparticles, were attributed to surface plasmon polariton resonance excitation in the Ag nanoparticles combined with interparticle coupling effects that increase the spectral range and spatial extent of the increased electromagnetic field arising from the plasmonic excitations.[17]

11.2.1.2 Application to Inorganic Solid-State Semiconductor Photovoltaics

While these studies focused primarily on the influence of plasmonic field localization and amplitude enhancement in organic semiconductors, presumably due to the very limited spatial extent of significant amplitude increases (typically ~10 nm or less), related effects can give rise to photocurrent response enhancement in solid-state inorganic semiconductor photodiodes and solar cells. However, bulk absorption coefficients for solid-state inorganic semiconductors are typically ~10^5 cm^{-1} or smaller at most wavelengths, corresponding to absorption lengths that are very large compared to the distances over which plasmonically enhanced fields are present. Because of this disparity and other factors, such as surface and interface recombination effects in these materials, the plasmonic field localization and amplitude enhancement effects exploited in organic materials and molecular spectroscopy are likely to be effective in only limited circumstances, e.g., for high-quality semiconductor nanostructures in extremely close proximity to a metal nanoparticle. However, strong optical scattering associated with metal (and dielectric) nanostructures can enable local field enhancement to be combined with scattering to engineer photon propagation into and within a solid-state semiconductor device for improved performance; indeed, optical scattering by nanoparticles is likely in many circumstances to dominate over local field enhancement in influencing solid-state semiconductor photodetector or solar cell performance.

Elementary calculations reveal a strong dependence of nanoparticle cross-sections for optical absorption and scattering on particle size. The cross-sections for absorption and scattering of incident radiation by a spherical nanoparticle, C_{abs} and C_{sca}, respectively, are given (assuming the particle size is very small compared to the incident wavelength) by[18]

$$C_{abs}(\lambda) = \pi a^2 \, 4\left(\frac{2\pi n_m a}{\lambda}\right) \text{Im}\left(\frac{\varepsilon - \varepsilon_m}{\varepsilon + 2\varepsilon_m}\right) \tag{11.3}$$

$$C_{sca}(\lambda) = \pi a^2 \frac{8}{3}\left(\frac{2\pi n_m a}{\lambda}\right)^4 \left|\frac{\varepsilon - \varepsilon_m}{\varepsilon + 2\varepsilon_m}\right|^2 \tag{11.4}$$

The relative importance of absorption and scattering processes induced by the nanoparticle, as quantified by the size of the scattering cross-sections C_{abs} and C_{sca} given by Equations 11.3 and 11.4, can play a substantial role in determining the efficacy of these processes in increasing optical absorption in solid-state semiconductor structures and devices. For $C_{abs} \gg C_{sca}$, the dominant process induced by the nanoparticles is absorption of photons and excitation of surface plasmon polaritons in a metal nanoparticle, accompanied by a large increase in the electromagnetic field amplitude in a volume within a distance of ~10 nm or less of the nanoparticle. This effect results in an increase in electron-hole pair photogeneration, but only in semiconductor material that is within a few tens of nanometers or less of the nanoparticle. Thus, it is likely to be of limited effectiveness in most solid-state semiconductor photodetector or solar cell structures, which have considerably larger dimensions both laterally and in depth. For $C_{sca} \gg C_{abs}$, the dominant process induced by metal nanoparticles is strong forward scattering of incident radiation at wavelengths near and somewhat longer than the surface plasmon polariton resonance wavelength. This forward scattering can increase the electromagnetic field amplitude for a considerable distance, i.e., a micron or more, below the particle, and also increases the overall transmission of electromagnetic energy into semiconductor material beneath the metal particle—an effect similar to that of an antireflection coating. This effect increases the

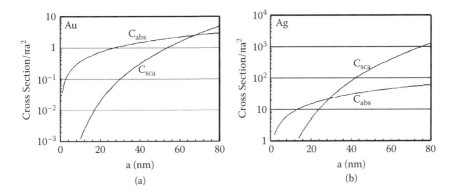

FIGURE 11.2

Computed values for scattering and absorption cross sections for (a) Au and (b) Ag nano-particles in vacuum, as functions of particle radius *a*, at the Fröhlich (resonant) wave-length. Values are normalized to the physical particle cross-section πa^2.

electron-hole pair generation rate per unit volume to a more moderate degree than the absorption and surface plasmon polariton excitation process, but within a much larger volume of semiconductor material—potentially extending a few micrometers or more below the metal nanoparticle. Thus, for most solid-state inorganic semiconductor devices it is likely to be desirable to operate in a regime of particle size and with particle materials for which the scattering cross-section C_{sca} is comparable to or larger than the absorption cross-section C_{abs}.

Figure 11.2 shows values of C_{abs} and C_{sca}, normalized to the particle cross-sectional area πa^2, computed as a function of particle radius for Au and Ag nanoparticles using Equations 11.3 and 11.4 combined with values for the dielectric function, ε, obtained from the published literature.[19] All values are computed at the Fröhlich frequency in vacuum, corresponding to the condition $\mathrm{Re}(\varepsilon) = -2$. From the figure we see that C_{sca} becomes dominant for larger nanoparticles, typically a few tens of nanometers or larger in radius. Thus, it is expected to be advantageous for many solid-state inorganic semiconductor device structures to employ particles in this size regime. However, it is also necessary to avoid particles that are substantially larger in size—~100 nm or more in radius—as such large particles typically suffer from the presence of multiple nearly degenerate modes that reduce the strength of the fundamental surface plasmon polariton resonance. As noted above, the scattering amplitudes tend to be largest in the forward direction, i.e., for propagation of the scattered wave in the same direction as the incident wave. Furthermore, the forward scattering peak tends to become more tightly focused as the size of the scattering nanoparticle increases, due to the inclusion of higher-order spherical harmonics in the scattered field amplitudes for larger particles.

The nature of localized field enhancement and scattering effects arising from plasmonic excitations in metal nanoparticles on the behavior of inorganic, solid-state semiconductor photodetector and solar cell device structures has been elucidated in a series of experiments in which photocurrent response in Si pn junction photodiodes has been characterized with and without Au nanoparticles of various sizes deposited on the device surface. On the basis of the preceding discussion, it would be anticipated that effects related to localized field enhancement in the vicinity of the metal nanoparticles would be most prominent in devices on which smaller nanoparticles were deposited, and for which photocurrent response arises predominantly from the semiconductor region near the device surface, e.g., for a device with a shallow pn junction and high dopant concentrations, which would tend to minimize minority carrier diffusion lengths, and therefore confine contributions to photocurrent response to the near-surface volume of the device. For devices with deeper junctions (~100 nm or deeper) and lower dopant concentrations, which should correlate with increased minority carrier diffusion lengths, the forward scattering effects described above would be expected to dominate.

Initial experiments in this regard were performed using Si pn junction photodiodes on which colloidal Au nanoparticles of varying diameter were deposited from solution.[20] Figure 11.3 shows a schematic diagram of the device structure and experimental configuration, scanning electron micrographs of Au nanoparticles, with diameters of 50, 80, and 100 nm, deposited on the Si photodiode surfaces, and photocurrent response spectra for a reference Si pn junction photodiode without nanoparticles, and for photodiodes functionalized with nanoparticles of varying diameter. The pn junction photodiode was fabricated by diffusion of B into an n-type Si wafer with donor concentration $\sim 5 \times 10^{18}$ cm^{-3}, resulting in an estimated junction depth of 80 nm and a B concentration at the device surface of $\sim 1.1 \times 10^{20}$ cm^{-3}. Due to the high dopant concentrations and relatively shallow junction depth in this device structure, one would expect that localized field enhancement effects arising from the presence of Au nanoparticles on the device might play a significant role in alteration of the device photocurrent response. In addition, field enhancement effects would be expected to be more prominent, relative to scattering effects, for

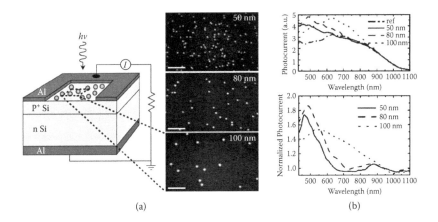

FIGURE 11.3

(a) Schematic diagram of device geometry and experimental configuration for measurement of nanoparticle-induced photocurrent enhancement in Si pn junction photodiodes, and scanning electron micrographs of 50, 80, and 100 nm diameter Au nanoparticles deposited on device surface. The estimated pn junction depth is ~80 nm. All scale bars are 1 μm. (b) Photocurrent response spectra for a reference Si pn junction photodiode without surface nanoparticle functionalization, and for photodiodes upon which 50, 80, and 100 nm diameter Au nanoparticles have been deposited (upper plot), and ratio of photocurrent response for nanoparticle-functionalized devices to that for the reference device (lower plot).

the smallest (50-nm diameter) nanoparticles, than for the larger (100-nm diameter) nanoparticles.

Extinction spectra measured for 50, 80, and 100 nm diameter Au colloidal nanoparticles in solution confirm the existence, and particle size dependence, of the surface plasmon polariton resonances in the nanoparticles. Figure 11.4 shows extinction spectra measured for particles of these sizes suspended in aqueous solution. The spectra have been normalized to account for the extinction response of the aqueous environment and the rectangular optical glass cell containing the solution. The extinction efficiency is defined as $\log(T_c/T_{Au})$, where T_c and T_{Au} are the light intensities transmitted through cells containing de-ionized water and the colloidal Au solution, respectively. Maxima in the extinction spectra are clearly visible at wavelengths corresponding to surface plasmon polariton excitations in the Au nanoparticles, with negligible extinction activity at wavelengths above 800 nm. As expected, the surface plasmon polariton resonance peak increases in wavelength and broadens with increasing particle size.

As shown in Figure 11.3a, deposition of colloidal Au nanoparticles results in the presence of predominantly isolated nanoparticles on the

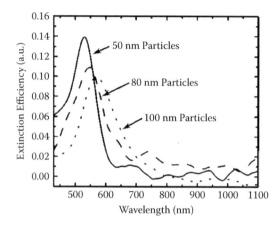

FIGURE 11.4

Extinction efficiency as a function of diameter for Au colloidal nanoparticles 50, 80, and 100 nm in diameter suspended in aqueous solution, clearly exhibiting peaks in extinction associated with surface plasmon polariton resonances. As expected, the resonance peaks increase in wavelength and width with increasing nanoparticle diameter.

device surface, at particle densities of approximately 6.6×10^8, 1.6×10^8, and 7.7×10^7 cm^{-2} for, respectively, 50-, 80-, and 100-nm diameter particles. These particle densities correspond to surface coverages of ~0.6 to 1.3%. The photocurrent response spectra shown in Figure 11.3b clearly reveal the influence of the Au nanoparticles on optical absorption and photocurrent generation in the device. For 50- and 80-nm diameter particles, a pronounced enhancement in photocurrent response is evident at wavelengths from below 450 nm to approximately 600 nm, suggesting that nanoparticle absorption and surface plasmon polariton excitation, accompanied by localized field enhancement in the immediate vicinity of each Au nanoparticle, play a significant role. For the 100-nm diameter Au nanoparticles, the maximum magnitude of photocurrent enhancement is reduced, but the enhancement extends to substantially longer wavelengths than for the smaller nanoparticles. This is consistent with an increased role, for the 100-nm diameter nanoparticles, of forward scattering, and consequently increased transmission of electromagnetic power into the semiconductor, as this effect is expected to increase in prominence for larger nanoparticles and to extend to substantially longer wavelengths than the field localization effects, consistent with the experimental observations shown in Figure 11.3.

These effects have also been extended to Au nanoparticles deposited on a-Si:H p-i-n thin-film solar cell devices.[21] Based on the results shown in

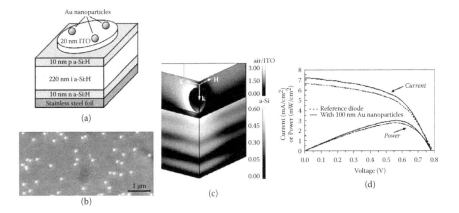

FIGURE 11.5

(a) Schematic diagram of a-Si:H p-i-n thin-film solar cell device with transparent ITO contact, functionalized with Au nanoparticles. (b) Scanning electron micrograph of 100 nm diameter colloidal Au nanoparticles deposited on device surface. (c) Electric field amplitude for Au nanoparticle atop ITO contact on a-Si:H, computed using finite-element numerical simulation for electromagnetic plane wave at 600 nm wavelength. (d) Current density-voltage and power output curves for a-Si:H photovoltaic reference device (without nanoparticles) and with 100 nm diameter Au nanoparticles on device surface.

Figure 11.3, colloidal Au nanoparticles 100 nm in diameter were employed. Figure 11.5 shows a schematic diagram of the experimental device geometry employed, a scanning electron micrograph of colloidal Au nanoparticles deposited on the device surface, results of a numerical simulation of the electric field amplitude in the vicinity of the nanoparticle, and current density-voltage and power output characteristics for a-Si:H devices with and without Au nanoparticle functionalization. The nanoparticle densities in these devices were, as may be deduced from Figure 11.5b, typically in the range of 2 to 4×10^8 cm^{-2}.

Finite-element numerical simulations of field distributions for electromagnetic plane waves incident on an Au nanoparticle atop the ITO/a-Si:H device structure, as shown in Figure 11.5c, confirm the increased field amplitude below the nanoparticle arising from scattering of the incident wave, which is expected to result in increased optical absorption and photocurrent generation. As is evident from Figure 11.5d, the incorporation of Au nanoparticles into the device does indeed result in experimental observation of substantially higher short-circuit photocurrent and also higher maximum power output. Specifically, for the device results shown in Figure 11.5, the short-circuit photocurrent is increased by 8.1% in the devices functionalized with Au nanoparticles, while the maximum power output is increased by 8.3%. The measured energy conversion efficiency of devices separately fabricated from the same amorphous Si:H material was ~5%. Furthermore, detailed diagnostic studies conducted on these devices have shown that the process used to deposit Au nanoparticles degrades the characteristics of reference devices, resulting in a decrease in short-circuit photocurrent typically of 3 to 5%. Thus, it is reasonable to conclude that the actual increase in short-circuit photocurrent and maximum power output arising from the presence of the Au nanoparticles is significantly higher than 8.1% and may approach 13% or more, even at the relatively low particle concentrations employed in these experiments.

The basic analysis of absorption and scattering effects associated with metal nanoparticles is sufficient to explain the major features observed in studies of photoresponse enhancement via integration of metal nanoparticles with semiconductor photodiode structures, and photodetectors generally. However, additional aspects of the observed device behavior can be accounted for only in a more detailed examination of the interaction between the electromagnetic field distribution arising from scattering or absorption by the nanoparticle and that arising from the presence of the semiconductor below. Figures 11.6a and b show a Si pn junction device structure, and 100-nm diameter Au nanoparticles deposited on the device surface, for a study in which some of these aspects have been elucidated.[22] The pn junction depth in the devices employed in these investigations was large, at ~0.5 μm, and the dopant concentration in the n-type Si substrate was quite low, at ~10^{15} cm^{-3}, resulting in significantly reduced photocurrent enhancement effects compared to those shown in Figure 11.3. Figure 11.6c shows the ratio of photocurrent response, as a function of wavelength, for a device upon which Au nanoparticles were deposited at a density of ~3.5×10^8 cm^{-2}, relative to that for a reference device without nanoparticles. Calculated ratios of photocurrent response, obtained via finite-element numerical simulations of electromagnetic field distributions for the device and experimental geometry shown in Figure 11.6a, are shown in Figure 11.6d. It is evident from the figure that the simulations are in very good, quantitative agreement with the experimental results, allowing the former to serve as a sound foundation for interpretation of our experimental observations.

In both the experimental and simulated photocurrent response ratios, we observe enhanced photocurrent response at wavelengths of ~600 to ~1,100 nm, as expected based on the forward scattering cross-section for Au nanoparticles in the size range employed here. Simulations confirm

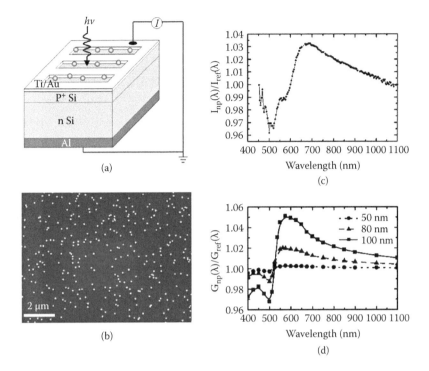

FIGURE 11.6

(a) Schematic diagram of Si pn junction photodiode and experimental geometry for detailed studies of nanoparticle-semiconductor scattering and photocurrent effects. The estimated pn junction depth is ~0.5 µm. (b) Scanning electron micrograph of 100 nm diameter Au nanoparticles deposited on device surface. (c) Experimental and (d) numerically simulated ratios of photocurrent response for Si pn junction photodiodes functionalized with Au nanoparticles, relative to that for a reference Si pn junction device without nanoparticles.

that this effect should be diminished for smaller nanoparticles, as can be seen in Figure 11.6d. At shorter wavelengths, however, a reduction in photocurrent response is observed. Specifically, substantially reduced photocurrent response occurs for wavelengths of ~430 to 500 nm and below, such that the presence of the Au nanoparticles leads to a decrease, rather than an increase, in photocurrent response at these wavelengths. This behavior is present in both the experimental data and numerical simulation results shown in Figure 11.6.

We attribute this behavior to the properties of the Au nanoparticle polarizability at these wavelengths, specifically a pronounced change in the phase of the polarizability, leading to a phase shift in the wave scattered by the nanoparticle, at ~500 to 550 nm. We note that, in the limit of a nanoparticle with a small radius a compared to the wavelength, the incident electric field $\mathbf{E}$ induces a dipole moment $\mathbf{p}$ in the nanoparticle given by

$$\mathbf{p} = \varepsilon_m \alpha \mathbf{E}$$

$$\alpha = 4\pi a^3 \frac{\varepsilon - \varepsilon_m}{\varepsilon + 2\varepsilon_m} \tag{11.5}$$

where ε_m is the dielectric function of the medium surrounding the nanoparticle, ε is the dielectric function of the nanoparticle, and α is the nanoparticle polarizability. The wave component scattered by the nanoparticle may be considered to be that radiated by the dipole moment given by Equation 11.5. A simple analysis shows that the imaginary component of the polarizability is maximized near the surface plasmon polariton resonance condition ($\varepsilon = -2\varepsilon_m$), so that the phase of the polarizability should be maximized near this condition as well.

Figure 11.7a shows a plot of the Au nanoparticle polarizability magnitude and phase, normalized to the particle volume and computed in the limit in which the particle radius is much smaller than the wavelength. For

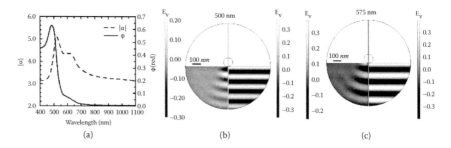

(a)   (b)   (c)

FIGURE 11.7
(a) Magnitude and phase of Au nanoparticle polarizability, computed for a particle radius a small compared to wavelength, normalized to the particle volume $4\pi a^3/3$. At right are shown electric fields obtained via numerical simulation for a plane wave incident on a 100-nm diameter Au nanoparticle on Si at wavelengths of (b) 500 nm and (c) 575 nm. In the lower left quadrant of each plot is the component of the incident wave scattered by the nanoparticle, and in the lower right quadrant is the component of the incident wave transmitted across the air-semiconductor interface. A clear phase shift between the two components, which we attribute to the nonzero phase of the particle polarizability, is apparent at 500 nm, but absent at 575 nm. This phase shift, and the resulting destructive interference between the two components, is responsible for the reduction in photocurrent response observed at shorter wavelengths.

wavelengths longer than the surface plasmon polariton resonance wavelength at ~550 nm, the phase of the Au nanoparticle polarizability is very small, ~0.1 rad or less. Thus, within the semiconductor, the electromagnetic wave arising from scattering by the nanoparticle is approximately in phase with that arising from direct transmission of the incident wave across the semiconductor-air interface, and these components will interfere constructively in the vicinity of the nanoparticle. For shorter wavelengths, however, the phase of the nanoparticle polarizability increases to a maximum of ~0.6 rad, resulting in a substantial phase shift, and partially destructive interference, between the electromagnetic field components arising from nanoparticle scattering and from direct transmission into the semiconductor.

The behavior of these phase shifts as a function of wavelength has been confirmed in finite-element numerical simulations, as shown in Figure 11.7b and c. Shown in each figure are electric field amplitudes for the incident wave component transmitted directly across the air-semiconductor interface (lower right quadrant in each plot), and for the component of the incident wave scattered by the nanoparticle (lower left quadrant in each plot). At a wavelength of 500 nm, for which the phase of the polarizability shown in Figure 11.7a differs substantially from zero, we see from Figure 11.7b that the wave component scattered by the nanoparticle is significantly out of phase with the directly transmitted wave. This leads to partial destructive interference between the scattered and directly transmitted wave, and consequently to a reduction in electromagnetic field intensity, optical transition rate, and photocurrent response at this wavelength. At a wavelength of 575 nm, for which the polarizability is close to zero, we see from Figure 11.7c that the wave component scattered by the nanoparticle and that directly transmitted across the interface are nearly in phase with each other, leading to constructive interference between these wave components, an increase in electromagnetic field intensity, and enhanced photocurrent response at this wavelength. As can be seen from Figure 11.6, the resulting suppression in photocurrent response at short wavelengths predicted from these simulations, and explained in terms of interference effects between the scattered and directly transmitted waves, matches very well with experimental observations.

The existence of such interference effects for wavelengths at or below the surface plasmon polariton resonance, and their undesirable consequences for photocurrent response, suggests that it would be beneficial to implement the concept of nanoparticle-induced photocurrent response enhancement in semiconductor photodiodes using either metal nano-particles for which the surface plasmon polariton resonance is shifted to wavelengths shorter than those at which response is desired, or to employ dielectric nanoparticles, for which the desired scattering effects are still present (although with smaller cross-sections than for metal nanoparti-cles), but for which surface plasmon polariton resonances do not occur. This concept has been explored in detail for silica nanospheres on Si pho-todiodes both through near-field optical studies of photocurrent response associated with the presence of individual nanoparticles[23] and for large area Si photodiode devices through characterization and analysis of cur-rent-voltage characteristics in Si pn solar cell structures functionalized with silica nanoparticles.[24] In the latter studies, increases in short-circuit current density of ~8.8% were observed in commercial Si solar cell struc-tures upon introduction of silica nanoparticles on the device surface.

Specifically, the utility of these effects for optimized crystalline Si pn junction solar cells and the effectiveness of both dielectric and metal nanoparticles in producing the desired scattering effects have been inves-tigated. Figure 11.8a shows a schematic diagram of a commercial Si solar cell device obtained from Silicon Solar, Inc. with front-side surface textur-ing, from which the preexisting antireflection coating and metal contacts were stripped. Ohmic contacts were then refabricated on the device sur-face, and then current-voltage characteristics and photocurrent response spectra measured on devices with and without nanoparticles deposited on the surface. Figure 11.8a also shows a scanning electron micrograph of Au nanoparticles, 100 nm in diameter, deposited on the device surface at a concentration of 1.4×10^9 cm^{-2}. The effects of both Au and silica nano-particles were studied: while, as noted above, silica and other dielectric nanoparticles do not support surface plasmon polariton excitations, Mie scattering behavior characteristic of all nanoparticles can still give rise to pronounced forward scattering effects and hence photocurrent enhance-ment in semiconductor photodiodes and photodetectors with dielectric nanoparticle functionalization.

Figure 11.8b shows photocurrent response spectra measured for these commercial Si solar cell devices functionalized with 150 nm diameter silica nanoparticles at concentrations of 1.03×10^9 cm^{-2} (dashed line) and 1.92×10^9 cm^{-2} (solid line), shown as ratios relative to reference response of the same devices prior to nanoparticle deposition. A clear increase in pho-tocurrent response is evident over wavelengths ranging from ~500 to over 1,000 nm. The photocurrent enhancement effect is not as sharply peaked

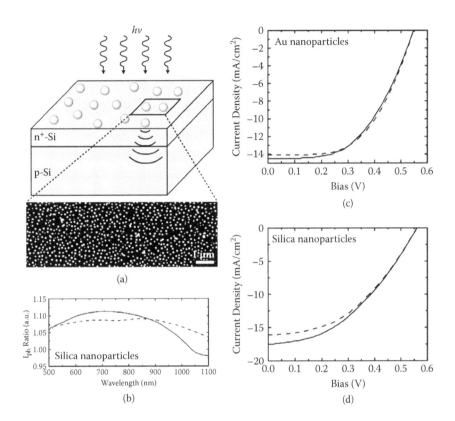

FIGURE 11.8

(a) Schematic diagram of crystalline Si photovoltaic device and experimental geometry, along with a representation of the forward scattered wave produced by nanoparticles on the device surface, and a scanning electron micrograph of 100 nm diameter Au nanoparticles deposited on the device surface at a concentration of 1.4×10^9 cm^{-2}. (b) Photocurrent response spectra for two devices with silica nanoparticles at concentrations of 1.03×10^9 cm^{-2} (dashed line) and 1.92×10^9 cm^{-2} (solid line), shown as ratios relative to reference response of the same devices prior to nanoparticle deposition. At right are shown plots of current density vs. voltage, under AM1.5 filtered illumination, for crystalline Si photovoltaic devices before (dashed lines) and after (solid lines) deposition of (c) 100 nm diameter Au nanoparticles at a density of 9.9×10^8 cm^{-2} or (b) silica nanoparticles at a density of 1.9×10^9 cm^{-2}. Enhancements in short-circuit current density of 2.8 and 8.8% are observed for Au and silica nanoparticles, respectively.

as that for Au nanoparticles, as shown in Figure 11.3b, which we attribute to the lack of a surface plasmon polariton resonance excitation in the silica nanoparticles. The size of the enhancement is also smaller, which may be due to the combination of the smaller scattering cross-section of the silica nanoparticles compared to that of an Au nanoparticle, and the different device structure employed in these experiments compared to those of Figure 11.3.

Figures 11.8c and d show current density vs. voltage, under AM1.5-filtered illumination, for these commercial crystalline Si photovoltaic devices before (dashed lines) and after (solid lines) deposition of either 100 nm diameter Au nanoparticles at a density of 9.9×10^8 cm^{-2} or 150 nm diameter silica nanoparticles at a density of 1.9×10^9 cm^{-2}. Enhancements in the short-circuit current density of 2.8 and 8.8% are observed for the Au and silica nanoparticles, respectively. The particle concentrations employed are close to the optimal values, based on results of numerical finite-element simulations of electromagnetic field distributions, and consequently optical transition rates, for various particle densities on Si semiconductor substrates in the presence of normally incident electromagnetic radiation. Indeed, numerical simulations have been found to yield quantitatively accurate predictions of observed enhancements in photocurrent,[24] and have confirmed the validity of the understanding of these and related effects as described here.

It is anticipated that these concepts can be readily extended beyond Si and a-Si:H to other photovoltaic materials. In general, nanoparticles can be considered to act as coupling devices to scatter incident light both in the forward direction and, for suitable device geometries, laterally into thin-film optical modes. While the demonstrations described above employ crystalline or amorphous Si solar cells and photodiodes, the universal nature of Mie scattering with small particles implies that particles of a wide range of materials can be engineered to increase the optical transmission of light regardless of the underlying absorbing substrate. Because this increased forward transmission is similar in its effect on overall photocurrent response to the effect of a conventional antireflection coating, it is worthwhile to assess its utility in this context. For Si solar cells, and especially those based on bulk crystalline Si, the increase in photocurrent attainable via nanoparticle scattering effects is smaller than that achievable with conventional antireflection coating technologies. However, numerical simulations we have performed indicate that nanoparticle-based scattering for light trapping may be beneficial for lower-index absorbers, such as organic semiconductors, where traditional antireflection technologies are difficult to apply. For example, the projected short-circuit current enhancement for silica nanoparticles on an organic semiconductor substrate, specifically poly-3-hexylthiophene, at concentrations of 4×10^8 and 2×10^9 cm^{-2} are 1.87 and 7.86%, respectively. Furthermore, nonabsorbing particles, such as silica, should scatter light more efficiently within the AM1.5 spectrum both in the forward direction and into lateral waveguide modes in thin-film organic photovoltaic cells when compared with most metallic nanoparticles.

11.3 NANOPARTICLE SCATTERING EFFECTS IN THIN-FILM WAVEGUIDE STRUCTURES

11.3.1 Application to Silicon-on-Insulator Device Structures

Optical scattering by nanoparticles atop a semiconductor photodetector device structure can also lead to coupling of photons incident normal to the device surface into lateral optical propagation paths, i.e., paths parallel to the device surface, within appropriately designed semiconductor thin-film device structures, due to the introduction of a lateral wave vector component in the scattered wave. Figure 11.9 shows a schematic diagram of a basic device geometry in which scattering of incident light by nanoparticles (or other nanostructures) on the surface of the device leads

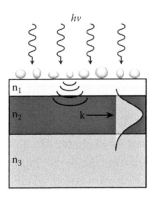

FIGURE 11.9
Schematic diagram of scattering of normally incident light into optical waveguide modes within an underlying device structure. The resulting lateral propogation of photons within a thin waveguiding layer can dramatically increase the efficiency of photon absorption.

to coupling of normally incident light into lateral optical propagation paths associated with a slab waveguide structure formed by an appropriately designed device. The resulting lateral propagation of photons within a thin waveguiding layer can dramatically increase the efficiency of photon absorption: normally incident photons may have a low probability of absorption while passing through the thin waveguiding layer, whereas lateral propagation provides a large increase in path length within that layer, and a corresponding increase in photon absorption probability. This concept can be particularly effective in photovoltaic or other photodetector devices in which thin layers are required, for example, to enable efficient collection of photogenerated carriers, satisfy critical-thickness constraints in epitaxial growth, or enable integration of devices with other materials or in form factors dictated by system-level requirements.

The validity and effectiveness of this concept was initially demonstrated over a decade ago in studies of photocurrent response in silicon-on-insulator photodetectors on which metallic nanoparticles were synthesized by thin-film deposition on a low-surface-energy dielectric followed by thermal annealing.[25] For Ag nanoparticles with a mean diameter of 108 nm on a silicon-on-insulator structure, an increase in photocurrent response by nearly a factor of 20 was observed at a wavelength of approximately 800 nm, for which a reference silicon-on-insulator device without nanoparticles would absorb approximately 1% of incident light.[26] Subsequent analysis indicated that coupling of radiation from a particle-like dipole at the device surface into optical modes confined by the silicon-on-insulator layer, which forms a slab waveguide structure due to the high refractive index of Si relative to the surrounding material, can exceed 80%.[27] Recent studies, combining experiment and numerical simulation, have also resulted in demonstration of nanoparticle-induced enhancement of both absorption and electroluminescence in silicon-on-insulator structures, with experimentally observed absorption enhancement at long wavelengths (1,050 nm) reaching a factor of 16 for Ag nanoparticles on a silicon-on-insulator photodetector structure with a 1.25 µm thick Si layer.[28,29]

The aforementioned studies exploited the presence of metal nanoparticles randomly distributed over the device surface, and with a substantial variation in size. A periodic arrangement of identical nanoparticles, however, can enable constructive interference among wave components scattered by individual nanoparticles in close proximity, thereby leading to substantially larger photocurrent enhancements, but over relatively narrow ranges of wavelength. Such behavior could be useful in photodetectors

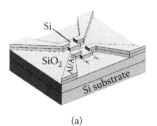 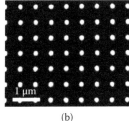 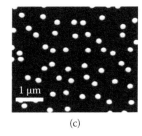

(a)   (b)   (c)

FIGURE 11.10

(a) Schematic diagram of silicon-on-insulator photodetector device structure. Also shown are scanning electron micrographs of (b) periodic and (c) random arrays of Au nanodots patterned by electron beam lithography on the surface of the active device region. Patterned dots enable constructive interference among wave components scattered by the nanoparticle, leading to large enhancements on photocurrent response in specific wavelength ranges.

for which high sensitivity is desired over a relatively narrow range of wavelengths, e.g., ~50 to 100 nm in width or less, or for photovoltaic devices employed in systems for which only similarly narrow ranges of wavelength response are required for a given device.

We have employed silicon-on-insulator photodetector structures to investigate the influence of nanoparticle periodicity on coupling of normally incident light into the silicon-on-insulator waveguide. Figure 11.10a shows a schematic diagram of the basic device structure employed, and Figure 11.10b and c shows scanning electron micrographs of Au nanodots, created by electron beam lithographic patterning and a liftoff process, arranged in either a periodic or a random array atop the active region of the silicon-on-insulator photodetector. For the random array, enhancements in photocurrent response are observed that agree well with previously published reports; enhancement in photocurrent response by a factor of up to ~2.5 is observed in a broad enhancement peak centered at a wavelength of ~1,000 nm, consistent with the relatively low nanodot density of ~2.8×10^8 cm^{-2} in these structures. For the periodic array, much sharper peaks in photocurrent enhancement, with increases in photocurrent response by factors as large as 5 to 6, are observed at wavelengths corresponding to confined waveguide modes of the silicon-on-insulator structure and constructive interference of electromagnetic wave components scattered by adjacent or nearby nanodots. These results indicate that pronounced enhancements in photocurrent response at specific wavelengths can be engineered by appropriate design of metal nanodot arrays combined with thin-film semiconductor structures with corresponding waveguiding behavior.

11.3.2 Application to Quantum Well Solar Cell Device Structures

We have also extended this concept to engineer, and improve, photocurrent response and photovoltaic power conversion efficiency in quantum well solar cell devices, in which low-bandgap quantum well layers are incorporated into a p-i-n photovoltaic device structure constituted primarily of material with a larger bandgap.[30–33] Incorporation of the quantum well layers enables absorption of long-wavelength photons to which the larger bandgap material would be transparent. However, efficient absorption of photons necessitates incorporation of a large number of quantum well layers constituting a thick (>1 μm) multiple-quantum-well layer, while efficient collection of the resulting photogenerated carriers mandates use of a thin (~300 nm or less) multiple-quantum-well layer to facilitate carrier escape from the quantum wells. These conflicting requirements have hindered the realization of quantum well solar cells with the very high power conversion efficiencies theoretically attainable by such devices: quantum well solar cells and related device concepts, such as intermediate band solar cells, have been predicted to yield maximum theoretical power conversion efficiencies ranging from 44.5%[32] to over 63%,[34] well in excess of the maximum theoretical single-junction device power conversion efficiency of ~31%.[35] To address this conflict, it is possible to exploit the fact that introduction of the multiple-quantum-well region, in addition to enabling long-wavelength photon absorption, also results in formation of a slab waveguide structure due to the higher refractive index of the quantum well layers relative to the surrounding material, in the manner indicated schematically in Figure 11.9.

The concept of nanoparticle-induced scattering into waveguide modes of a quantum well solar cell structure has been demonstrated in lattice-matched $InP/In_{1-x}Ga_xAs_yP_{1-y}$ ($y \approx 2.2x$) multiple-quantum-well p-i-n solar cell structures,[36] illustrated schematically in Figure 11.11a. Devices incorporated ten-period multiple-quantum-well structures within the intrinsic region of the epitaxial layer structure, with each period consisting of 10 nm $In_{0.91}Ga_{0.09}As_{0.2}P_{0.8}$ barriers alternating with 10 nm $In_{0.81}Ga_{0.19}As_{0.4}P_{0.6}$ quantum wells. Incorporation of the multiple-quantum-well layer enables the absorption spectrum of the device to extend to longer wavelengths

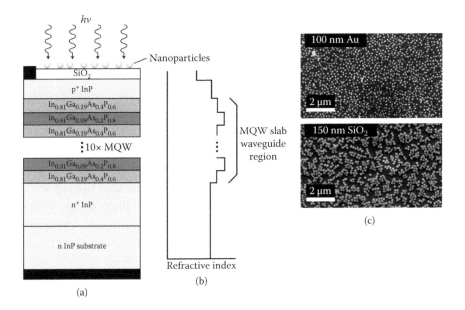

FIGURE 11.11
(a) Schematic diagram of InP-based quantum well solar cell device with nanoparticles on the device surface. (b) Refractive index profile of the quantum well solar cell device, with a slab waveguide region, 225 nm thick in this device structure, defined by the refractive index contrast between the material of the multiple-quantum-well region and the surrounding InP layers. (c) Scanning electron micrographs of 100 nm diameter Au nanoparticles (top) and 150 nm diameter SiO_2 nanoparticles (bottom) deposited on a quantum well solar cell device surface, at densities of 2.7×10^9 cm^{-2} and 2.1×10^9 cm^{-2}, respectively.

than that of a homojunction cell consisting entirely of the electrode material (here, InP), yielding an increase in short-circuit current density J_{sc}.[37] The resulting maximum power delivered for a quantum well solar cell device can therefore exceed that of a corresponding homojunction device,[37,38] despite a drop in open-circuit voltage (V_{oc}) resulting from less than unity efficiency in collection of photogenerated carriers from the multiple-quantum-well layer.

However, optimization of the collection of photogenerated carriers from the multiple-quantum-well region requires that the intrinsic region of the device be sufficiently thin that a substantial electric field, typically ~30kV/cm or more, be maintained even at the maximum power operating point of the device, corresponding to a moderate forward bias. This requirement severely restricts the acceptable range of thicknesses of the intrinsic multiple-quantum-well region, limiting it to ~250 nm or less for InP-based devices. Intrinsic layer thicknesses sufficiently small to ensure efficient carrier collection are generally also small compared to the optical absorption length at wavelengths between the quantum well and barrier/electrode absorption edges, resulting in poor efficiency in photon absorption at these wavelengths.

This intrinsic conflict between the necessity for thin multiple-quantum-well regions to enable efficient photogenerated carrier collection and thick multiple-quantum-well regions for efficient photon absorption can be resolved by exploiting the nanoparticle scattering behavior illustrated schematically in Figure 11.9. Scattering of incident light by the nanoparticles enables both improved transmission of photons into the semiconductor active layers and coupling of normally incident photons into lateral optically confined paths within the multiple-quantum-well waveguide layer, resulting in increased photon absorption, photocurrent, and power conversion efficiency.

Figure 11.12a shows the photocurrent response spectrum measured for InP homojunction, p-InP/i-$In_{0.91}Ga_{0.09}As_{0.2}P_{0.8}$/n-InP barrier only, and p-i-n quantum well solar cell devices. We see clearly in the figure that the photocurrent response for the InP control sample extends only to the InP absorption edge at 960 nm, while for the barrier-only control and quantum well devices the photocurrent responses extend to the absorption edges of $In_{0.91}Ga_{0.09}As_{0.2}P_{0.8}$ (1,040 nm) and $In_{0.81}Ga_{0.19}As_{0.4}P_{0.6}$ (1,160 nm), respectively. Figure 11.12b shows the maximum-power curves for a very similar set of InP homojunction, p-InP/i-$In_{0.91}Ga_{0.09}As_{0.2}P_{0.8}$/n-InP barrier-only, and p-i-n quantum well solar cell devices. Despite the drop in V_{oc} from 0.63 to 0.64 V for the homojunction and barrier-only devices to 0.53 V for the quantum well devices, the quantum well devices exhibit an increase in maximum power output of 7.4 and 4.6% relative to the homojunction and barrier-only devices, respectively. These results confirm that incorporation of lower-bandgap material in the quantum well solar cell structure indeed enables absorption of photons, and consequently generation of photocurrent, at wavelengths beyond the absorption edge of the InP electrode regions, thereby improving power conversion efficiency.

The effectiveness of nanoparticle scattering in improving photocurrent response over a broad range of wavelengths is illustrated in Figure 11.12c. Shown in the figure are photocurrent response spectra for quantum well solar cell devices with either 100-nm diameter Au or 150-nm diameter SiO_2 nanoparticles deposited on the surface, plotted as ratios relative to the spectrum for the same device without nanoparticles. Surface particle densities of ~2.7×10^9 cm^{-2} and ~2.1×10^9 cm^{-2} were employed for Au and SiO_2

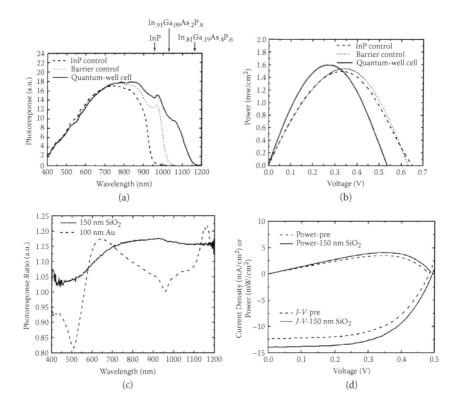

FIGURE 11.12

(a) Photocurrent response spectra for InP p-i-n control device (dashed line), p-InP/i-In$_{0.91}$Ga$_{0.09}$As$_{0.2}$P$_{0.8}$/n-InP barrier material control device (dotted line), and p-i-n ten-well In$_{0.91}$Ga$_{0.09}$As$_{0.2}$P$_{0.8}$/In$_{0.81}$Ga$_{0.19}$As$_{0.4}$P$_{0.6}$ quantum well solar cell device (solid line). The expected absorption edges for InP, In$_{0.91}$Ga$_{0.09}$As$_{0.2}$P$_{0.8}$, and In$_{0.81}$Ga$_{0.19}$As$_{0.4}$P$_{0.6}$ are indicated by the arrows. (b) Power output curves for InP p-i-n control device (dashed line), p-InP/i-In$_{0.91}$Ga$_{0.09}$As$_{0.2}$P$_{0.8}$/n-InP barrier material control device (dotted line), and p-i-n ten-well In$_{0.91}$Ga$_{0.09}$As$_{0.2}$P$_{0.8}$/In$_{0.81}$Ga$_{0.19}$As$_{0.4}$P$_{0.6}$ quantum well solar cell device (solid line). The quantum well device yields increases of 7.4 and 4.6% over the homojunction and barrier-only control devices, respectively. (c) Photocurrent response spectra of p-i-n In$_{0.91}$Ga$_{0.09}$As$_{0.2}$P$_{0.8}$/In$_{0.81}$Ga$_{0.19}$As$_{0.4}$P$_{0.6}$ quantum well solar cells with either 100-nm diameter Au nanoparticles (dashed line) or 150-nm diameter SiO$_2$ nanoparticles (solid line) deposited on the surface, plotted as ratios relative to the spectrum for the same devices without nanoparticles. (d) J-V and output power curves measured for quantum well solar cell devices without SiO$_2$ nanoparticles (dashed lines), and for the same devices after deposition of 150 nm diameter SiO$_2$ nanoparticles (solid lines). Short-circuit current density and maximum power increase by 12.9 and 17.0%, respectively, for devices incorporating SiO$_2$ nanoparticles.

nanoparticles, respectively. Scattering of incident light by Au nanoparticles yields a reduction in photocurrent response at wavelengths of ~570 nm and below, and a broad increase from ~570 nm to >900 nm. This behavior arises from strong forward scattering of light by the nanoparticles combined with interference between the scattered wave and the waves directly transmitted across the semiconductor device surface, as described in the context of nanoparticle-induced forward scattering for photocurrent enhancement in crystalline Si pn junction photodiodes and illustrated in Figures 11.6 and 11.7.[22] Scattering of incident light by SiO_2 nanoparticles yields increased transmission and photocurrent response over the entire 400 to 1,200 nm range of wavelengths, as no surface plasmon polariton resonance is present. This is consistent with the behavior illustrated in Figure 11.8.

For devices functionalized with Au nanoparticles we also observe a pronounced increase in photocurrent response between 960 nm and cutoff at ~1,200 nm. We attribute this behavior to the scattering of incident radiation into optical propagation modes associated with the slab waveguide formed by the multiple-quantum-well region and surrounding layers. A textbook analysis[39] of the slab waveguide structure shown in Figure 11.11a reveals that two confined modes are supported at wavelengths of 960 to 1,200 nm. The improved photocurrent response in this wavelength range is interpreted as arising from coupling of incident photons into the guided and substrate radiation modes of the waveguide structure, enabling the marked increase in photon path lengths within the multiple-quantum-well region associated with lateral, rather than vertical, photon propagation paths to be exploited to achieve greater efficiency in photon absorption.

Figure 11.12d confirms the effectiveness of nanoparticle scattering effects in improving short-circuit current density and power conversion efficiency in quantum well solar cell devices. Shown in the figure are current density-voltage characteristics and the corresponding power output for a quantum well solar cell device before and after deposition of SiO_2 nanoparticles on the device surface. For a SiO_2 nanoparticle surface density of ~2.1 × 10^9 cm^{-2}, a 12.9% increase in J_{sc}, an increase in the fill factor from 58 to 59%, and a 17.0% increase in maximum power conversion efficiency were observed relative to the same device prior to nanoparticle deposition. For an Au nanoparticle surface density of ~2.7 × 10^9 cm^{-2}, a 7.3% increase in J_{sc} and an increase of 1% in maximum power conversion efficiency relative to the same device prior to nanoparticle deposition were measured. The more modest improvements in performance associated with the presence of Au rather than SiO_2 nanoparticles are taken to be indicative of the detrimental effect of interference between scattered and directly transmitted waves on propagation of photons into the active semi-conductor device region. This factor appears to outweigh the increased scattering cross-section of the Au nanoparticles relative to SiO_2.

A key issue in such structures is the effectiveness of coupling between the scattered wave and the waveguide modes within the device structure. For a nanoparticle that is small compared to the wavelength, the scattered wave can be approximated as radiation from the dipolar resonant mode excited in the nanoparticle, and coupling of this scattered wave into device waveguide modes can be modeled by considering the coupling of radiation from a dipole at the surface of the device into the waveguide. The electric field for dipole oscillations corresponding to light at normal incidence is in the horizontal direction, i.e., parallel to the device surface. Calculations of this coupling can be performed by computing the dipole's total dissipated power in terms of the vector field amplitudes of the dipole moment and local electric field, modified due to the presence of the device structure below.[40] The influence of the device layers, expressed in terms of Fresnel reflection coefficients, can be computed using a transfer-matrix method.[39] It is then possible to calculate the resulting power spectrum, $S(u)$, for the dipole above the device structure, where u is a normalized wave vector component in a direction parallel to the waveguide layers and to the oscillating dipole moment.

For a waveguide structure such as that shown in Figure 11.9, three types of optical modes, corresponding to differing values of u, can be considered. For $u \in [0,1]$, the optical mode is propagating both above the waveguide (into air) and below (into the substrate with refractive index n_3). For $u \in [1,n_3]$, the optical mode is evanescent above the waveguide, and propagating into the substrate; these are so-called leaky or substrate radiation modes. For $u \in [n_3, \infty]$, the optical mode is evanescent both above and below the waveguiding layers; these correspond to fully guided modes.

Figure 11.13 shows the power spectrum $S(u)$ computed in the manner described above for the quantum well solar cell structure shown in Figure 11.11a. The portion of the spectrum corresponding to $u > 1$ constitutes the fraction of power radiated by the dipole that couples into either leaky or guided modes of the quantum well solar cell device structure, with $u > n_{InP}$ corresponding to coupling into true guided modes. While coupling into guided modes for this structure occurs with an efficiency typically less than 10%, coupling into the combination of leaky and guided modes occurs with efficiency in the vicinity of 90%, even for the low level

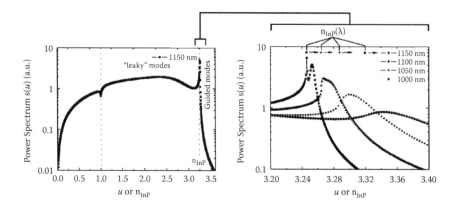

FIGURE 11.13

Calculated power spectrum, $S(u)$, for a horizontal electric dipole located above a multiple-quantum-well solar cell. In the plot at left, computed for a wavelength of 1,150 nm, the vertical dashed lines delineate the three regions that light can couple into (from left to right): propagating, leaky, and guided. The magnified view in the plot at right shows $S(u)$ for wavelengths ranging from 1,000 to 1,150 nm. The symbols near the top of the upper plot indicate the value of n_{InP} and consequently the onset of guided mode coupling (for $u > n_{InP}$) at wavelengths of 1,150 nm (squares), 1,100 nm (triangles), 1,050 nm (stars), and 1,000 nm (circles). The peaks present in the plots for $u \geq n_{InP}$ indicate strong coupling into the guided optical modes of the multiple-quantum-well waveguide region. The greatest coupling occurs at 1,150 nm as indicated by the sharp peaks on the left (TM0 mode) and right (TE0 mode). As the wavelength decreases, the modal peaks merge together due to increasing absorption in the material.

of refractive index contrast present in the quantum well solar cell device structures. While the effectiveness of propagation within leaky modes in improving photon absorption efficiency in the multiple-quantum-well region remains an open question, these analytical approaches provide useful guidance in the design of device structures in which nanoparticle scattering into lateral optical propagation paths can be most effectively exploited. Indeed, earlier studies of silicon-on-insulator structures provide proof of principle that very high coupling efficiencies, in the range of 80%, to guided modes in appropriately designed devices can be realized.[27]

11.4 SUMMARY

A variety of approaches have been explored, and continue to emerge, for coupling the optical properties of metal and dielectric nanoparticles to the behavior of semiconductor photodetectors and photovoltaic devices to improve broad-spectrum photosensitivity or wavelength-specific photoresponse. For applications involving optical absorption in molecules or organic semiconductors, approaches based on exploitation of surface plasmon polariton excitation in metal nanoparticles, and associated increases in local field amplitudes, have dominated. In contrast, for devices based on solid-state inorganic semiconductors, the disparity between the length scales for plasmon-induced field amplitude increases in the vicinity of metal nanoparticles, and those for optical absorption in bulk semiconductors suggest that greater emphasis on nanoparticle-induced optical scattering is likely to be most effective.

Studies of plasmonic field enhancement and forward scattering effects arising from the presence of metal nanoparticles on the surface of crystalline Si and a-Si:H pn junction photodiodes and solar cell structures have provided insight into the relative significance of local field enhancement effects and scattering, the very significant role of phase shifts in wave components scattered by nanoparticles, and the potential for, and limitations of, approaches for improving photodetector and photovoltaic device performance based on these effects realized using both metal and dielectric nanoparticles. Additional approaches for improving performance of photodetectors and photovoltaics based on nanoparticle-induced scattering of incident light into lateral optical propagation paths offer the potential to realize large increases in efficiency of photon absorption by dramatically increasing the optical path length for photons in thin layers, which in more conventional device geometries would yield only very low photon absorption efficiency. Very high efficiencies in coupling of incident radiation into waveguide modes in very thin device layers or complete device structures could enable not only very high power conversion efficiencies in photovoltaics, but also increased flexibility in device design and material utilization over a range of efficiencies, and in design of photovoltaic devices compatible with a variety of system-driven form factors.

ACKNOWLEDGMENTS

Part of this work was supported by a grant from the UCSD Von Liebig Center, the Air Force Office of Scientific Research (FA9550-07-1-0148), and the Department of Energy (DE-FG36-08GO18016).

REFERENCES

1. Moskovits, M. 1985. Surface-enhanced spectroscopy. *Rev. Mod. Phys.* 57:783–826.
2. Kneipp, K., Wang, Y., Kneipp, H., Perelman, L. T., Itzkan, I., Dasari, R. R., and Feld, M. S. 1997. Single molecule detection using surface-enhanced Raman scattering (SERS). *Phys. Rev. Lett.* 77:1667–70.
3. Nie, S. M., and Emery, S. R. 1997. Probing single molecules and single nanoparticles by surface-enhanced Raman scattering. *Science* 275:1102–6.
4. Schultz, S., Smith, D. R., Mock, J. J., and Schultz, D. A. 2000. Single molecule detection with nonbleaching multicolor optical immunolabels. *Proc. Natl. Acad. Sci. USA* 97:996–1001.
5. Taton, T. A., Mirkin, C. A., and Letsinger, R. L. 2000. Scanometric DNA array detection with nanoparticle probes. *Science* 289:1757–60.
6. Li, K., Stockman, M. I., and Bergman, D. J. 2003. Self-similar chain of metal nanospheres as an efficient nanolens. *Phys. Rev. Lett.* 91:227402-1–4.
7. Kalkbrenner, T., Ramstein, M., Mlynek, J., and Sandoghdar, V. 2001. A single gold particle as a probe for apertureless scanning near-field optical microscopy. *J. Microsc.* 202(Pt 1):72–76.
8. Zhang X., and Liu, Z. W. 2008. Superlenses to overcome the diffraction limit. *Nature Mater.* 7:435–41.
9. Ditlbacher, H., Krenn, J. R., Schider, G., Leitner, A., and Aussenegg, F. R. 2002. Two-dimensional optics with surface plasmon polaritons. *Appl. Phys. Lett.* 81:1762–64.
10. Engheta, N. 2007. Circuits with light at nanoscales: Optical nanocircuits inspired by metamaterials. *Science* 317:1698–702.
11. Bohren, C. F., and Huffman, D. R. 1983. *Absorption and scattering of light by small particles*. New York: John Wiley & Sons.
12. Kelly, K. L., Coronado, E., Zhao, L. L., and Schatz, G. C. 2003. The optical properties of metal nanoparticles: The influence of size, shape, and dielectric environment. *J. Phys. Chem. B* 107:668–77.
13. Yu, P. Y., and Cardona, M. 2001. *Fundamentals of semiconductors*. 3rd ed. Berlin: Springer-Verlag.
14. Stenzel, O., Stendal, A., Voigtsberger, K., and von Borczyskowski, C. 1995. Enhancement of the photovoltaic conversion efficiency of copper phthalocyanine thin film devices by incorporation of metal clusters. *Sol. Energy Mater. Sol. Cells* 37:337–48.
15. Stenzel, O., Wilbrandt, S., Stendal, A., Beckers, U., Voigtsberger, K., and von Borczyskowski, C. 1995. The incorporation of metal clusters into thin organic dye layers as a method for producing strongly absorbing composite layers: An oscillator model approach to resonant metal cluster absorption. *J. Phys. D Appl. Phys.* 28:2154–62.
16. Westphalen, M., Kreibig, U., Rostalski, J., Lüth, H., and Meissner, D. 2000. Metal cluster enhanced organic solar cells. *Sol. Energy Mater. Sol. Cells* 61:97–105.
17. Rand, B. P., Peumans, P., and Forrest, S. R. 2004. Long-range absorption enhancement in organic tandem thin-film solar cells containing silver nanoclusters. *J. Appl. Phys.* 96:7519–26.
18. Bohren, C. F., and Huffman, D. R. 1983. *Absorption and scattering of light by small particles*. New York: John Wiley & Sons.
19. Johnson, P. B., and Christy, R. W. 1972. Optical constants of the noble metals. *Phys. Rev. B* 6:4370–79.

20. Schaadt, D. M., Feng, B., and Yu, E. T. 2005. Enhanced semiconductor optical absorption via surface plasmon excitation in metal nanoparticles. *Appl. Phys. Lett.* 86:063106-1–3.
21. Derkacs, D., Lim, S. H., Matheu, P., Mar, W., and Yu, E. T. 2006. Improved performance of amorphous silicon solar cells via scattering from surface plasmon polaritons in nearby metallic nanoparticles. *Appl. Phys. Lett.* 89:093103-1–3.
22. Lim, S. H., Mar, W., Matheu, P., Derkacs, D., and Yu, E. T. 2007. Photocurrent spectroscopy of optical absorption enhancement in silicon photodiodes via scattering from surface plasmon polaritons in gold nanoparticles. *J. Appl. Phys.* 101:104309-1–7.
23. Sundararajan, S. P., Grady, N. K., Mirin, N., and Halas, N. J. 2008. Nanoparticle-induced enhancement and suppression of photocurrent in a silicon photodiode. *Nano Lett.* 8:624–30.
24. Matheu, P., Lim, S. H., Derkacs, D., McPheeters, C., and Yu, E. T. 2008. Metal and dielectric nanoparticle scattering for improved optical absorption in photovoltaic devices. *Appl. Phys. Lett.* 93:113108-1–3.
25. Stuart, H. R., and Hall, D. G. 1996. Absorption enhancement in silicon-on-insulator waveguides using metal island films. *Appl. Phys. Lett.* 69:2327–29.
26. Stuart, H. R., and Hall, D. G. 1998. Island size effects in nanoparticle enhanced detectors. *Appl. Phys. Lett.* 73:3815–17.
27. Soller, B. J., Stuart, H. R., and Hall, D. G. 2001. Energy transfer at optical frequencies to silicon-on-insulator structures. *Optics Lett.* 26:1421–23.
28. Catchpole, K. R., and Pillai, S. 2006. Absorption enhancement due to scattering by dipoles into silicon waveguides. *J. Appl. Phys.* 100:044504-1–8.
29. Pillai, S., Catchpole, K. R., Trupke, T., and Green, M. A. 2007. Surface plasmon enhanced silicon solar cells. *J. Appl. Phys.* 101:093105-1–8.
30. Barnham, K. W. J., Ballard, I., Connolly, J. P., Ekins-Daukes, N. J., Kluftinger, B. G., Nelson, J., and Rohr, C. 2002. Quantum well solar cells. *Physica E* 14:27–36.
31. Johnson, D. C., Ballard, I. M., Barnham, K. W. J., Connolly, J. P., Mazzer, M., Bessière, A. Calder, C., Hill, G., and Roberts, J. S. 2007. Observation of photon recycling in strain-balanced quantum well solar cells. *Appl. Phys. Lett.* 90:213505.
32. Wei, G., Shiu, K. T., Giebink, N. C., and Forrest, S. R. 2007. Thermodynamic limits of quantum photovoltaic cell efficiency. *Appl. Phys. Lett.* 91:223507.
33. Anderson, N. G. 2002. On quantum well solar cell efficiencies. *Physica E* 14:126–31.
34. Bremner, S. P., Corkish, R., and Honsberg, C. B. 1999. Detailed balance efficiency limits with quasi-Fermi level variations. *IEEE Trans. Electron. Devices* 46:1932–39.
35. Henry, C. H. 1980. Limiting efficiencies of single and multiple energy gap terrestrial solar cells. *J. Appl. Phys.* 51:4494–500.
36. Derkacs, D., Chen, W. V., Matheu, P., Lim, S. H., Yu, P. K. L, and Yu, E. T. 2008. Nanoparticle-induced light scattering for improved performance of quantum-well solar cells. *Appl. Phys. Lett.* 93:091107-1–3.
37. Barnham, K. W. J., Braun, B., Nelson, J., Paxman, M., Button, C., Roberts, J. S., and Foxon, C. T. 1991. Short-circuit current and energy efficiency enhancement in a low-dimensional structure photovoltaic device. *Appl. Phys. Lett.* 59:135–37.
38. Raisky, O. Y., Wang, W. B., Alfano, R. R., Reynolds, C. L. Jr., Stampone, D. V., and Focht, M. W. 1998. $In_{1-x}Ga_xAs_{1-y}P_y$/InP multiple quantum well solar cell structures. *J. Appl. Phys.* 84:5790–94.
39. Yeh, P. 1988. *Optical waves in layered media*, 319. New York: John Wiley & Sons.
40. Soller, B. J., Stuart, H. R., and Hall, D. G. 2001. Energy transfer at optical frequencies to silicon based waveguiding structures. *J. Opt. Soc. Am. A* 18:2577–84.

Epilogue: Future Manufacturing Methods for Nanostructured Photovoltaic Devices

Loucas Tsakalakos and Bas A. Korevaar

Nano is a beautiful new science and technology world. We have seen exciting new physics at play in generation III band structures, novel device architectures employing nanostructures, as well as the application of nanostructures to new module concepts (e.g., luminescent solar concentrators) and even conventional modules. These potential future applications are all enabled by the fascinating world of structures that are only visible by scanning electron microscopy (SEM) and transmission electron microscopy (TEM). However, going from very small to very large is a huge step. For example, from single nanowire solar cell efficiencies to 1 cm² devices, and from 1 cm² devices to modules takes significant effort. Quarterly, the journal *Progress in Photovoltaics: Research and Applications* publishes a list of record efficiencies at the cell level and at the module level, and even on those scales large differences in performance can be observed. Larger differences will be present when discussing the quantum efficiency in a single wire vs. an array of wires, or charge transport in a sheet of quantum dots vs. a fully integrated quantum dot device.

The environment in which one does his or her research drives choices, and both basic and applied aspects of the work have enormous value. Basic mechanisms need to be understood to successfully develop and design high-efficiency cells, and costly technology should not be avoided during this stage. However, in the back of one's mind one always has to think about manufacturability on larger scale, particularly for those colleagues who happen to work in an industrial research setting.

Solar is driven by a huge energy demand, from both a resource and a global environmental perspective. Many nano-features will fit in 1 m², whereas many square meters are needed for a terawatt energy generation. In this concluding section we summarize various manufacturing technologies and their applicability to making nanoscale devices over large areas. We will discuss both technologies that are currently being used and technologies that might have future applicability within this area. The list is not assumed to be complete; future technology development can also change this picture a lot, and of course, we are well aware that many basic science and engineering discoveries are required to bring the various approaches discussed in the preceding chapters to full fruition. This discussion is therefore a snapshot in time, trying to discuss the state of the art, which might be considered old for future generations in this fast-developing technology area.

Prior to discussion of various potential nanostructure-specific manufacturing technologies, it is instructive to review the major criteria that are often used in selecting a manufacturing approach. These are almost always accompanied by detailed cost analyses that are beyond the scope of this book. A major criterion is the capital expenditure (CAPEX) cost, which is paid up front and depreciates over time. A high CAPEX can have a strong impact on the decision of whether to pursue a given technology. The overall manufacturing/process cost is also critical, as it will decide the direct $\$/W_p$ cost of a technology, as well as impact the levelized cost of electricity (LCOE) in \$/kWh. Many factors impact the manufacturing cost, including process size, throughput, yield, materials costs, electrical power, waste treatment, and others. We will highlight some of these dependent features in our discussion, with a focus on scalability (to large areas, e.g., >1 m²), throughput, and the environmental friendliness of the process. The latter aspect can have a direct cost impact since less environmental control measures will be required. Throughput may be associated with a batch process or a continuous roll-to-roll process, each having advantages and disadvantages. Another key feature of a process is the ability to control the structures of interest at the nano (e.g., diameter distribution, doping, etc.) and macro (e.g., thickness, area uniformity) scales, with technologies requiring minimal process controls (sensors/feedback) being preferred.

The decision to choose a specific manufacturing technology is not an easy one, yet it can often be a critical deciding factor in the ultimate success or failure of a product or company. It is perhaps apparent that each manufacturing technology has its advantages and disadvantages. Of course, it is early in the field of nano-PV to select a leading manufacturing approach, since this must be considered in light of a specific PV approach that provides high performance. Nevertheless, it may be instructive to consider the major approaches known today and their attributes with regard to the manufacturing requirements discussed previously, as an abbreviated example of how one makes such decisions in an industrial setting. Table 1 shows the major processing approaches that are possible for various classes of nanostructures relevant to PV and provides such an analysis at a cursory level. This is based not only on our experience in studying nanoscale process technologies from an industrial perspective, but also on detailed analysis of numerous thin-film and silicon manufacturing approaches in the PV industry. This is by no means meant to be authoritative, and many of the points are open to debate, but it is exactly such debate we wish to stimulate such that in coming years, as nano-PV technologies mature and reach more advanced stages, such factors are considered by researchers in academia and industry alike. Our objective is not to select a leading process approach, or condemn others, since we are well aware that advances will be made in the future. Rather, we wish to highlight the thought processes often used in industry. We now briefly summarize our thoughts behind such a table.

NANOCOMPOSITES

Nanocomposite solar cells, including dye-sensitized solar cells and organic solar cells, rely on a nano-featured network of intertwined continuous nanoscale features. When considering fabrication methods, one has to make sure that the conducting part, in most cases the semiconducting polymer or TiO_2, is continuous while at the same time highly porous such that a large surface area is obtained. Various techniques may be used to deposit TiO_2 and ZnO, which are widely used materials in nanocomposite hybrid organic-inorganic solar cells. The main target is a high porous network of TiO_2 with a large surface area, enclosing continuous porosity, accessible by dye or another secondary phase. One of the important aspects of choosing a certain technology for the deposition is the choice of the right precursor that allows for large area deposition, high throughput, and low CAPEX. This may entail forming small particles in a solution, which then are placed on a substrate. Potential methods include tape casting, spray coating, spin coating, or using a thermal/plasma spray method. Spray coating has many advantages, including scalability and low cost,

TABLE 1

Example of a Trade-off Analysis Matrix for Leading Process Techniques for the Major Classes of Nanostructured Photovoltaics with Regard to the Critical Manufacturing Requirements*

Technology/Process	CAPEX	Scalability (Large Areas)	Structure Flexibility	Environmentally Friendly	Throughput	Combined Manufacturing Cost
Nanocomposites						
Spray coating	++	++	+	+	++	++
Tape casting	+	+/−	+	+	+	++
Spin coating	++	−	−	+	−	+
Thermal spray/flame hydrolysis	−	+	−	−	+	+
Quantum Wells						
MOCVD	− −	+/−	+	− −	+/−	−
MBE	− −	− −	++	−	− −	− −
Nanowire/Tubes						
CVD/PECVD	−	++	+	+/−	+	+
Templated electrochemical deposition	+	+	++	+/−	−	+
Solution deposition	++	+/−	+/−	+/−	+/−	+
Quantum Dots—Nanoparticles						
Sputtering	+	++	+/−	+/−	++	+
Plasma processing	+/−	+	+/−	−	+/−	+/−
Dip coating/Langmuir-Blodgett	++	+	+	+/−	+/−	+

* Relative comparisons are made between processes for each type of nanostructure, not by comparing all processes in the table.

with perhaps the one drawback being the need for an additional annealing step. Tape casting has similar advantages, though thickness control to less than 10 microns is difficult to achieve. Spin coating is also a low-cost process, yet generally not feasible for a large area substrate or a continuous process. Thermal/plasma spray is also of great interest for nanocomposite manufacturing, since the desired sintering between the particles will either happen within the heat zone or on the substrate. However, the choice for the one-step process like thermal spray or flame hydrolysis may not be obvious, as the multistep process using a cold process can be less capital intensive from an equipment perspective and furnaces are typically capable of very high throughput.

QUANTUM WELLS

Quantum well solar cells are almost exclusively fabricated using III-V materials by metal-organic chemical vapor deposition (MOCVD) or molecular beam epitaxy (MBE) methods. MBE has the advantage that one can control the structure of matter at the submonolayer and atomic level. However, scale-up, throughput, and cost continue to be challenges for MBE. These issues must be addressed for future large-scale manufacturing in the major terrestrial markets. MOCVD is more amenable to scale-up, yet still allows for good control of thickness and stoichiometry. Cost associated with process gases is of concern with MOCVD, however. Throughput is also usually low due to reactor designs limited to smaller-area substrates in batch mode, as well as the need to grow slowly to minimize defect formation in heteroepitaxial materials systems. At present these processes are well suited for space, defense, and terrestrial concentrator applications.

NANOWIRES/TUBES

Solar cells based on nanowires are a very new area of research and many new groups are entering this field. Most groups have started working on Si nanowires, demonstrating both single-wire efficiencies and even large area functionality. More exotic materials are also investigated, with groups working on InGaN nanowires and even CIGS nanowires. Many manufacturing methods are possible for future nanowire/tube solar cells, including plasma-enhanced chemical vapor deposition (PECVD), electrodeposition, vapor-liquid-solid (VLS)/CVD, sputtering, etching, etc.

Chemical vapor deposition and related processes (e.g., plasma-enhanced CVD) are perhaps the most widely used method for growing nanowires and tubes. Advantages of such methods, particularly when combined with catalytic growth processes (e.g., VLS), include the fact that such manufacturing equipment has a well-established base in the electronics industry, is readily scalable, and has orders of magnitude higher growth rates for catalytic growth relative to noncatalyzed thin-film growth. Disadvantages include the high CAPEX required for high throughput in-line CVD systems, as well as relatively hazardous process gases that are often utilized. CVD of carbon nanotubes suffers from a lack of chirality control, as discussed previously.

Templated electrodeposition (ECD) is another potential manufacturing route to nanowire/tube-based devices. This method allows for good control over key geometrical parameters such as diameter and length, low CAPEX, and low manufacturing costs. ECD suffers from relatively low throughput (deposition rates in nanopores) and difficulty in scale-up due to electric field and diffusion nonuniformity within nanopores across a large area substrate. Surfactant-mediated solution-based growth has also been used to grow certain types of nanowires (e.g., ZnO) and has many of the same advantages of ECD. This process is easy to scale up and requires low CAPEX, yet to date also requires long reaction times to achieve long nanowire arrays. It is also possible to grow nanowire/tubes by a vapor process, harvest the wires in solution, and then deposit them on a solid substrate.

Processing challenges that must be addressed for all nanowire/tube manufacturing processes include cleaning and handling of nanostructures from a mechanics perspective, oxidation, the ability to dope nanowires/tubes reproducibly, physical strength, junction direction, and orientation control (physical and crystallographic), among others. More environmentally friendly processes should be developed for all nanowire/tube structure fabrication.

QUANTUM DOTS/NANOPARTICLES

Nanoparticles can be formed in many different ways. One can form them separately and then store them in solution, or one can form them directly, where one needs them within a device structure. Methods vary from plasma dust to annealing of continuous films. For example, an equilibrium plasma like the expanding thermal plasma technique developed at the Eindhoven University of Technology generates charged particles containing up to ten atoms. Based on mass and charge, these particles could be extracted from the plasma into a second chamber, where they could, for example, be incorporated in a growing film. The yield of generating these particles within the plasma would need to be increased significantly, though, to make it useful for manufacturing. On the other hand, thin layers of gold, silver, aluminum, or copper could be annealed such that the films break up and a surface covered with particles is left, which could, for example, be used for plasmonic effects.

When considering nanoparticles based on precious metals such as gold or exotic materials compositions, a first question to address is the cost impact of such structures. A simple cost calculation is illustrative: assuming a 1 m^2 module and, for example, a 50-nm-thick gold film, we multiply by the density (assuming 10 g/cm^3 for this example) to obtain a required mass of 5×10^{-4} kg/m^2. At a gold price of $6,000/kg, that means a cost of $3/$m^2$. An efficiency of 10% would result in $0.03/$W_p$. This is well within the ranges of the materials cost of the other components of the solar cell. While this serves as an illustrative example, it is exactly this type of first-order analysis, coupled with more extensive cost models, that is required when considering such technologies.

Most quantum dot particle systems are synthesized by solution-based colloidal methods that are readily amenable to mass production. Dispersion of such particles into functional layers may be achieved by dip coating and related methods (e.g., Langmuir–Blodgett deposition). These processes are relatively low in required CAPEX and manufacturing costs, though they may require additional engineering to improved large area deposition/uniformity and throughput if one wishes to achieve technologically relevant film thicknesses.

SUMMARY AND FUTURE OUTLOOK

In summary, there are several manufacturing approaches that one can implement depending on the desired nanostructure type and device structure. Each of these has advantages and disadvantages that with proper analysis and engineering may be alleviated through further research and development. It is early in the field of nanotechnology for photovoltaics, and many exciting basic science and technological developments remain to be demonstrated and ultimately implemented. While we are cautiously optimistic that these developments will find their way into PV products, we believe the future of nanophotovoltaics is bright!

Index